TECHNOLOGY OF METAL POWDERS

TECHNOLOGY OF METAL POWDERS

Recent Developments

Edited by L.H. Yaverbaum

NOYES DATA CORPORATION
Park Ridge, New Jersey, U.S.A.
1980

Library of Congress Catalog Card Number: 80-31
ISBN: 0-8155-0794-1
Printed in the United States

Published in the United States of America by
Noyes Data Corporation
Noyes Building, Park Ridge, New Jersey 07656

Library of Congress Cataloging in Publication Data

Yaverbaum, Lee.
Technology of metal powders.

(Chemical technology review ; no. 153)
Includes index.
1. Powder metallurgy--Patents. I. Title.
II. Series.
TS245.Y38 671.3'7'0272 80-31
ISBN 0-8155-0794-1

Foreword

The detailed, descriptive information in this book is found in U.S. patents issued since March 1974 that deal with the technology of metal powders.

This book is a data-based publication, providing information retrieved and made available from the U.S. patent literature. It thus serves a double purpose in that it supplies detailed technical information and can be used as a guide to the patent literature in this field. By indicating all the information that is significant, and eliminating legal jargon and juristic phraseology, this book presents an advanced commercially oriented review of the technology of metal powders. This title contains all recent developments since our previous title *Powdered Metals Technology*, published in 1974.

The U.S. patent literature is the largest and most comprehensive collection of technical information in the world. There is more practical, commercial, timely process information assembled here than is available from any other source. The technical information obtained from a patent is extremely reliable and comprehensive; sufficient information must be included to avoid rejection for "insufficient disclosure." These patents include practically all of those issued on the subject in the United States during the period under review; there has been no bias in the selection of patents for inclusion.

The patent literature covers a substantial amount of information not available in the journal literature. The patent literature is a prime source of basic commercially useful information. This information is overlooked by those who rely primarily on the periodical journal literature. It is realized that there is a lag between a patent application on a new process development and the granting of a patent, but it is felt that this may roughly parallel or even anticipate the lag in putting that development into commercial practice.

Many of these patents are being utilized commercially. Whether used or not, they offer opportunities for technological transfer. Also, a major purpose of this book is to describe the number of technical possibilities available, which may open up profitable areas of research and development. The information contained in this book will allow you to establish a sound background before launching into research in this field.

Advanced composition and production methods developed by Noyes Data are employed to bring these durably bound books to you in a minimum of time. Special techniques are used to close the gap between "manuscript" and "completed book." Industrial technology is progressing so rapidly that time-honored, conventional typesetting, binding and shipping methods are no longer suitable. We have by-passed the delays in the conventional book publishing cycle and provide the user with an effective and convenient means of reviewing up-to-date information in depth.

The table of contents is organized in such a way as to serve as a subject index. Other indexes by company, inventor and patent number help in providing easy access to the information contained in this book.

15 Reasons Why the U.S. Patent Office Literature Is Important to You –

1. The U.S. patent literature is the largest and most comprehensive collection of technical information in the world. There is more practical commercial process information assembled here than is available from any other source.
2. The technical information obtained from the patent literature is extremely comprehensive; sufficient information must be included to avoid rejection for "insufficient disclosure."
3. The patent literature is a prime source of basic commercially utilizable information. This information is overlooked by those who rely primarily on the periodical journal literature.
4. An important feature of the patent literature is that it can serve to avoid duplication of research and development.
5. Patents, unlike periodical literature, are bound by definition to contain new information, data and ideas.
6. It can serve as a source of new ideas in a different but related field, and may be outside the patent protection offered the original invention.
7. Since claims are narrowly defined, much valuable information is included that may be outside the legal protection afforded by the claims.
8. Patents discuss the difficulties associated with previous research, development or production techniques, and offer a specific method of overcoming problems. This gives clues to current process information that has not been published in periodicals or books.
9. Can aid in process design by providing a selection of alternate techniques. A powerful research and engineering tool.
10. Obtain licenses–many U.S. chemical patents have not been developed commercially.
11. Patents provide an excellent starting point for the next investigator.
12. Frequently, innovations derived from research are first disclosed in the patent literature, prior to coverage in the periodical literature.
13. Patents offer a most valuable method of keeping abreast of latest technologies, serving an individual's own "current awareness" program.
14. Copies of U.S. patents are easily obtained from the U.S. Patent Office at 50¢ a copy.
15. It is a creative source of ideas for those with imagination.

Contents and Subject Index

Introduction

Powder metallurgy is one of the earliest "technologies" to appear in the evolution of man. It is a process that reduces metals to powdered form and presses the powder into defined shapes, rather than using molten metal. Sometimes melting is impractical because of the very high melting point of the metal, e.g., tungsten, with a melting point of 6100°F which is too high for ordinary melting and casting methods, or sometimes an alloy is desired of two metals that cannot fuse. Thus, with powder metallurgy a single metal or a mixture of metals may be used. After compaction, a heating and cooling process usually follows, adding strength to the finished product.

Metals are made into powders in many ways, i.e., chemically, where a reducing agent breaks the metal into particles and by-products which can be removed; mechanically, where the metal can be milled by several techniques; and electrolytically. Related techniques include atomization, vapor condensation, precipitation, and use of a magnetic field.

The growth of powder metallurgy as an industrial technology has created a growing demand for "tailor-made" powders having carefully checked granulometric and morphological characteristics. For example, when a greater uniformity in processing an end product is required, sphericity of the particles is most desirable. Submicronic powders also have significant effects in decreasing the temperature and/or the sintering time of powder metallurgical processes.

Other advantages of powder metallurgy include better homogeneity of the end products, processes that cannot be duplicated by other methods, rapid production, low scrap loss, dimensional accuracy, energy conservation, use of unskilled labor in most processes, better economy, mixing of normally unmixable powders, and greater porosity for permeation of liquids or gases, e.g., self-lubricating gears.

The broad significance of the technology is indicated by the worldwide effort which has been devoted to powder metallurgical processes over the recent years. This book describes over 200 patented processes relating to all phases of production. Over one-third of these were developed outside the United States. Thus, Germany, France, Italy, Japan, Canada, and the Scandinavian countries have all developed technology in these areas.

The book describes all aspects of powder metallurgy, from powder preparation and recovery through applications, including special apparatus. In some cases a process may be of value for the production of more than one type of metal powder. Where possible, the book is arranged to reflect the utility of a given process to a specific metal or family of metals. However, in some cases, the composite is so large or the application so broad, that the last chapter has been devoted to these processes rather than identifying them within a limited grouping.

Iron and Iron Processes

POWDER METALLURGICAL TECHNIQUE

Selectively Carburized Forged Powder Parts

M.C. Sarnes and R. Pathak; U.S. Patent 4,165,243; August 21, 1979; assigned to Federal-Mogul Corporation describe a powder metallurgical process for producing dense low-alloy ferrous parts having selected carburized surface portions in which a metal powder of the desired composition is briquetted into a shape-retaining preform; the preform is masked, sintered, carburized whereafter the mask is removed and the resultant preform is forged to a desired final shape. The forged part is then quenched immediately following the forge operation for hardness or if desired it can be further heat treated in another manner to obtain other mechanical properties.

Figure 1.1 is a flow diagram sequentially illustrating the preferred steps in the practice of the process.

In Figure 1.1, the steps in the preferred practice of this process include providing a metal powder of the desired composition which is briquetted into a green briquette of a desired configuration and which thereafter is masked over selected surface portions thereof. The masked briquette thereafter is sintered whereafter it is carburized over the exposed surface portions thereof to provide a case depth of the desired magnitude.

At the completion of the carburizing treatment, the mask is removed and the sintered and selectively carburized preform is forged into a part of the desired final configuration and density. The resultant forged part is preferably quenched after stabilization of temperature to further enhance its mechanical properties. The quenched part optionally, can be subjected to further heat treatment to

Figure 1.1: Flow Diagram for Selective Carburization of Forged Powder Parts

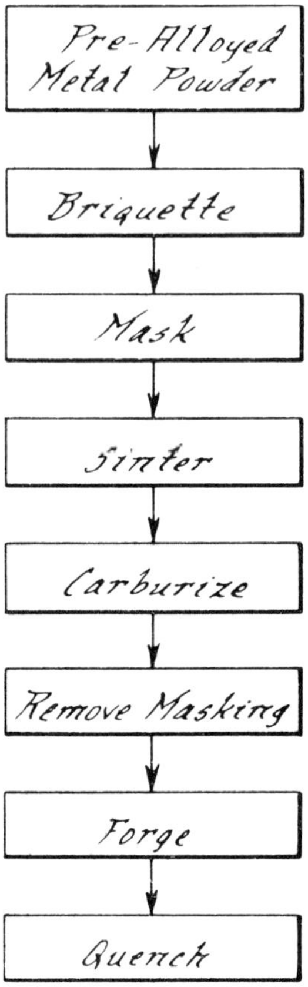

Source: U.S. Patent 4,165,243

further optimize its mechanical properties and thereafter it is subjected to final machining or finishing operations as may be desired or required to produce a precision part.

The metal powder can comprise a powder of AISI No. 4600 series which nominally contains from zero up to about 0.5% manganese, about 0.25% to about 2.25% nickel, about 0.25% to about 0.70% molybdenum, up to about 1.25% carbon with the balance consisting essentially of iron along with conventional impurities and residuals. While the powder may comprise a mechanical mixture of the several alloying constituents to provide the desired final alloy chemistry, it is preferred that the powder consist of prealloyed powder particles in which each particle comprises an alloy of the desired composition.

Such prealloyed powders of the desired alloy chemistry can readily be produced by any one of a variety of techniques including water atomization of a melt of the alloy which can be screened and classified to extract particles of the desired size range. While the particle size and configuration of the metal powder is not generally thought to be critical in the practice of the present process, it is preferred to employ particles of fairly even size distribution within a range of about minus 60 mesh (U.S. standard sieve size) to about minus 325 mesh, with no more than 10% by weight being greater in size than minus 60 mesh and no more than 40% by weight being finer in size than minus 325 mesh.

The powder alloy preliminary to briquetting, is admixed with a suitable die lubricant and binder, of any of the types well known in the art. In addition, the metal powder may also be admixed with graphite to supplement and/or adjust the carbon content thereof to within a desired range.

A premeasured quantity of the metal alloy powder is placed in a die cavity of a desired configuration and is compacted in a manner well known in the art into a green briquette or preform of a size and configuration suitable for subsequent forging into a finished part of the desired configuration. The briquetting operation is performed in a manner to produce a preform of a density preferably at least 75 of 100% theoretical density, and preferably from about 85% to about 90% of theoretical density. The briquetting operation preferably employs a ram and die-set.

The resultant green briquette or preform is thereafter masked employing masking means subsequently to be described such that only selected portions of the preform are exposed. After masking, the preform is placed in a sintering furnace in which it is preliminarily heated to an intermediate temperature effecting thermal decomposition and/or volatilization of the lubricant employed followed by a final sintering step which for 4600 series steel alloys preferably ranges from about 2000° to about 2100°F. The sintering is performed in a batch or continuous manner in accordance with techniques known in the art employing a protective atmosphere to avoid undesirable oxidation and scale formation on the surfaces of the powder particles. The sintering operation is carried out for a period of time sufficient to effect a diffusion bonding of the powder particles together at their points of contact forming an integral sintered matrix.

Following the sintering step, the masked and sintered preform is transferred to a caburizing furnace provided with a carbon atmosphere of any of the types known in the art including those disclosed in *Metals Handbook,* vol. 2, 8th ed., pp 67-117, published by the American Society for Metals. Preferably, class 302 endothermic base carburizing gases are employed and the carburizing treatment is carried out at a temperature generally ranging from about 1500° to about 1900°F for a period of time sufficient to form a high carbon casing of the desired depth on the unmasked portions of the preform. Generally, case depths of the required carbon levels of about 0.4% are formed on the unmasked portion of the preform ranging from about 0.030 inch up to about

0.080 inch. The specific depths will vary depending upon the specific type of metal alloy employed, including its initial carbon content, the geometry of the preform, the final case hardness and core hardness desired, and the particular type and surface conditions to which the final part is to be subjected during service, and other considerations well known in the art.

Upon removal of the sintered preform from the carburizing furnace which may be of a batch-type or, preferably, of the continuous type, the mask is removed from the sintered, and now selectively carburized, preform and coated with a lubricant and the part thereafter is subjected to forging. The forging temperature can vary from as low as about 1600°F up to about 1850°F, beyond which high oxidation occurs and die life is detrimentally affected. These problems can not be tolerated in production of forged parts of the type described herein, namely precision forged parts being of final dimension following the forging step and wherein no machining other than minimal finish grinding is required.

The foregoing preferably is carried out in a manner such that the preform is within a protective atmosphere to avoid oxidation of the surfaces thereof and is further preferably performed in a single step employing forging pressures of about 60 to about 80 tons per square inch. Following the forging operation, the forged part is permitted to stabilize in temperature whereafter it is quenched employing a liquid quenchant such as oil or a water-glycol mixture. The resultant forged part is characterized as having a density of at least about 99.6% of theoretical to a density approaching substantially that of the density of a wrought part.

IRON OXIDE REDUCTION PROCESSES

Spray Calcination

K.Grebe, H. de Haas, N. Dautzenberg and J. Hewing; U.S. Patent 3,975,186; August 17, 1976; assigned to Mannesmann AG, Germany provide a method for making iron powder having a low bulk density, e.g., not more than about 1 g/cm^3 and being at least substantially free from internal pores.

Since the powder particles should have fiber-like, i.e., long and thin configuration, because such particles have a kind of bulky configuration, then upon random positioning of each particle in relation to those of its environment a rather loose packing is inevitable. If the particles are, in fact, free from pores, additional material may be included in the powder without danger of retaining voids.

Fine grain iron oxide can be obtained from waste and otherwise exhausted or used steel or iron pickling solution which has been subjected to spray-roasting or spray calcination. The iron oxide is made available either as bulk-powder or at grainy consistency and in loose fill or as pellets. This material is reduced at a temperature from 800° to 1000°C in an atmosphere that consists of or includes CO, but is at least substantially free from SO_2. The reduction product is then cooled in a still reducing or inert atmosphere until e.g., reaching room temperature, and the cooled product is ground, i.e, the cake is at least broken up.

Iron oxides made by spray calcination yield an iron powder, which, unlike powder reduced from iron oxides obtained otherwise, does not have porous particles, but nonporous elongated particles of fiber-like configuration. The SO_2 content of the reduction gas should not exceed 1% in relation to the CO content, as a higher SO_2 content interferes with the formation of the fibers.

Generally speaking, oxidizing calcination enhances subsequently the fibrous consistency of the iron powder particles which, of course, means that the bulk density is lower. The presence of H_2 in and as a reducing agent generally has a counteracting effect here, which is the reason for limiting its content in the reducing atmosphere. If the tendency to form fibers is reinforced by the prior heat treatment, it was found that precalcination for 8 to 20 minutes at about 1100°C readily suffices.

The tendency to form fibers can also be enhanced by seeding the iron oxides with an additive. A seeding material which can be used with advantage is, for example, CaO at a ratio from 0.2 to 2%. A higher content of such additives will not yield any significant improvement in the iron powder.

In practicing the process, the CO-atmosphere may be produced in situ by adding, e.g., a carbon carrier such as charcoal to the iron oxide to obtain the CO in a manner known per se. The reduction generally is carried out by passing the reducing gas over the iron oxide.

Hydrogen has a limited use as reducing agent here. The method can be improved if the reduction is interrupted after about 80% of the oxide-oxygen has been extracted. The remaining oxygen is reduced further continuing the reducing process in hydrogen. One should proceed in this manner if the danger exists that towards the end of the reduction, carbon that is released, e.g., from the CO atmosphere contaminates the iron powder. Completing the reduction in hydrogen obviates that danger and, as stated, has little effect on the fiber structure of the particles, if used to a minor extent while, on the other hand, more extensive use of hydrogen impedes to some extent fiber formation which in turn is a parameter for control of the resulting bulk density.

The oxide used as raw material may contain chlorine and sulfates which could corrode the reduction vessel. Therefore, these contaminants should be removed by scrubbing the oxide in water and filtering prior to the reducing.

The bulk density of the iron powder as product may actually be too low for some uses. Therefore, one can grind the powder subsequently which is a convenient way of controlling the final bulk density. Thus, one should distinguish between grinding merely to break up the cake from the reducing chamber and further grinding to reduce particle size, e.g., fiber length.

Example: Pellets at a size somewhat in excess of 3.15 mm are made from iron oxide made by spray calcinating a pickling solution for steel. As per an analysis, the pellets consist of:

SiO_2	0.05%
CaO	0.2%
MgO	0.2%
Al_2O_3	0.2%
TiO_2	0.01%
Cr_2O_3	0.1%
Mn	0.27%
P	0.04%
S	0.06%
Na_2O	0.09%
K_2O	0.01%
Ni	0.02%
FeO	0.41%
Fe_2O_3	95.8%

whereby the iron content in the two oxides amounted to 67.4% by weight.

1,000 grams of such raw material were reduced in a rotating or revolving tube-like reduction vessel or chamber in the following manner. The oxide pellets were heated in N_2 up to 900°C in 45 minutes, whereupon pure CO (no SO_2) was introduced for 180 minutes and at 900°C. The amount of gas throughput was 1,000 Nl per hour (1 Nl = 22.414 liters). Thereafter, the reduction product was cooled in N_2 and 685 grams reduced iron cake could be removed from the chamber. The cake was then ground (broken up) in a toothed disk mill of the type traded under the name Condux. More than 95% of the powder particles that resulted therefrom had a diameter smaller than 200 μ, and the powder had a bulk density of 0.53 g/cm^3. The powder consistency was 98.6% Fe (total) of which 98.4% was metallic iron and the remainder was contained in the 0.13% FeO and 0.13% Fe_2O_3. Further analysis yielded 0.04% P, 0.01% S and 0.35% C (all percentages by weight).

Perforated Belt Support

In the production of iron powder from iron oxide, normally it is difficult and costly to obtain complete reduction by conventional techniques.

W. Whigham; U.S. Patent 3,957,482; May 18, 1976 has found in carrying out the reduction of powdered metal oxides that if the starting material to be reduced is formed into a cohesive layer without introducing harmful contaminants or impurities, which when applied onto a belt perforated with small closely-spaced apertures does not flow through the aperatures, then the disadvantages of the prior art may be substantially reduced in that the heavy inert gases produced during reduction escape readily through the perforations in the belt, thus facilitating the continuous penetration of the oxide layer by hydrogen or reducing gas.

In one embodiment, the iron oxide starting material to be reduced is formed into a cohesive layer or cake, which undergoes the formation of additional internal void space under reducing conditions, to permit an increased rate of penetration of the reducing gas therethrough. In this embodiment, the iron oxide starting material is preferably formed into a cake or layer; in such a case, the cake or layer of the starting material initially formed may not have sufficient strength or cohesiveness to be completely self-sustaining or -supporting, but once the cake or layer is introduced into a reduction furnace, the effects of heat render the cake or layer cohesive and self-supporting to withstand the reducing conditions and to permit the surfaces of the cake or layer to be subjected to the action of the reducing atmosphere without reverting into small particles which may fall through the belt perforations and be lost to the process. In such an embodiment where the initial formation of the cake or layer does not have sufficient strength to be self-supporting, the cake may be fed to the reducing furnace while being supported, following which the non-self-supporting cake may be rendered self-supporting due to the conditions existing in the reduction furnace to provide a self-supporting cake or layer with an increased void space which may then be reduced in the above-described manner.

Also, the iron oxide starting material may be reduced in two stages; the first stage involving the reduction of finely ground iron oxide powder to obtain only a partially reduced product, following which the reduction is completed by forming the partially reduced powder into a self-supporting form, and subjecting that self-supporting material to reducing conditions to complete the reduction.

A practical form of the process, involves reducing the starting material in a belt furnace in which is provided a perforated belt capable of transporting the iron oxide material through the furnace. To this end, various types of perforated belts may be employed, for example, perforated or apertured sheet material is preferably used to form the belt; alternately, woven or mesh type belts may also be employed. In these instances, and with any other type of belt employed, it is essential that its porosity be very high to permit the gravity flow of the spent gases, which are denser than the reducing gas, through the oxide layer and through the belt so that the oxide particles are constantly exposed to the reducing action of fresh dry reducing gas.

By way of typical examples, perforated belts of suitable material may have apertures ranging up to ½ inch or more, but preferaby about ¼ inch, with adjacent apertures being spaced approximately ½ inch (this may be more or less, as desired) from each other. In the case of mesh, since the material being reduced is self-supporting, there is no criticality in the configuration of the wires forming the mesh with the exception that the mesh must of course be strong enough to transport the material through the furnace. In general, the type of belt employed should have porosity such that it permits the reducing atmosphere to come into contact with the maximum surface area of the material being reduced and also permits the ready escape of the inert gases formed during reduction.

Example: A belt furnace which operated with an unperforated sheet metal belt, carrying finely ground iron oxide into which no additives had been introduced had a productive capacity of about five tons per day of iron powder product with a typical residual oxygen content after reduction of about 0.5% by weight.

Upon the installation of a belt perforated according to one embodiment of this process, and with the oxide feed to the process converted to a cohesive form by the addition of 1% starch, calculated on the dry weight of the oxide, in the form of an aqueous solution or dispersion, it was found that a production rate in excess of 15 tons of iron powder product per 24-hour day, with a residual oxygen content of about 0.3% by weight could be maintained on a continuous basis. In other words, by following this process the capacity of the process was increased by about 200%, and the quality of the product, with respect to completeness of reduction, also improved significantly.

Shaft Furnace

C.M. Jaco, Jr.; U.S. Patent 4,054,443; October 18, 1977; assigned to Midrex Corporation produces finely divided highly metallized iron powder by conducting a shaft furnace direct reduction process at certain selected feed choices, temperature ranges, and throughput speeds.

This process includes the steps of selecting ores and other components of the feed material from which iron powder is to be prepared, reducing the particle size of the feed material, chemically reducing the feed material to iron having a metallization in excess of about 94% in a direct reduction furnace, controlling the reducing gas temperature between 1550° and 2050°F in the furnace, controlling the pressure between 3 and 15 pounds per square inch in the furnace, and removing the reduced iron powder from the furnace.

Feed materials are selected from iron ores, both particulate and lump, iron oxide as pellets or other agglomerated form or even as fines, partially reduced iron from the reduction of iron ores or iron oxides in both pellet, lump and finely divided forms and mixtures of any or all of the foregoing. Feed material includes lump iron ore or agglomerated iron oxide pellets which, for predictable geological and physical reasons, are relatively easy to crumble. Suitable ores include hematite, magnetite and limonite as well as those having geological origins which result in a laminar or sedimentary structure. Hard, dense, massive lump ores having low porosity and characterized by a blue, shiny, specular appearance tend to fragment on heating.

The fragmentation can be violent, occurs in the 700° to 900°F temperature range, and is accompanied by a loud, audible cracking or popping sound. It is relatively independent of heating rate; the total amount of fines produced when a sample is heated in 10 minutes is about the same as the quantity produced when the sample is heated to the same temperature over a six or eight hour period. Cooling and reheating, even at extremely high rates, does not cause further thermal fragmentation of such materials.

Samples taken from deeper parts of the ore body fragment more severely on heating than those mined nearer the surface. Stresses and strains induced into the material as it was being formed might be a strong contributing factor.

The softer, red appearing ores generally tend to reduce more easily and be less subject to thermal fragmentation. Those which have a pronounced laminar structure may tend to fracture at the lamination on heating, but when this does occur such a lump tends to break into smaller pieces. Other suitable ores include those whose formulations incorporate bound water which can give them a layered, foamy or interstitial structure; and which release the bound water under the influence of heat or pressure causing the ores to decrepitate. Materials that retain organic matter during formation are also suitable.

Fine materials produced in a direct reduction furnace according to this process have a metallization always in excess of 80%, but preferably at least 94% and normally on the order of 96%. In order to achieve a purity sufficient for blending with powders made by other more expensive techniques the fines product may be recycled through the reduction furnace to upgrade its average metallization to as high as 98%.

PREPARATION AND PURIFICATION PROCESSES

Silanes as Passivating Agents

J.E. Ehrreich and A.R. Reti; U.S. Patent 3,846,118; November 5, 1974; assigned to Graham Magnetics, Inc. have developed a process for making small, uniform, metallic particles of controlled pyrophoricity, in substantially discrete form as opposed to highly agglomerated form, without the use of highly expensive apparatus or the use of toxic gases.

A soluble metallic salt is reacted with a strong reducing agent in a solvent medium which, in addition to the reactants, comprises a silane solute which has a strong chemical affinity for the newly-formed metallic surface.

The reduction is carried out in a liquid phase in which both the metallic salt and passivating agent have substantial solubility. The products are free-flowing powders having a bulk density of about a magnitude greater than materials prepared by analogous processes but without silane.

The reducing agent which is used can be any strong reducing agent permitting the metal salt to be reduced to the metal itself. Metal borohydrides are particularly useful in aqueous media at moderate temperatures. Among such metal borohydrides are the alkali metal borohydrides such as sodium and potassium borohydride, the alkaline earth borohydrides such as calcium borohydride, magnesium borohydride, and the like.

The precise particle size of the metallic particles being formed will, in part, depend on the concentration, temperature, and degree of agitation of the fluid mass in which the reduction is accomplished. However, the particle size is also influenced by the choice of a proper passivating agent which must be characterized by a moderate activity in solution whereby it does not react so quickly with newly-formed metal that one obtains little more than a organometallic compound, or metallic oxide dust, yet does combine quickly enough with active-sites on a newly-forming metal surface to prevent the growth of metal particles of excessive size.

The particles formed by this process become more magnetic rather than less magnetic when subjected to elevated temperatures, e.g., in the 300°-650°F range in a nonoxidizing atmosphere.

Silanes, especially those having molecular weights of from about 100 to 500, are particularly desirable passivating agents when one wishes to produce, at moderate temperatures, dispersions of metallic particles in the submicron range, for example, those having maximum average diameters below 0.5 microns.

Example 1: A 0.5 molar, aqueous solution of cobalt chloride was prepared. A quantity of 25 drops of a vinyl silane, vinyl triacetoxysilane having a molecular weight of 232 (Z6075, Dow Corning Corporation) was added to 100 milliliters of this aqueous solution using a conventional eye dropper.

In a separate vessel 2.0 grams of sodium borohydride was dissolved in 50 milliliters of water. Thereupon, using a magnetically actuated mixing bar to maintain adequate agitation, the borohydride solution was added dropwise to the cobalt chloride solution. Care was taken to add the borohydride slowly enough to allow the substantially complete reaction of the borohydride in each drop before adding a subsequent drop.

The reaction resulted in a very fine dispersion of cobalt which was relatively free from agglomeration of individual metal particles. When the same reaction was run without the use of the silane a highly conglomerated cobalt dispersion was obtained which settled rapidly rather than remained in dispersion.

Example 2: A quantity of 9.9 grams of $FeCl_2 \cdot 4H_2O$ was dissolved in 100 milliliters of water. Two grams of sodium borohydride and five drops of amino silane were dissolved in 50 milliliters of water in a separate vessel. The iron-bearing solution was agitated with a magnetic stirrer and the borohydride solution was slowly added thereto. A fine dispersion of iron powder was obtained.

After being filtered and washed four times with distilled water, the iron powder was dispersed in acetone to aid the drying thereof. The powder was then separated from acetone by use of magnet and air dried at 30°F. When the foregoing procedure was followed without use of passivating agent, an iron powder was obtained which had agglomerated into a fluffy material of relatively low apparent density.

The bulk density of the material prepared with the passivating agent was compared to the bulk density of the material prepared without the passivating agent: with, 0.53 g/cc and without 0.06 g/cc.

Direct Reduction of Ore in Arc Heater

M.G. Fey and E.A. Dancy; U.S. Patents 3,992,193; November 16, 1976 and 3,989,511; November 2, 1976; both assigned to Westinghouse Electric Corporation have developed a process for the production of metal powder from an ore by direct reduction in an arc heater characterized by the steps of introducing a finely divided ore into an arc heated plasma gas to effect reduction of the ore to small liquid droplets of elemental metal in a reducing atmosphere and quenching the metal droplets to form solid metal powder. The metal has a melting point of not more than the temperature of the chemical reduction reaction, such metal being iron, chromium, vanadium, manganese, cobalt or nickel, in the first process, and molybdenum, tungsten; tantalum, or niobium in the second.

The process may be carried out in a structure characterized by that shown in Figure 1.2a in which a reactor or collecting vessel **5** is provided at the lower end. The structure also comprises arc heater means including at least one and preferably three arc heaters **11, 13, 15** (Figure 1.2b). The reactor or vessel **5** is composed of a suitable material, such as metal, the upper end of which includes a reduced cylindrical portion or chamber **17** with which the inlet **7** communicates. The exit ends of the arc heaters **11, 13, 15** (Figure 1.2b) likewise communicate with the plenum chamber **17** so that similar plasma jet streams **19** extend from each of the arc heaters into the plenum chamber. In addition, quenching means such as spray nozzles **21** are disposed around the vessel **5** and below the positions of the arc heaters **11, 13, 15,** whereby a fluid of quenching material, such as an inert gas, is injected into the vessel **5**.

The arc heaters **11, 13, 15** (Figure 1.2b) are each a single phase, self-stabilizing ac device capable of power levels up to about 3500 kilowatts for up to about 10,000 kilowatts for a three phase plant installation. It is preferred that three arc heaters be provided, one for each of the three phases of the ac power supply. Two arc heaters **11** and **15** are shown in Figure 1.2a.

During operation of the arc heaters **11, 13, 15,** a reducing gas, such as methane,

Figure 1.2: Apparatus for Direct Reduction of Ore

a.

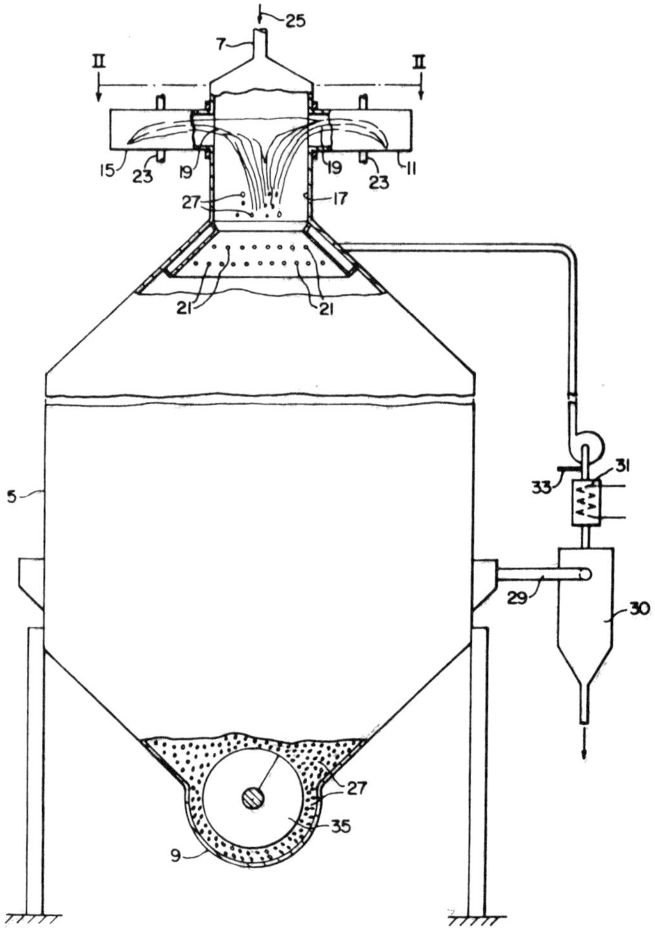

b.

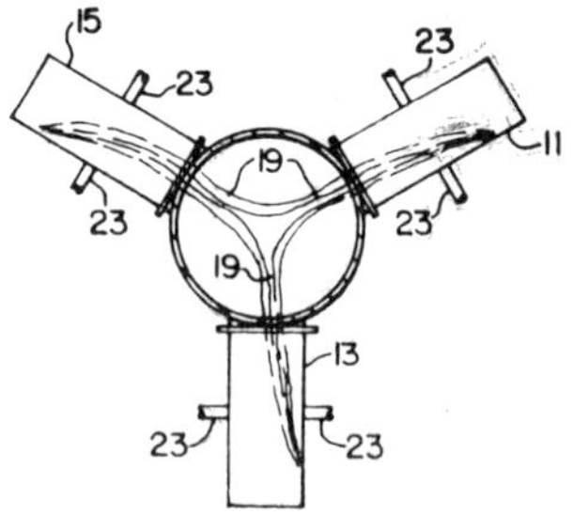

Source: U.S. Patent 3,992,193

is introduced into the arc heaters through peripherally disposed inlets **23** which gas comprises a greater portion of the plasma jet streams that enter the chamber **17**.

Finely divided ore **25**, such as iron oxide (Fe_2O_3 and Fe_3O_4), or molybdenum oxide ($Mo\ O_3$) as an example of other metals having melting points greater than the temperature of the chemical reduction reaction is introduced into the plenum chamber **17** via the inlet **7**. In the plenum chamber **17**, the ore **25** enters the plasma jet stream **19** where, in the presence of the reducing gas atmosphere (methane), the ore is reduced to elemental metal, for example, iron. Upon reduction of the ore to the elemental metal state, small droplets of the liquid metal form and drop from the plenum chamber to the lower end of the vessel **5**.

The droplets **27** of metal pass through a quenching zone which comprises jets of reducing gas, such as a mixture of CO and H_2, emitted into the vessel **5** through the spaced nozzles **21** which reducing gas is introduced through the nozzles at temperatures considerably below the melting point of the elemental metal. Additional quenching occurs by radiation of the hot molten particles **27** to the cold wall of the vessel **5**. For example, the melting point of iron is 2795°F so that the temperature of the coolant reducing gas is considerably below 2795°F in order to accomplish solidification of the droplets **27** before they drop to the bottom of the vessel **5**.

As an alternative the droplets **27** of metal may be quenched by providing for cooling of the gas as it passes through an expansion nozzle (not shown) at the chamber exit. As a further alternative, the droplets may be quenched by passing through a nozzle and then through the gas atmosphere as set forth above. The gas solids pass through the expansion nozzle. As the gas is cooled by expansion, the entrained particles transfer their latent heat of fusion to the gas and become solidified.

The reaction of the metal oxide or ore with a reductant, such as methane (CH_4), is shown in the following formula:

$$Fe_2O_3 + 3CH_4 \rightarrow 2Fe + 3CO + 6H_2$$

The carbon monoxide and hydrogen gas mixture resulting from that reaction is removed from the vessel **5** through an outlet conduit **29** which conduit passes through a cyclone particle separator **30** and then a coil **31** of the heat exchanger for withdrawing the heat from the gases, a portion of which gases are then returned to the nozzles **21** and the balance is removed from the system through a vent **33**. As the solidified metal particles collect at the bottom of the vessel **5**, they may be withdrawn through the outlet **9** in a suitable manner such as by a screw conveyor **35**.

Agglomeration of Carbonyls in Magnetic Field

W. Dawihl and W. Eicke; U.S. Patent 3,955,962, May 11, 1976; assigned to

Klockner-Werke AG, Germany have developed an apparatus wherein a magnetic field is generated whose flux lines extend normal to a planar supporting surface which is located in an enclosed annular space. Carbonyls of metallic material, e.g., iron, are fed into the enclosed space so that they travel lengthwise of the flux lines and normal to the plane of the supporting surface. The space is then heated to a decomposition temperature at which the carbonyls undergo thermal decomposition and liberate metal atoms which agglomerate under the influence of the flux lines on the supporting surface and form on the latter metal fibers which project from the supporting surface in a direction normal thereto, i.e., opposite to the direction of travel of the carbonyls themselves.

The supporting surface, or rather the element on which it is provided or of which it forms a part, can be shifted axially of the enclosed space, which is of annular configuration, and the shifting can be regulated in accordance with the speed at which the metal fibers grow on the supporting surface. The fibers begin to grow at the plane of the supporting surface, and grow outwardly away from the supporting surface in direction oppositely to the travel of the incoming carbonyl vapors. Heating gas is supplied into the annular reactor chamber from the center of the same, and a temperature gradient is established from the center towards the outer wall bounding the chamber.

The inner wall bounding the chamber is heated to a temperature substantially higher than the decomposition temperature of the carbonyls, e.g., to approximately 500°C if iron carbonyls are to be decomposed. The temperature of the outer wall bounding the reaction chamber, on the other hand, is substantially below the decomposition temperature of the carbonyls, e.g., approximately 120°C when iron carbonyl is to be decomposed.

The carbonyls themselves are vaporized outside the reaction chamber and the vapor is then admitted into the reaction chamber from the exterior. It is possible to provide two or more separate vaporizing devices located exteriorly of the reaction chamber, and this is particularly advantageous if two or more different types of carbonyls are to be vaporized, for example to be vaporized simultaneously. The vapors of these different carbonyls can then be introduced together or simultaneously into the reaction chamber and this makes it possible to produce metal fibers which are composed of metal alloys.

A further possibility is to first produce metal fibers in the reaction chamber and to thereupon introduce carbonyls of a metal or metals different from the one of the metal fibers e.g., of nickel, molybdenum or chromium, which liberate metal atoms that then form a coating or plating on the already produced metal fibers. To obtain advantageous flow conditions and agitation of the vapor or vapors in the chamber, the gaseous vaporized carbonyl is preferably introduced into the annular reaction chamber in tangential direction.

The apparatus of the process comprises a reaction chamber, magnetic means for producing in the reaction chamber magnetic flux lines extending in a predetermined direction, feeding means for feeding into the reaction chamber vaporized metal carbonyls for travel in the chamber along the flux lines in the aforementioned direction, and heating means for heating the interior of the chamber to a temperature at which the carbonyls undergo thermal decomposition and liberate metal atoms.

According to the process intercepting means is provided in the chamber for intercepting the metal atoms; this intercepting means has a supporting surface extending normal to the direction of travel of the incoming carbonyl vapors, and on which the metal atoms agglomerate under the influence of the flux lines and form metal fibers which project from the supporting surface normal thereto and in direction opposite to the direction of flow of the incoming metal carbonyl vapor.

The intercepting means may be provided at or in the region of its surface of a nonferromagnetic material, for example a nonferromagnetic wire net. The planar supporting surface constitutes the stabilizing or crystallizing plane, at which the metal atoms crystallize and become stabilized to begin forming the metal fibers which it is desired to produce.

As mentioned before, the metal carbonyls are vaporized exteriorly of the reaction chamber in vaporizing devices. A single such device may be provided, or if it is desired to be able to vaporize a plurality of metal carbonyls, for example different metal carbonyls in each device and at the same time, a plurality of such devices may be provided; the devices may be arranged annularly about the reaction chamber and/or may be offset axially of the reaction chamber relative to one another.

Control of Bulk Density

G. Findeisen, N. Dautzenberg and H. Gesell; U.S. Patent 3,909,239; September 30, 1975; assigned to Mannesmann AG, Germany have found that the bulk density of ferrous powder made by atomizing molten steel can be controlled by adding oxygen to the atomizing process and by reducing subsequently any resulting oxides to obtain a spongy surface layer of the particles, having thickness in dependence upon the added oxygen and reducing bulk density accordingly.

A conventional atomizer can be used, wherein a stream of molten iron or steel pours from a bottom opening of a ladle and passes a nozzle or nozzles directing pressurized jets of water against the down pouring metal stream. The metal stream as such disintegrates and multiple droplets coagulate, falling to the bottom while solidifying, at least at their surface before piling on the bottom. The atomizing water is provided under pressure to a chamber with a nozzle or nozzles from which the water jets are ejected. Pure oxygen or oxygen enriched air should be fed to that chamber so that a water-oxygen mixture is ejected through the nozzle or nozzles. Additionally or alternatively, the pressurized water is mixed with hydrogen peroxide, which gives off the oxygen at the point of atomization.

As a consequence, a raw powder is produced with strong oxidation at the surface of the individual particles. The subsequent reduction converts, so to speak, this oxidation into a spongy coating on and as part of the surface of the particles. That coating is thicker, the thicker an oxidation layer has formed on the particles prior to reduction, and that thickness, in turn, is the greater the more oxygen was added and/or developed. The apparent density and bulk density of the final powder decreases with increasing dimension of that spongy coating so that accordingly the bulk density increases with decreasing oxidation.

It follows that the amount of oxygen added under controlled conditions to the water jet atomizing process controls the resulting bulk density, and variations here permit production of iron powder with bulk density as low as 2.5 g/cm^3 or even lower or as high as 3.2 g/cm^3. The oxygen added under controlled conditions has a higher pressure than the partial pressure of oxygen in water as occurring normally. The added oxygen may have a pressure of 2 to 6 atmospheres.

It is uneconomical to reduce the raw powder by means of adding, e.g., hydrogen. Rather, internal reaction will be used. For example, molten steel to be atomized is enriched with carbon to such an extent that the carbon and the added oxygen have stoichiometrical relation for formation of carbon dioxide during malleablizing in its own atmosphere.

Atomized Powder with Low Oxygen Content

In order to manufacture atomized metal powder with extremely low oxygen content *U.R. Larson; U.S. Patent 4,124,377; November 7, 1978; assigned to Rutger Larson Konsult AB, Sweden* has developed a process which comprises the steps of providing a substantially closed granulation chamber, providing a reducing liquid in the lower part of the chamber and a reducing gaseous atmosphere above the reducing liquid, melting the metal of which the powder is to be formed, producing a casting stream of the molten metal in the reducing gaseous atmosphere of the substantially closed vessel, subjecting the stream to pressurized atomizing agent to atomize the stream, cooling the metal particles at least partly in the reducing gaseous atmosphere and collecting the resulting powder in the reducing liquid.

In the process, a casting stream is subjected to a reducing atomizing medium, preferably gaseous or liquid hydrocarbon or a mixture thereof, for instance petroleum products such as liquified petroleum, oil, benzene or the like. In order to protect the powder against oxidation, the actual pulverization process is performed in a closed granulation chamber which is partially filled with liquid medium and is under pressure from gaseous reducing agent. This also avoids any risk of explosion. One advantage of this method of manufacture is also that by regulating the quantity of atomizing medium, such as oil, in relation to the quantity of metal, the carbon content in the finished powder can be regulated.

Figure 1.3 is a schematic view of one embodiment of apparatus.

Figure 1.3: Apparatus for Producing Atomized Powder with Low Oxygen Content

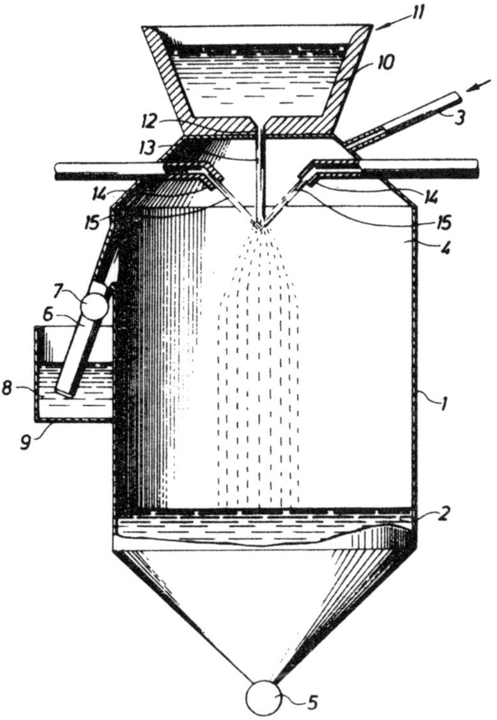

Source: U.S. Patent 4,124,377

In the drawing, a granulation chamber **1** is partly filled with a reducing liquid **2**, for instance oil, preferably fuel oil comprising 86.8% carbon, 12.5% hydrogen, 0.58% sand and the remainder including ash 0.12%. The chamber **1** is provided with a bottom teeming aperture **12** in communication with a casting vessel **11** containing a metal melt **10**. An inlet **3** is provided in the upper part of the chamber **1** for reducing gas and nozzles **14** protrude into the chamber for the supply of a reducing atomizing agent **15**. In the embodiment shown, a liquid lock, in the form of a channel **9**, is provided. The channel **9** cooperates with tube **6** communicating via valve **7** with the chamber **1**, the open end of the tube being the liquid level of a liquid **8** in the channel **9**.

Before pulverization commences, the valve **7** and bottom valve **5** are closed, after which the granulation chamber **1** is filled with a reducing liquid up to the draw hole **12**. When the granulation chamber is completely filled, a reducing gas is supplied through the tube **3** at the same time as the liquid level is lowered to the level desired for the pulverization process. The valve **7** is then opened, whereupon the reducing gas **4** in the upper part of the granulation chamber **1** will maintain a super-atmospheric pressure corresponding to the length of the pipe **6** which is immersed in the liquid **8** in the liquid lock channel **9**. The actual pulverization process may now be performed. Molten metal **10** from the casting vessel **11** runs down through the draw hole **12** in the form of a metal stream **13** which is hit by the reducing atomizing agent **15** expelled from the nozzles **14**.

Catalytic Decomposition of Iron Pentacarbonyl

W.S. McEwan and H.B. Jonassen; U.S. Patent 4,056,386; November 1, 1977; assigned to The U.S. Secretary of the Navy have found that ammonia catalyzes the decomposition of iron pentacarbonyl. Additionally, it has been found that several other compounds are useful in promoting the decomposition of iron pentacarbonyl, e.g., H_2, NO, PF_3, PH_3 and halogens such as I_2. According to this process, the rate of decomposition of gaseous iron pentacarbonyl into iron and carbon monoxide at any given temperature (such as 280° to 350°C) is increased about five to seven fold by utilizing one of the abovenamed compounds as a catalyst.

Caustic Leach for Silica Removal

C.D. Carey, M.C. Kuhn and M.K. Witte; U.S. Patent 4,154,608; May 15, 1979; assigned to UOP Inc. have developed a method for obtaining high purity iron powder by treating an iron bearing source which contains impurities such as silica along with minor amounts of aluminum, calcium and magnesium.

The iron bearing source such as hematite or magnetite is subjected to a grinding process whereby the iron ore is reduced to a desired particle size. The particle size will be such that a major portion of the silica is liberated from the iron and therefore the desired particle size should be about −200 mesh, this size being sufficient to effectively separate the silica particles from the ore. Thereafter the ground ore is subjected to a caustic leach at an elevated temperature and pressure. The leach solution which is utilized to effect a separation of silica comprises an aqueous sodium hydroxide solution which contains from about 100 to 200 g/l of sodium hydroxide, although a lesser amount or greater amount of sodium hydroxide may be employed if so desired. The leach is effected at temperatures which may range from about 300° to 400°F at an applied pressure in the range of from about 200 to 400 psi.

After treating the ore with the caustic solution for a period of time which may range from about 0.5 up to about 4 hours or more in duration, the silica particles in the leach head material will form sodium silicate of varying composition which will be soluble in nature. This caustic treatment of the ore results in the reduction of the silica content which was present in the original ore down to less than 1%. Following the treatment of the ore within the reaction parameters set forth above, the solid ore which may also contain insoluble silicates of aluminum, calcium, and magnesium, is separated from the leach liquor by conventional means such as filtration, centrifugation, decantation, etc. The residue may then be subjected to a reducing process in which the iron is treated with hydrogen at an elevated temperature and pressure for a predetermined period of time in either a one or two step reduction process.

In one embodiment the iron ore is reduced by treatment with hydrogen at a pressure in the range of atmospheric to about 500 psi in a two step process, the first step being effected at a temperature in the range of from about 700° to about 1300°F. The oxygen level in the iron will be reduced to less than 5%; however, the iron oxide particles at this point in the process are highly pyrophoric and therefore must be protected by an inert atmosphere which is provided for by the presence of nitrogen before being subjected to the second stage of the reduction step.

The second stage reduction is effected at temperatures ranging from about 1600° to about 1900°F and serves to completely reduce the iron oxide to iron powder. In addition to reducing the iron oxide to metallic iron, the second stage of the reduction process also serves to sinter the surface of the iron particles in such a manner so that the pyrophoricity of the particles is eliminated and therefore the iron particles may be exposed to the atmosphere with the concurrent elimination of the necessity of an inert gas such as nitrogen being present.

Example: An iron ore concentrate which contains 6.3% silica was reduced by treatment with hydrogen at a temperature of about 850°F and atmospheric pressure until the hematite was reduced to magnetite.

The magnetite product was separated from the silica gangue material by three passes through a permanent magnetic separator, the original 6.3% silica in the starting material being reduced by means of the magnetic separation to about 0.75% silica.

The magnetite product which was recovered from the magnetic separator was ground in a ball mill to a particle size which ranged from −100 mesh to +270 mesh. Following this the ground magnetite was leached in a pressure autoclave with a caustic solution which contained about 150 g/l of sodium hydroxide, the leach operation being effected at a temperature of 400°F and a pressure of 275 psi, the silica particles in the leach head material reacting with the caustic to form soluble sodium silicate. The caustic leach slurry was filtered using a plate and frame filter and the filtration residue which contained from about 0.2 to about 0.3% impurities including aluminum silicate, calcium silicate, magnesium silicate and silica was washed with hot water, which had a temperature of about 150°F, to remove any traces of caustic leach liquor which may still have been present on the solids.

The solids from the caustic leach were then treated with a weak hydrochloric acid solution containing about 5 weight percent, the washing being effected at room temperature for a period of about 30 minutes. The insoluble silicates which had precipitated from the lixiviant during the leaching and filtration steps were dissolved in the hydrochloric acid along with about 2% of the iron. At the end of the 30 minute period the hydrochloric acid leach slurry was filtered and the residue again washed with hot water which had a temperature in the range of about 150° to about 200°F. At this point, the silica content of the magnetite comprised 0.08%.

The washed magnetite was then treated with pure hydrogen at a temperature in the range of from about 700° to about 1300°F to lower the oxygen level in the ore to less than 5%. Upon completion of the reduction, the temperature was increased to a range of from about 1600° to about 1900°F whereby the iron oxide was completely reduced to an iron powder.

After cooling, the iron powder which resulted from the reduction was in an agglomerated form and was ground to the desired particle size in a ball mill, the resultant particles being in a range of from about −40 mesh to about +270 mesh. After reaching the desired particle size, the iron powder was subjected to three passes through a permanent magnetic separator whereby any foreign particles which may still have been present in the iron powder and freshly liberated impurities were separated. The iron powder which was recovered had a silica content of from about 0.03 to about 0.06% and was suitable for use in powder metallurgical applications.

Continuous Passivation of Sponge Iron Particles

D. Beggs; U.S. Patent 3,844,764; October 29, 1974; assigned to Medrex Corporation describes a process for passivating active sponge iron particles by maintaining a continuously gravitationally descending column of sponge iron particles within a vertically oriented housing or vessel. A gaseous medium is continuously recirculated through the descending column in counterflow relationship to the descending particles. The gaseous medium is maintained at a predetermined temperature at its introduction to the column of descending particles. Additionally oxygen is added to gaseous medium at a predetermined rate and the flow of the gaseous medium in the column is likewise maintained at a predetermined rate.

Figure 1.4 is a diagrammatic representation of apparatus for carrying out the passivation process.

It shows an enclosed vessel or housing **10** for containing a gravitationally descending bed of sponge iron particles **11**. A feed hopper **12** for receiving active sponge iron particles connects to a feed pipe **13** for feeding particles into vessel **10**. Feed pipe **13** extends below the top of the vessel to create a gas exit chamber **14** between the top of the vessel and an angle of repose of the particles indicated at **15**. A belt feeder or other type discharge feeder **16** serves to withdraw passivated sponge iron particles from vessel **10** through a discharge

Figure 1.4: Apparatus for Continuous Passivation Process

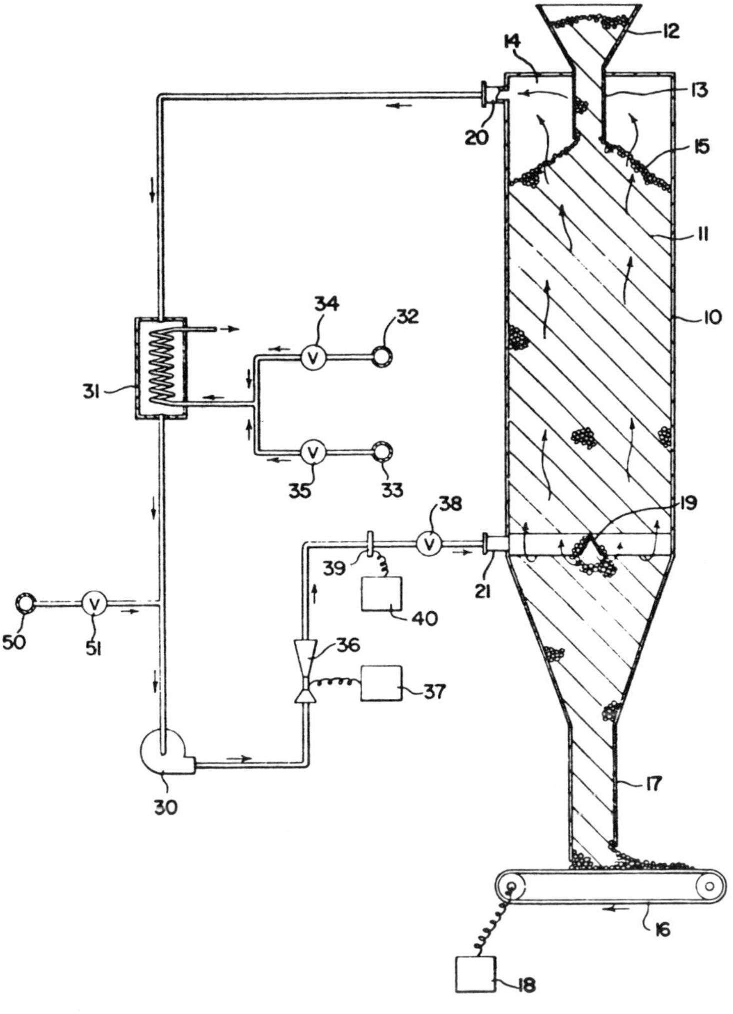

Source: U.S. Patent 3,844,764

pipe **17**. The feeder **16** connects to a particle discharge rate measuring device **18** such as would be conventional for a belt type weigh scale.

A gas introduction member **19**, shown as an inverted V cross, serves to provide means for introducing gas into the bed of particles in the lower region of vessel **10**. A gas offtake **20** connects to gas exit chamber **14** and a gas intake **21** connects to gas introduction member **19**. A recirculating fan **30** serves to recirculate gas from gas offtake **20** to gas intake **21** to cause generally upward gas flow in vessel **10** in countercurrent relationship to the gravitational descent of the sponge iron particles.

The gas recirculation system shown includes an indirect heat exchanger **31** for heating or cooling gas withdrawn from offtake **20**. This heat exchanger can, for example, be a conventional fin tube type supplied either with a heating medium such as steam from a source **32** or a cooling medium such as water from a source **33**, with associated control valves **34** and **35**. A recirculating gas flow metering device **36** such as a venturi meter is provided with a flow measuring device **37**.

A recirculating gas flow control valve **38** serves to control the gas flow in a manner hereinafter described. A gas temperature sensor **39** such as a thermocouple is connected to a temperature controller **40** which serves to control the action of valves **34** or **35** depending on whether the recirculating gas needs to be heated or cooled in heat exchanger **31**, in order to maintain a desired gas temperature at sensor **39**. An oxygen containing gas such as air is admitted to the recirculating gas stream from a source **50** through a control valve **51** in a controlled manner described later.

Sponge iron particles in the form of reduced pellets or lump made from oxide pellets or lump ore, as produced and when initially exposed to air will oxidize at different rates depending upon the physical and chemical properties of the oxide material, such as its porosity and gangue content. In many laboratory and field tests it has been discovered that active sponge iron particles can be effectively passivated with minimum oxidation by exposure to a gas containing oxygen for a one to two day period under controlled oxygen and temperature conditions. During such exposure, a very thin skin of magnetite is formed on the sponge iron surface throughout the interior of the particles which inhibits further oxidation. During the exposure, the particles progressively passivate and consume oxygen at a declining rate which approaches zero after one to two days.

It has been found that effective passivation can be accomplished by maintaining the particles at a temperature in the range of 125° to 175°F during the exposure period. At 100°F, exposure to oxygen does not effectively passivate in a two day period. At 200°F and higher, the passivation is effective but there is a tendency to oxidize more of the iron than is desired in order to achieve effective passivation.

During exposure in the above preferred temperature range of 125° to 175°F, the oxygen consumption necessary to achieve effective passivation is in the range of 0.004 and 0.01 cubic feet of oxygen per pound of sponge iron particles. This is equivalent to oxidation of from about 0.1 to 0.25% of the iron in the particles.

During the exposure period it has been found that oxygen concentrations in the range of 500 ppm to 20,000 ppm are effective in achieving passivation.

However, the oxidation reaction is exothermic and oxygen concentrations in the range of 500 to 2,000 ppm are preferred in order to limit the initial reaction rate and resultant temperature rise.

In order to establish and maintain the temperature of the particles in the preferred range without dependence upon incoming particle temperature, it is necessary that not only the temperature of the incoming gas be controlled within the range but also the thermal capacity of the gas must be at least equal to the thermal capacity of the particles. With an equal thermal capacity, a 1°F change in gas temperature is the same amount of heat as 1°F change in particle temperature.

The specific heat of sponge iron is approximately 0.12 Btu/lb/°F and the specific heat of all gases is approximately 0.0185 Btu/ft^3/°F, in a temperature range of 125° to 175°F. Thus, approximately 0.12 divided by 0.0185, or 6.5 ft^3 of gas will have the same thermal capacity as one pound of sponge iron.

Referring again to Figure 1.4, the preferred method of operation will now be described. The temperature of the recirculated gas being introduced to the bed of particles through member **19** is controlled by the action of temperature control **40** acting upon heating control valve **34** or cooling control valve **35** in a conventional manner to maintain the temperature of introduction in the preferred range of 125° to 175°F.

The admission of oxygen to the recirculated gas from source **50** is controlled by the action of particle discharge rate measuring device **18** acting upon control valve **51** in a conventional manner to admit between 0.004 and 0.01 ft^3 of oxygen per pound of particles being discharged by discharge feeder **16**. Alternatively, control valve **51** can be acted upon by an oxygen analyzer-controller (not shown) to maintain the oxygen level in the range of 500 to 2,000 ppm in the recirculated gas. The fan **30**, in addition to providing the means for gas recirculation also serves to mix the freshly admitted oxygen with the recirculated gas.

The flow rate of the recirculating gas is controlled by the action of particle discharge measuring device **18** acting upon flow measuring device **37** which in turn acts upon flow control valve **38** in a conventional manner to maintain a flow of at least 6.5 ft^3 of recirculated gas per pound of particles being discharged by discharge feeder **16**.

In another system *W.B. Pietsch; U.S. Patent 4,033,559; July 5, 1977; assigned to Midrex Corporation* provides a hot sponge iron compacting apparatus followed by a shearing apparatus. The hot compacting apparatus rolls the sponge iron into an elongated mass. The shearing apparatus cuts the mass across its longitudinal dimension, compacting the newly exposed edges, and creating small "compacts" which are easily handled, transported, and used in subsequent processes.

Where the compacting apparatus produces a relatively wide strip, apparatus may be provided for slitting the elongated mass longitudinally to produce a plurality of elongated masses.

Figure 1.5a illustrates apparatus for continuously passivating hot reactive particulate metal material consisting essentially of three basic parts, a material accumulator **10**, a compactor **12** and a divider-densifier **14**.

Figure 1.5: Continuous Passivation Apparatus

a.

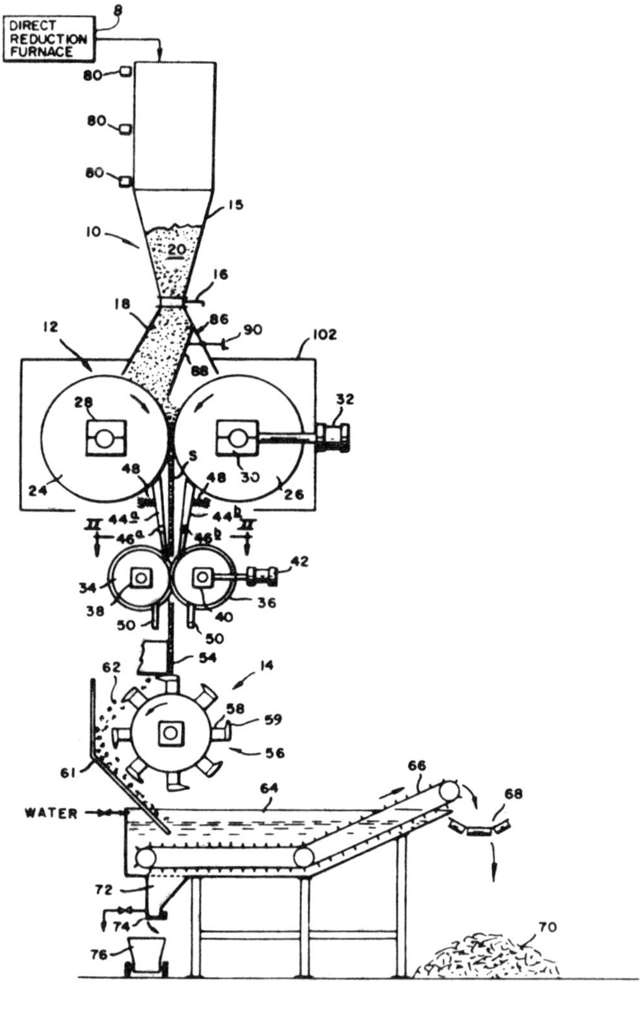

(continued)

Figure 1.5: (continued)

b.

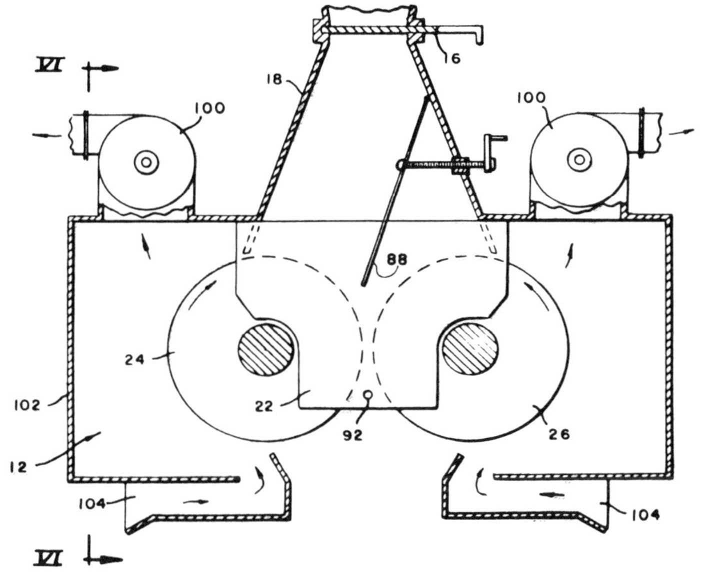

c.

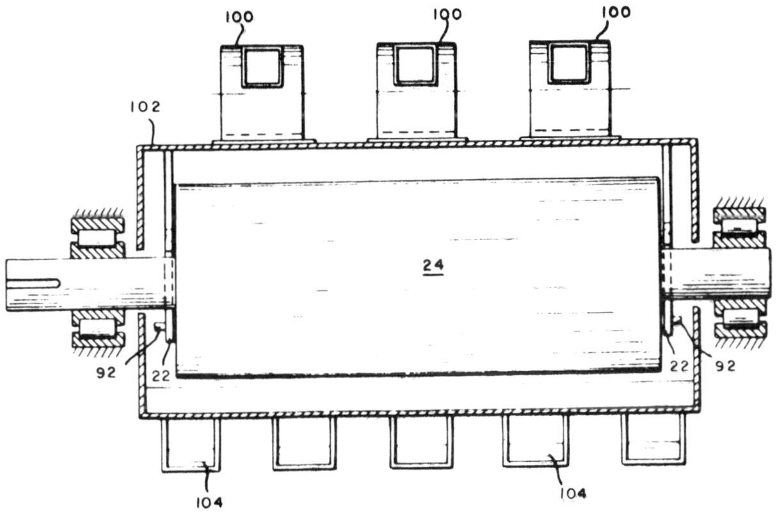

Figure 1.5a: Elevational view of the preferred embodiment of the process.

Figure 1.5b: End view of the compactor rolls and their associated parts.

Figure 1.5c: Sectional view taken along the line VI-VI of Figure 1.5b.

Source: U.S. Patent 4,033,559

Accumulator **10** may take the form of a surge bin **15** having a slide valve **16** at the bottom thereof positioned above a feed hopper **18**, which hopper is adapted both to contain hot feed material **20** and to control the rate of feed of the material to the compactor apparatus **12**. The feed material **20** may be sponge iron, pellets or lump ore, or a combination thereof, which material has been reduced in an associated direct reduction furnace **8**. Feed hopper **18** includes cheek plates **22**, shown in Figure 1.5b, which bear against the end of the bodies of power driven, large diameter compactor rolls **24** and **26** to seal the interior of the hopper **18** against atmospheric air. As shown, roll **24** is fixed in its chocks **28** whereas roll **26** is movably mounted in horizontally movable roll chocks **30**, the movement of which is controlled by hydraulic cylinders **32**. Rolls **24** and **26** may have a flat contour or may have a small collar at each end of the roll body.

The region from slide valve **16** to the nip of rolls **24** and **26** is designated below as the feeding and compaction zone.

Scrapers **44a** and **44b** are pivotally mounted beneath the compacting rolls at pivot points **46a** and **46b** respecitvely, each of which is well below the center of gravity of its respective scraper. Thus, the upper edge of each scraper tends to maintain contact with its respective compacting roll. A tension spring **48** biases each scraper against its compacting roll to ensure sure contact.

Beneath the compactor rolls and aligned with the pass-line thereof are a pair of horizontally opposed shearing or slitting rolls **34** and **36** for longitudinal slitting of wide strip. Roll **34** is mounted in fixed chocks **38** while roll **36** is mounted in horizontally movable chocks **40** the motion of which is controlled by hydraulic cylinders **42**.

A doctor device **50** which may also be known as a roll stripper or guide is located on the discharge side or bottom of each roll **34** and **36**. Such a guide may have the full contour of the roll, extending completely across the roll body. Alternatively, the device may consist of a number of guides each of which fits into a recess or onto a collar of the roll, respectively.

Aligned with the pass-lines of the compactor rolls and the slitter rolls is anvil **54**, which likewise may have a horizontal contour matching that of roll **34**. A rotating shear **56**, containing a number of blade holders **58** with their associated blades **59**, is mounted beneath the anvil **54** in such a manner that the shear blades transversely cut the workpieces by shearing them against the anvil. Blades **59** preferably have a flat shearing face **60**. A feed chute **61** may be provided beneath the rotary shear to collect the metallized iron compacts **62** and direct them into a cooling tank **64** which is filled with water.

An apron conveyor **66** removes the compacts from the cooling tank, transferring them to a belt conveyor **68** for transshipment to a stockpile **70**. The small amount of fine material that may be produced by the shearing and abrasion of the compacts falls through the perforations in the apron conveyor and collects in a funnel **72** which has a lock valve **74** at its lower extremity. The funnel can be emptied into a trough car **76** periodically to remove the fines from system.

In operation, hot sponge iron, pellets and/or lump material, including fines, is charged into surge bin **15**. Level control devices **80** (such as a C-E Invalco Nuclear Level Control, Combustion Engineering, Inc.) may be interlocked with the speed control on the roll drives for rolls **24** and **26** to ensure sufficient volume of material to maintain a constant width of the elongated strips.

The hot sponge iron has a temperature of at least 600°C preferably 700° to 800°C and an average metallization of at least about 75%, but normally at least 85%, and preferably at least 90% metallized. The hot sponge iron passes from the surge bin **15** through the region of the slide valve **16** into feed hopper **18**. The slide valve is in the open position during operation of the apparatus, and is closed only when the machinery is shut down.

The feed hopper, slide valve and surge bin must be gas tight to prevent ambient air from contacting the hot, extremely reactive material. The flow rate of hot sponge iron to the nip of the two compacting rolls **24** and **26** may be controlled by a movable feed tongue mechanism **86**. This controls the volume of sponge iron reaching the rolls, and thus the thickness and density of the strip produced by compaction. If too great a volume of material is fed into the rolls, they will open, producing a thicker strip having an unfavorable density distribution or gradient. The feed tongue **88** is pivotally attached to the feed hopper, and has an adjusting arm **90** which extends outside the hopper. The tongue may extend the full width of the hopper, or a number of narrower tongues may be employed.

Hot sponge iron is fed to the compacting rolls by gravity. An alternative feeder arrangement such as a screw feeder may be used which will exert a positive feed pressure on the material entering the roll nip. This will control both the rate and volume of flow of the feed material.

After entrainment of the hot sponge iron into the roll nip, the sponge iron is continuously densified by the counterrotating rolls **24** and **26** which exert large compressive forces on the sponge iron causing formation of an increasingly compacted iron mass until it reaches the narrowest gap at the horizontal centerline of the rolls. During this densification procedure, the individual pellets or lumps are deformed and the spongy structure of the iron is destroyed by the pressure of compaction exerted by the compacting rolls, and the fines are assimilated into the densified mass.

The gas which has been in the interstices between the hot pellets as well as in the pores of the pellets, if forceably expelled therefrom and escapes from the hopper **18** through gaps between the ends of the compacting rolls and the cheek plates **22** as well as through gaps between the roll body and the base of feed hopper **18**. This gas, which remained in and around the pellets on discharge from the direct reduction furnace **8** is reducing in character, and provides a steady stream of nonoxidizing gas to protect the feeding and compaction zone against contact by the surrounding atmospheric air.

Thus it is unnecessary to provide a sophisticated sealing system. At startup a nitrogen purge is used. Nitrogen or other nonoxidizing gas is introduced to the feed hopper through orifice **92** (Figure 1.5b and Figure 1.5c) in the cheek plates **22** near the nip of the rolls. After a few feet of compacted strip has been formed, the nitrogen purge is stopped, as the reducing gas forced out of the pores will displace the nitrogen and maintain a reducing atmosphere.

Compaction of the hot sponge iron forms a continuous strip or sheet **S** (Figure 1.5a) having such high density that the formerly very high affinity of the iron for oxygen is so far reduced that it is no longer subject to catastrophic reoxidation. In fact oxidation of the surface of this extremely dense material will now result in a very small loss of metallization.

The mean density of the strip depends on the thickness of the strip. A pronounced density gradient toward the less dense center of the strip reduces the mean density with increasing strip thickness. The mean density must be at least 4.5 g/cc. Below a density of 4.5 g/cc, passivation is insufficient for long-term open bulk storage without significant loss of metallization. Note that while the mean density of a thick strip may be only 4.5, the surface contacted by the surrounding atmosphere has a very high density with an attendant high degree of passivation.

During compaction, the compacting rolls **24** and **26** become heated due to conduction of heat from the hot feed material **20**, the temperature of which has been increased by the extremely great amount of energy input that has been transformed into heat and is absorbed by the feed material. Exhaust fans **100** (Figure 1.5b and Figure 1.5c) may be employed to remove excess heat from the rolls. For the exhaust fans to work efficiently, the compacting rolls **24** and **26** should be surrounded by an enclosure **102** best shown in Figure 1.5b. Enclosure **102** has at least one, but preferably a multiplicity of air intakes **104** along each side of its bottom face. The suction created by exhaust fans **100** will circulate ambient air through intakes **104** around compacting rolls **24** and **26** and out of enclosure **102** through the exhaust fans **100**. These fans may be associated with a dust collector, bag house or precipitator to remove particulate material from the exhaust air.

Scrapers **44** act as roll guides for the compacting rolls **24** and **26** to prevent the elongated strip from wrapping around one of the rolls. In addition the scrapers assist in guiding the strip into the slitting rolls **34** and **36**. The scrapers **44** are biased against the compacting rolls by springs **48** and guide the strip into the pass-line of the slitting rolls.

Batch Passivation of Bulk Metallized Particles

D. Beggs and W.A. Ahrendt; U.S. Patent 3,926,617; December 16, 1975; assigned to Midrex Corporation have developed a process for passivating bulk masses of metallized pellets having a volume for the most part of at least 0.02 cubic inch, so that oxidation thereafter may be substantially eliminated while the pellets are being shipped, stored, handled or conveyed.

The term "bulk mass" as used here is intended to define a sufficient quantity of pellets oriented in such a geometrical arrangement so that heat generated during passivation will be retained. The thermal conductivity of a bulk mass of metallized pellets was found to be approximately 4 Btu/ft^2/hr/°F/in. By comparison, the thermal conductivity of insulating block is 1 Btu/ft^2/hr/°F/in and for fire brick it is 9. Thus, the pellets act as self-insulators thereby tending to retain the heat of reaction. This process incorporates this observation of an element in a passivating process.

Freshly reduced metallized pellets may be treated in a bin, silo, stockpile or the like. Preferably, the pellets are treated in a container, such as a bin, where the flow of gas containing oxygen through the bin may be uniformly distributed and controlled. The freshly reduced pellets may be at a temperature such that no further treatment is necessary before their introduction to the bin; however, it may be necessary to either heat or cool the pellets at a temperature between 100° to 250°F before, or after, they are placed in the bin.

It has been found that the bin should be completely enclosed with a solid base in order to prevent external air from flowing upwardly in uncontrolled fashion and a roof to protect the pellets from precipitation. Controlled means for supplying a flow gas containing oxygen uniformly throughout the bulk mass of metallized pellets is provided. The gas containing oxygen is ventilated upwardly through the pellets periodically and the temperature of the bed is checked to insure that the pellets have not overheated. Between the periodic venting of gas, the pellets are allowed to remain in a quiescent state, i.e., no air drafting at which time the metallized pellets will react with the oxygen in the vented gas which eventually, after periodic quiescence and venting, results in passivation.

Fast cooling may be achieved in order to reduce the amount of time the bins are full of pellets. After the pellets have been passivated, a cooling gas may be continuously passed through the bed in order to induce cooling of the pellets.

The bin has suitable controls to regulate the gas containing oxygen introduced in the bin so that the entire bin receives uniform exposure to the gas and simulates the quiescent characteristics of the bulk mass. This bin is shown generally at **10** in Figure 1.6 and has a circumferential wall **12**, a floor **14** and a roof **16**. The bin **10** serves as a housing for a bulk mass of pellets **17** to be passivated and, more specifically, in the test employed, had sufficient volume to hold approximately 12,700 cubic feet of pellets which had a weight of approximately 625 tons.

The roof **16** is placed at an angle which approximates the angle of repose of

Figure 1.6: Apparatus for Batch Passivation

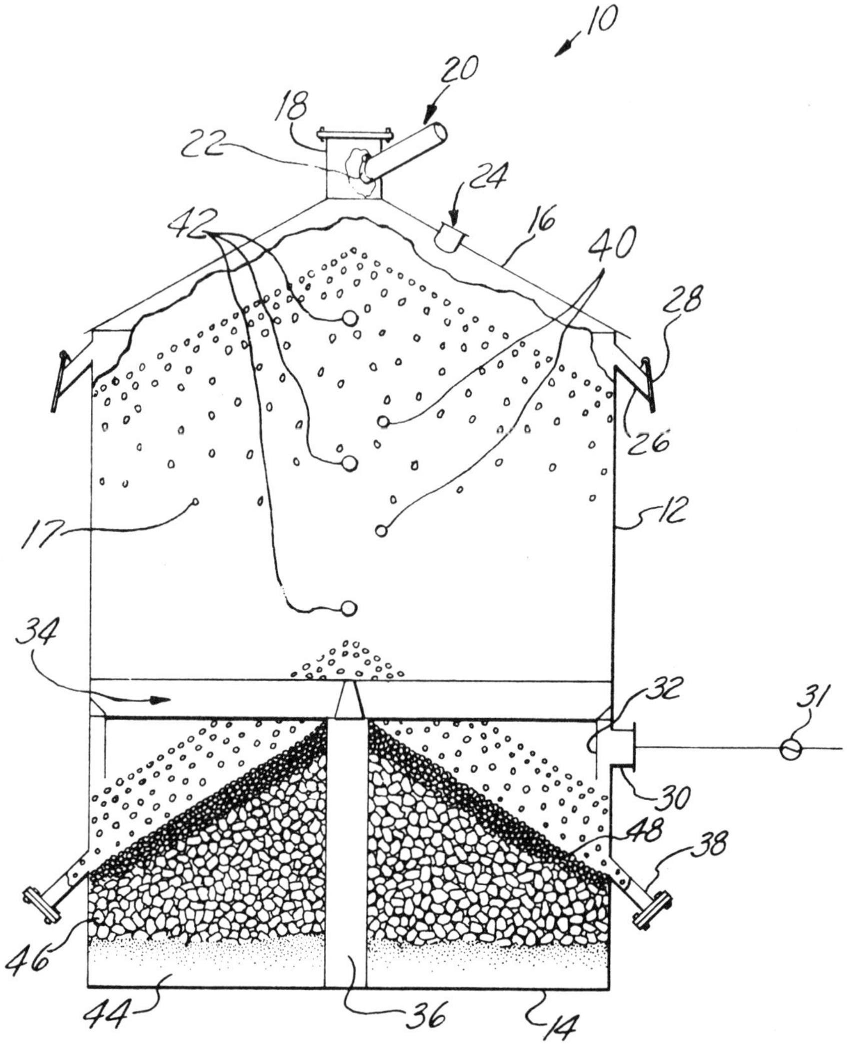

Source: U.S. Patent 3,926,617

the pellets **17** to be placed therein. Received within the roof **16** is an axially extending passageway **18** that is closed at its top and which receives a charge pipe **20** through which the metallized pellets may be introduced into the bin **10**. The interior end of the charge pipe **20** has a flap **22**, made of a material such as neoprene, so that the charge pipe may be closed to the atmosphere during periods other than loading. Located within the roof **16** is a level indicator that indicates the quantity of pellets within the bin. Immediately below the roof **16**, and received within the upper portion of the wall **12**, are a plurality of air vents **26**, each of which is supplied with a vent flap **28** which also allows air to flow outwardly from the bin **10**.

Intermediate the ends of the walls, an air line **30** is received which is connected to a fan **31** that is selectively operative to push air into the bin **10**. A bustle **32** is located immediately inside the wall **12** and is closed at the top but open at the bottom. In this way, the air line **30** will not become plugged with pellets and the air is able to flow into the pellet bed. Located immediately above the bustle **32** is a pair of horizontal, inverted V-shaped cross members **34** which are at a right angle to one another and each of which is in open communication at its ends with the bustle **32**. Air is received within the cross members **34** from the bustle **32** and is distributed equally throughout the area of the bin immediately below the cross members. The cross members **34** are provided with a support member **36** at their junction.

Located at the lower portion of the wall **12** are a plurality of discharge pipes **38** through which the pellets may be discharged after processing. Located at various points within the bin are gas sample outlets **40** through which gas may be removed from the bin to determine the level of oxygen present therein. Also received within the bin is a plurality of thermocouples **42**, only one being shown, which measure the temperature at various locations.

Located within the bottom of the bin **10** is a layer of sand **44**. This sand **44** assures the sealing of the bottom of the bin and eliminates the possibility of external air drafting in from below the pellet bed. Immediately above the sand is a layer of crushed stone **46**, the crushed stone having an upper surface level which approximates the angle of repose of the metallized pellets. Immediately above the crushed stone and below the discharge pipes **38** is a section wherein "dead" pellets **48** remain.

The bin **10** used for the final investigations was 26 feet in diameter, 24 feet high and was instrumented with thermocouples **42** which were located 2 feet from the center line, 3 feet from the wall, and 1 foot from the wall. Six couples **42** were located 4 feet apart vertically at each radius. Two gas sample **40** points were located just off the center line near the top and bottom of the bed.

Iron Alloys and Steels

IRON ALLOYS AND OTHER COMPOUNDS

Oxidation of Ferrochromium Powders

K. Yamagishi, T. Banba, T. Nakajima; U.S. Patent 3,821,032; June 28, 1974; assigned to Nippon Kokan KK, Japan have developed a method for oxidizing ferrochromium powders which comprises establishing and maintaining a fluidized bed of high-carbon ferrochromium powder having a sieve size of less than 300, blowing upwardly through a body of such powder an oxygen containing gas at a rate of 11 to 40 cm/sec, heating the bed to a temperature of from 800° to 1050°C so as to start the oxidation of the powder, terminating the heating step when the oxidation is started, and continuing to blow the oxygen containing gas at the same rate so the oxidation degree is restricted to lie within 1.10 to 1.20 O:C by mol.

This process can be distinguished from the conventionally used procedures for oxidation of high-carbon ferrochromium powders by the following characteristics:

1. The particle size of the starting ferrochromium powder falls within the ASTM sieve range number 100 to 300.
2. The heating supplied from an external source is carried out until the temperature within the range of 800° to 1050°C has been reached and thereafter the reaction is maintained by the heat of self-combustion of the ferrochromium, no heat from any external source being required.
3. The degree of oxidation is required to be kept within the range of 1.10 to 1.20 O:C by mol.

Example 1: The chemical composition of a finely divided high-carbon ferrochromium to be treated in accordance with the process and having a particle size of less than 0.149 mm is set out below.

Component	%
Cr	61.2
C	7.6
Si	0.8
P	0.029
S	0.030
Fe	30.341

The abovedescribed high-carbon ferrochromium is supplied at 120 kg/hr into the fluidized reactor described in detail in the patent. The ferrochromium used initially is partially oxidized up to 30%. There is supplied for fluidizing and oxidizing the ferrochromium 9.5 Nm^3/min of air which following startup is reduced to 2.5 Nm^3/min at 800°C. The reaction proceeds smoothly without any sintering taking place. The temperature existing in the reactor amounts to 870°±20°C in the fluidized zone and is controlled by introduction of waste gas or other suitable means. No heat from an external source is supplied. The amount of air supply for fluidization and oxidation amounts on the average to 2.0 to 3.0 Nm^3/min after startup has been completed.

The amount of oxidized ferrochromium discharged from the outlet and collected in a bag filter is 105 kg/hr and 25 kg/hr, respectively. The ferrochromium recovered has the composition set out below.

Component	%
Cr	53.4
C	6.0
Si	0.7
P	0.027
S	0.023
O	9.7
N	Trace
Fe	30.15

The above values have calculated mol ratios of O:C 1.21. In this connection, the chemical composition of the ferrochromium powder collected at both points was substantially the same.

The diameter of the oxidized ferrochromium particles was less than 0.149 mm (100%) indicating that no fusion or sintering has taken place. The thusly obtained oxidized ferrochromium (obtained with an oxidation ratio of 1.21) was briquetted and then charged into a vacuum deoxidizing furnace and there deoxidized at a temperature of from 1200° to 1400°C and a pressure of 3 to 0.1 mm Hg. The chemical compositions of the products thereby obtained are set out below. The values correspond to JIS (Japanese Industrial Standards) G 2303 for low-carbon ferrochromium.

Component	%
Cr	67.8
C	0.007
Si	0.7
P	0.027
S	0.011
Insoluble (acid)	2.407
N	0.005
Fe	29.05

Example 2: The same high-carbon ferrochromium as used in Example 1 was fed into the fluidized bed reactor at a rate of 100 kg/hr and oxidized. The reaction was self-maintaining at the reaction temperature of 970°±20°C with the same operating conditions as set out in Example 1. The amount of air supply for effecting fluidization and oxidation amounted to 3.3 Nm^3/min on the average. The reaction was carried out smoothly without any sintering or fusing taking place. The amount of oxidized ferrochromium recovered at the lower outlet and in the bag filter amounted to about 105 kg/hr and 25 kg/hr respectively. The composition of the collected powder is set out below.

Component	%
Cr	47.5
C	1.3
Si	0.6
P	0.022
S	0.020
O	27.2
N	Trace
Others	23.358

In view of the above-set-out chemical values, the oxidation rate was equivalent to about 90% of the theoretical.

Ferrosilicon Alloy Powder

J. Cziska, G. Strauss, W. Portz, K. Komorniczyk and J. Kandler; U.S. Patent 3,839,014; October 1, 1974; assigned to Knapsack AG, Germany have developed a ferrosilicon alloy in powder form consisting of smooth, spheroidally shaped particles and having a density of more than 7 g/cc. The alloy contains more particularly between 8 and 15 weight percent of silicon, between 0.5 and 5 weight percent of nickel, between 1.4 and 5 weight percent of copper and between 0.3 and 2.5 weight percent of phosphorus as additional ingredients.

As a further ingredient, the alloy may contain between 0.02 and 2 weight percent of carbon.

In addition to carbon, the iron-copper-nickel-silicon-phosphorus alloy may contain customary commercial contaminants including, for example, manganese, aluminum, titanium, chromium, molybdenum, vanadium or sulfur in maximum proportions of altogether 3 weight percent.

The alloys can be made by melting iron, copper, nickel, quartz gravel, carbon and ferrophosphorus in an electrothermal reduction furnace, or by melting iron, copper, nickel, ferrosilicon and ferrophosphorus in an induction furnace, at temperatures between 1200° and 1650°C, and atomizing the resulting melt in known manner under pressures between 2 and 30 atmospheres absolute using water, steam or air as the atomization-inducing agent.

The resulting finely divided and fused droplets are chilled in water, subjected to preliminary dehydration, dried and sieved. The powder particles so made substantially have a compact, smooth, spheroidal, drop-shaped or elongated surface. The alloy powder particles have a size substantially between 0.001 and 0.4 mm, the particle size distribution being very regular. It is therefore possible for the screen analysis curves to be plotted, practically as a straight line, in the Rosin-Rammler diagram.

In the following examples, the rust index was identified. To this end, a porcelain dish filled with 20 g of pulverulent alloy and 5 ml distilled water was placed in a drying cabinet, which had a temperature of 75°C, and allowed to remain therein over 75 minutes. Following a cooling period of 25 minutes in each particular case, the above impregnation and vaporization procedures were repeated an additional 3 times and the increase in weight percent was identified.

Example 1: 780 kg of iron turnings (shovel grade), 350 kg of quartz gravel (diameter between 5 and 45 mm), 175 kg of coke (diameter between 60 and 100 mm), 38 kg of waste copper, 38 kg of waste nickel and 75 kg of ferrophosphorus, which contained 20 weight percent of phosphorus (fist size) were melted in an electrothermal reduction furnace and the resulting melt, which had a temperature of 1550°C, was delivered to an atomization means. The melt was atomized through an annular slit nozzle with steam under a pressure of 12 atmospheres (gauge).

The resulting powder was collected in water. Following dehydration and drying, the ferrosilicon powder so made (referred to hereinafter as product B), was tested in the laboratory to determine its rust index and found to have a very good resistance to corrosion. A comparative test was made with product A. This was atomized ferrosilicon which was prepared in analogous manner but was free from phosphorus.

	. . . Product . . .	
	A	B
Pycnometer density (g/ml)	7.14	7.18
Apparent density (g/ml)	3.80	3.94
Rust index increase, wt %	1.50	0.04
Si (wt %)	12.10	12.20
Cu (wt %)	3.60	3.50
Ni (wt %)	3.70	3.60
P (wt %)	0.05	1.40
C (wt %)	1.21	1.20

The balance was iron and customary commercial impurities. Particle size distribution in starting material in weight percent is as follows:

Size (mm)	Weight Percent
>0.100	4.8-5.0
0.063-0.100	10.5
<0.063	84.7-84.5

Example 2: 780 kg of iron scrap, 180 kg of ferrosilicon, which contained 75 weight percent of Si, 38 kg of waste copper, 38 kg of waste nickel and 75 kg of ferrophosphorus, which contained 20 weight percent of phosphorus, were melted in an induction furnace. The resulting melt was atomized under conditions the same as those reported in Example 1 and tested in the laboratory (Product D). Comparative test C was made with an atomized ferrosilicon alloy which was prepared in analogous manner but was free from phosphorus.

	. . . Product . . .	
	C	D
Pycnometer density (g/ml)	7.17	7.21
Apparent density (g/ml)	3.81	3.90
Rust index increase, wt %	1.36	0.03
Si (wt %)	12.40	12.30
Cu (wt %)	3.70	3.60
Ni (wt %)	3.80	3.70
P (wt %)	0.05	1.24
C (wt %)	0.20	0.20

The balance was iron and customary commercial impurities. Particle size distribution in starting material in weight percent is as follows:

Size (mm)	Weight Percent
>0.100	5.4-5.2
0.063-0.100	10.6-10.5
<0.100	84.0-84.3

Iron-Chromium Sintering Alloy

F. Thummler, G. Zapf and M. Ahmed; U.S. Patent 3,853,537; December 10, 1974 have developed a sintering alloy powder for forming products by powder metallurgical techniques containing from 2 to 13% chromium, optionally up to 5% copper and up to 2.5% carbon and the remainder except for impurities iron. The alloy powder is preferably made by annealing an iron-chromium alloy powder having a particle size less than 150 μm for 2 hours at a temperature of between 850° and 950°C and subsequently mixing the annealed powder with iron powder with a particle size less than 400 μm to adjust the chromium content to the desired value. Alternatively an iron-chromium alloy powder with a particle size less than 150 μm is mixed with a fine iron powder with a particle size less than 40 μm, the mixture is annealed with exclusion of air for 2 hours at a temperature of from 850° to 950°C and after this the annealed powder is reduced in particle size and is mixed with iron powder to give the desired chromium content.

Example 1: Ferrochrome powder with a particle size not more than 50 μm was annealed for 2 hours at 850°C with the exclusion of air in an atmosphere of hydrogen. After cooling, the powder was reduced in particle size and a powder mixture prepared containing 1% zinc stearate, 6% chromium, 2% copper and 91% iron. The extra iron, apart from the iron in the ferrochrome, was added to the mixture in the form of electrolytic iron powder.

The powder mixture was prepressed under a pressure of 6 Mp/cm^2 to form a blank. The blank had a density of 6.8 g/cm^3. The blank was sintered for 2 hours in an atmosphere of cracked ammonia gas in a closed box sealed against the entry of water vapor and oxygen by means of a getter substance. The sintered product had a tensile strength of 51 kp/mm^2 and an elongation at rupture of 9%.

Example 2: A mixture containing 3% chromium in the form of ferrochrome in which the chromium concentration was 45% by weight was milled down to a particle size below 50 μm. The milled powder mixed with sponge iron of particle size below 100 μm was annealed for two hours in a closed box at 900°C. After cooling the powder was reduced in particle size and then mixed with 1% of zinc stearate.

The mixture was prepressed at 6 Mp/cm^2 to give a blank which was sintered at 1200°C for 2 hours in an atmosphere of inert gas containing 70% N_2 and 30% H_2. For sintering the blank was packed in a box sealed against direct penetration by the sintering atmosphere by means of a getter substance. The resulting sintered product had shrunk during the sintering to a density of 7.0 g/cm^3. The tensile strength was 30 kp/mm^2 and the elongation at rupture 11%.

Example 3: A mixture containing 4% of chromium in the form of finely ground ferrochrome powder with a chromium concentration of 42%, 4% of electrolytic copper powder, 1% of zinc stearate and the remainder iron powder in the form of a reduction powder with a maximum particle size of 150 μm was prepressed under a pressure of 4 Mp/cm^2 to give a blank which had a density of 6.4 g/cm^3.

The blank was sintered in a closed box, in which the blank was protected by a getter substance against penetration by the oxygen and moisture of the sintering atmosphere, in a travelling belt furnace at 1100°C. After cooling the product had a tensile strength of 33 kp/mm^2 and an elongation at rupture of 2.5%.

Master Alloy Powder

S. Mocarski; U.S. Patent 4,071,354; January 31, 1978; assigned to Ford Motor

Company has developed a master alloy powder for admixture to an iron based powder to provide liquid phase sintering and production of a substantially homogeneous product having the characteristics of a wrought alloy product. The master alloy powder contains at least two elements selected from the group consisting of manganese, nickel, molybdenum, chromium, copper, carbon and iron. The master powder may contain additions of silicon up to 5% and rare earth metals up to 2%, either of which assists to speed up diffusion and create a more favorable liquidus-solidus relationship within the master alloy powder.

A successful multicomponent master alloy mixture (Designated No. 342) derived from metal melted under inert gas, gas atomized, and screened to a −200 mesh size and having the following chemical analysis provided an initially satisfactory liquidus and melting range: nickel 28.20%, iron 10.52%, manganese 40.78%, molybdenum 5.37%, and chromimum 15.15%.

When this master alloy mixture was added into a base iron powder, the addition being 2½% by weight, together with natural graphite in four different proportions, and after being subjected to a conventional technique of precompacting, sintering in hydrogen atmosphere at 2250°F and hot forming at 1800°F (982°C) the resulting steels contained a final composition of 1.0% manganese, 0.03% copper, 0.82% nickel, 0.14% molybdenum, 0.42% chromium, the remainder iron. The master alloy mixture had a liquidus of 2140°F (1171°C) and a solidus of 1830°F (999°C) during heating, producing a 310°F (154°C) melting range which is deemed useable for commercial applications.

With two groups of master alloy powders available, one multialloy (Mo-Mn-Cr-Ni-Fe), the other binary (Ni-Mn with copper substituted for some of the nickel), the master alloy powders can be made easily diffusible by small percentage additions of silicon (about 1-5%), rare earth metals (about 0.5-1.5%), or about 0.1% yttrium (an element that acts like rare earth for purposes of this process). This makes it possible to provide a low alloy steel by liquid phase sintering responding to any hardenability requirement, either for quenched and drawn steel or for carburized parts. Diffusion of molybdenum, even in a small amount, increases significantly the hardenability of the case.

Molybdenum is also known to overcome the difficulties associated with temper embrittlement; upwards to 0.08% Mo in the final product should be used as an alloying addition for this purpose. The table below summarizes nominal compositions of some master alloys pertinent to this process.

In the table, Alloy No. 524 exhibits a very low liquidus and solidus, the respective values being 2065°F ((1129°C) and 1730°F (943°C), melting range being 335°F (168°C). Alloy No. 524 had five times deeper penetration into the iron than the alloy powders No. 342 and No. 400 during the liquid diffusion test run under the same conditions for all the alloy powders.

Master Alloy Mix No.	Chemical Composition, wt%								°F	°F	°F
	Mn	Ni	Cr	Mo	Fe	Cu	Si	R.E.	Liquidus	Solidus	Melting Range
342	40	30	15	5	10	—	—	—	2140	1830	310
400	44	25	—	11	19	—	—	—	2200	2130	70
524	55	18	3	8	14	—	2	—	2065	1730	335
533	56	24	3	6	11	—	—	—	2115	1890	225
533S			———*———				2.5	—	2130	1820	310
533M			———*———				2.5	1	2020	1850	270
534	52	22	8	6	12	—	—	—	2110	2070	40
534S		———*———				—	2.5	—	2070	1870	200
534M		———*———				—	2.5	1	2100	1860	240
535	47	20	13	6	14	—	2.5	—	2210	2130	80
535S		———*———				—	2.5	1.0	2100	1960	140
535M		———*———				—	—	—	2145	1930	215
528	75	25	—	—	—	—	—	—	1930	1800	130
527	74	12.5	—	—	—	12.5	1	—	1940	1700	240
344	36	30	18	6	10	—	—	—	2205	2005	200
345	41	25	18	6	10	—	—	—	2220	1970	250
346	38	23	18	6	15	—	—	—	2245	2000	245
506	64	16	0	10	10	—	—	—	2250	1955	290
508	56	14	0	15	15	—	—	—	2300	2000	300
509	56	14	15	5	10	—	—	—	2170	2070	100
510	56	14	10	10	10	—	—	—	2240	2015	225
511	59	11	15	5	10	—	—	—	2280	2040	240
512	53	17	15	5	10	—	—	—	2220	2000	220
513	56	14	22	8	—	—	—	—	2435	1920	515
514	50	20	15	5	10	—	—	—	2200	2090	110
515	46	24	15	5	10	—	—	—	2200	1990	230
531	72	14	—	—	—	14	2	—	1910	1770	140
532	72	14	—	—	—	14	—	1	2020	1790	230

*The same above percentages as immediately above except reduced proportionately for the presence of silicon and/or rare earths.

Boron Alloy for Filler Metals

G. H. von Scheele; U.S. Patent 4,049,436; September 20, 1977; assigned to Hoganas AB, Sweden adds an alloying element in the form of metal powder as a part of the flux composition (with regard to nucleation agents and/or nitrogen binding elements and strong deoxidizers alloyed iron powder) where the alloying content exists in an essentially more diluted state than normally used in commercially available alloying metals. By this a practical application of the microalloying technique is possible, which is used when producing fine-grained steels, also for a reproducible production of filler metal.

The iron-based weld filler powder to be used either in the as-is condition or diluted with a boron-free unalloyed iron powder as a raw material in the production of filler metals such as stick electrodes, agglomerated fluxes, and flux cored wires, contains boron within the limits 0.03% and 0.07%, one or more of the elements Al, Ti, Zr, V in a total content of up to 1%, and one or more of the elements Ni, Cr, Mo in a total content of up to 20%. The weld deposit is made by depositing an iron-based powder having boron within the limits of 0.005% and 0.5% in a weld cavity between the two metal members and applying a welding action to the powder to form a high-impact weld deposit between the two metal members.

Acicular Powder Containing Titanium and Tin

A.R. Corradi and G. Fagherazzi; U.S. Patent 4,056,410; November 1, 1977; as-

signed to Montedison, S.p.A, Italy have developed a process for preparing metallic powders based on iron and suitable for magnetic recording, starting from acicular particles of an iron compound selected from the class consisting of α-Fe_2O_3, α-FeOOH and Fe_3O_4, by reduction with a gas containing more than 50% by volume of hydrogen. The particles of the iron compound are treated with titanium, in the form of $TiO(OH)_2$ or TiO_2, and with tin, in the form of $Sn(OH)_2$ or $SnO(OH)_2$; after which they are subjected to a heat treatment in air of at least 400°-550°C; and finally to a reducing treatment at 340°-420°C.

The titanium is employed in an amount of 0.5-1% by weight, calculated as TiO_2, and the tin in an amount of 3-6% by weight, based on pure iron. Prior to reduction, the particles of the iron compound may be treated not only with titanium and tin, but also with cobalt and/or silica, the cobalt in an amount not exceeding 6% by weight with respect to pure iron and the silica in an amount not exceeding 5% by weight with respect to pure iron. The metallic powders based on iron are suitable for magnetic recording.

They contain 0.5-1% by weight, with respect to pure iron, of titanium calculated as TiO_2, and 3-6% by weight of Sn, with respect to pure iron, and have an average particle length ranging from 0.3 to 1 μ and an average particle diameter ranging from 0.1 to 0.4 μ, with or without a maximum of not more than 5% by weight of SiO_2 with respect to pure iron and not more than 6% by weight of Co with respect to pure iron.

Example 1: 6.3 g of α-FeOOH (obtained by precipitation from a $FeSO_4 \cdot 7H_2O$ solution), whose acicular particles had an average length of 0.8 μ and an average diameter of 0.1 μ were imbibed with 5.6 cc of a 0.2 M $Co(CH_3COO)_2$ solution and with 25 cc of H_2O, thus forming a fluid paste. Such paste was brought to 80°C, under intense stirring, until water evaporated so as to obtain a pasty mass that was successively placed into a furnace at 100°C, kept therein for 2 hours, and then treated at 495°C in air for 2 hours. The iron compound now contained 4.8% by weight of Co with respect to pure iron. 0.5 g of the so-treated iron compound was reduced for 5 hours and 30 minutes with H_2 at 351°C.

The resulting product was in the form of particles having an average length of 0.25 μ and an average diameter of 0.14 μ. The coercivity (H_C) = 440 Oe and the saturation magnetism at room temperature (σ_S) = 165 emu/g. The ratio of σ_R (residual magnetism) to σ_S = 0.55. Such product, after having been kept for 30 days in the air exhibited a reduction of the σ_S value equal to 19%.

Example 2: 5.77 g of the same α-FeOOH of Example 1 were suspended in H_2O and 15 cc of a $SnCl_2 \cdot 2H_2O$ solution acidified by HCl (0.1 M solution) were admixed therewith. The pH value was brought to 4.5 and stirring was conducted for 30 minutes at 45°C. It was filtered and washed by decantation until the filtrate reacted no longer positively to Cl^- with $AgNo_3$. It was dried at 100°C and then subjected to heat treatment in air for 2 hours at 510°C.

Such sample exhibited on analysis a Sn content of 4.83% with respect to pure iron. 0.5 g of the so-treated iron compound, reduced for 28 hours at 350°C with H_2, yielded a product in the form of particles having an average length of 0.5 μ and an average diameter of 0.14 μ and exhibiting the following characteristics: H_C = 850 Oe, σ_S = 132.5 emu/g, σ_R/σ_S = 0.32. Such product, after having been kept in the air for 30 days, exhibited a reduction of the σ_S value equal to 11%.

Example 3: 10.83 g of the same α-FeOOH of Example 1 were suspended in water. 4.4 cc of a Na_2SiO_3 solution containing 31.04 g/l of SiO_2 and so much HCl as to bring the pH value to 6 were admixed therewith. The resulting mixture was stirred at 50°C for 30 minutes, then filtered and the residue washed. It was dried during the heat treatment at 500°C for 2 hours in air. The sample obtained contained 1.9% by weight of Si, expressed as SiO_2, with respect to pure iron.

0.5 g of the so-treated iron compound was subjected to reduction in H_2 at 370°C for 13 hours and provided a product in the form of particles having an average length of 0.3 μ and an average diameter of 0.18 μ, and exhibiting the following characteristics: H_C = 586 Oe, σ_S = 172 emu/g, σ_R/σ_S = 0.33. The product, after a 30-day exposure to air, exhibited a reduction in the σ_S value of 10%.

Welding Filler Alloy by Atomization

J. Mikurak; U.S. Patent 3,881,912; May 6, 1975; assigned to Hoeganaes Corporation describes a method of making an iron alloy filler material for use in welding. The filler material has a chemical composition which includes between 0.01 and 0.15% carbon; 0.15 to 2.0% manganese; between 0.01 and 2.0% nickel; between 0.01 and 1.00% molybdenum; about 0.25% chromium with traces of silicon, sulfur, phosphorus, residuals normally found during conventional steelmaking procedures and with the balance being iron.

The filler material is produced through the consecutive steps of melting, atomization, annealing and particle size determination. The filler material is made such that it contains a maximum of 5.0% particles by weight which are smaller than the mesh size of a U.S. Standard No. 200 sieve. The filler is further made such that it contains a maximum of 5.0% particles by weight which are larger than the mesh size of a U.S. Standard No. 40 sieve.

Example: Filler material alloy iron particles were prepared having the following weight percentages:

	Molten Alloy Iron %	Annealed Alloy Iron %
Carbon	0.55	0.05
Manganese	2.0	1.99

(continued)

	Molten Alloy Iron %	Annealed Alloy Iron %
Silicon	0.01	0.01
Sulfur	0.020	0.018
Phosphorus	0.15	0.15
Chromium	0.25	0.25
Nickel	2.0	2.0
Molybdenum	1.0	0.97
Iron	Balance	Balance

The above-stated molten alloy composition was atomized to produce substantially composition-consistent particles. The atomized powder was annealed in an annealing furnace at a temperature of 1100°C for 1.5 hours. The annealing furnace atmosphere was disassociated ammonia. After annealing the mixture was mixed in an equalizer and screened through a No. 200 and a No. 20 sieve. The resulting filler material had a screen analysis weight percentage as follows:

Sieve No. (U.S. Standard)	Weight Percentage
+20	0.0
−32+40	5.0
−40+60	10.5
−60+80	44.5
−80+100	25.0
−100+200	10.0
−200	5.0

The filler material was once again applied to a submerged arc welding process using a basic flux composition. However, the basic filler composition was adjusted throughout the range of 2.9 g/cc to 3.85 g/cc. The following results were obtained:

No.	Apparent Density (g/cc)	Weld Comments
1	2.9	Large heat-affected zone
2	3.0	Excellent weld, no porosity, high strength
3	3.25	Excellent weld, no porosity, high strength
4	3.5	Good weld, some backflash and bubbling seen
5	3.85	Fair weld-backflash and bubbling observed

Utilizing substantially the same example composition, the percentage of powder fines having a particle size less than No. 200 sieve was increased to 8.0% of the filler composition by weight. Upon welding, it was observed that porosity existed at the weld interface resulting in a low weld strength capability.

Induction Heating for Low Oxygen Content

S. Ito, Y. Morioka, Y. Kajinaga, M. Nitta and I. Sakurada; U.S. Patent 3,966,454; June 29, 1976; assigned to Kawasaki Steel Corporation, Japan have developed a method of producing iron or iron alloy powders having a low oxygen content to be used for powder metallurgy by subjecting the powder materials to induction heating under a relative density of 5 to 65% based on the density of the molding steel of the powders.

Alternating current of 50 cycles-1 megacycle is used to heat the powders at a temperature of 750° to 1400°C to effect reduction rapidly. The starting iron or iron alloy powder materials have an oxygen content of not more than 8% by weight and a total carbon amount including carbon alloyed in the powders of not more than 6% by weight.

STEEL COMPOSITIONS

Prealloyed for Powder Forging

Y. Kondo, T. Kawamura, H. Aoyama and T. Araki; U.S. Patent 3,901,661; August 26, 1975; assigned to Toyo Kohan Co., Ltd., Japan have developed a prealloyed steel powder for formation of structural parts by powder forging. It comprises up to 0.5% by weight of carbon, 0.8 to 5.0% by weight of copper, 0.1 to 0.7% by weight of molybdenum, 0.3 to 1.3% by weight of nickel, and up to 0.6% by weight of manganese.

This prealloyed steel powder exhibits a good forgeability and is inexpensive. It can also give a powder forged article which is comparable or superior to structural wrought alloy articles manufactured from conventional wrought alloy steels in respect to heat treatability, carburizing, quench hardening properties and mechanical properties.

Example: Low carbon steel scrap was melted in a high frequency induction furnace, and prescribed alloying elements were added to the melt so as to attain a desired chemical composition. The molten steel was then allowed to fall from small holes, and the molten prealloyed steel streams were atomized under a pressure of 65 kg/cm^2 by means of high-pressure water projected from nozzles.

The resulting powder was dried and subjected to reduction annealing for two hours at 730°C. The annealed powder had a chemical composition comprising 0.02% of carbon, 2.32% of copper, 0.50% of molybdenum, 0.84% of nickel, 0.11% of manganese, 0.013% of phosphorus and 0.010% of sulfur, and was characterized by a micro-Vickers hardness of 175, an oxygen content of 0.15%, the following particle size distribution:

20 to 80 mesh	36.0%
80 to 100 mesh	17.7%
100 to 150 mesh	20.2%
150 to 200 mesh	17.2%
200 to 250 mesh	5.9%
250 to 325 mesh	1.0%
smaller than 325 mesh	2.0%

Its apparent density was 3.4 g/cm^3 and its flow rate, 18.5 sec/50 g. The above powder was mixed with graphite powder and incorporated with 1% zinc stearate, and the mixed powder was subjected to compaction forming under a pressure of 4.5 tons per square centimeter to have a cylindrical shape having a diameter of 58 mm and a height of 40 mm. The green density was 6.2 g/cm^3 at this time. The so-formed compact was dewaxed in a dissociated ammonia gas, and sintered for 30 minutes at 1120°C in a dissociated ammonia gas.

Then, it was heated in a nitrogen inert gas (NX gas) at 900°C for 30 minutes and forged under a pressure of about 13 tons per square centimeter with use of a mechanical press. The resulting powder forged article has a carbon content of 0.17% and a density of 7.87 g/cm^3. Then this powder forged article was heated at 900°C for 30 minutes in a nitrogen inert gas (NX gas), and quenched in oil and tempered at 600°C for 30 minutes. The resulting powder forged, heat-treated article was characterized by a tensile strength of 113.5 kg/mm^2, an elongation of 16.7%, a Rockwell C hardness of 31.3 and an impact strength with U notch of 8.6 kg/cm^2.

The optical microscopic structure of the product was a quenched and tempered one. When a specimen cut from this powder forged article was subjected to the Jominy test, it was found, that at a point of 2.5 mm from the quenched end the Rockwell C hardness was 46, at a point of 10 mm from the quenched end the Rockwell C hardness was 34, and at a point of 40 mm from the quenched end the Rockwell C hardness was 24. For comparison, a powder forged article was prepared in the same manner as above from a prealloyed steel powder comprising 0.19% of carbon, 1.85% of nickel, 0.30% of molybdenum and 0.6% of manganese. When this comparative article was subjected to the Jominy test, the quench hardenability of this comparative powder forged article was very poor.

Ferronitrogen Addition Agent

A. Sarian and A.G. Sarian; U.S. Patent 4,133,678; January 9, 1979 describe a process for making a ferronitrogen addition agent for steel making processes. The ferronitrogen agent is derived from waste materials of steel making and processing.

The ferronitrogen has the advantages of the conventional ferroalloys method but is not limited in its usage by the presence of manganese or chromium. It can be used in the fabrication of any type of steel in which the presence of nitrogen is necessary for the achievement of all the improvements the nitrogen conveys to the physical properties of the steel.

Raw materials which can be used to prepare the ferronitrogen include metallic ejections (grits, sludges, powder from exhausted fumes-flue dust) resulting from various steel making processes; iron oxides (powder and scale), steel powder and iron powder resulting from secondary steel processes such as forging, rolling, drawing, etc.; iron pellets and their residual powder; and iron powder.

The ferronitrogen is produced by treating the raw materials, under the following conditions:

1. Grinding the raw materials and separating (size selection) the fine powder with particles smaller than 30 microns,
2. Reducing the fine powder of iron oxides with hydrogen or carbon monoxide at a temperature of 1400° to 1600°F (760° to 870°C). To avoid the sticking of the iron powder an addition of fine silica (SiO_2) should be used. The final grading of the iron powder must be below 30 microns.
3. Diffusing atomic nitrogen into the iron or steel powder in order to obtain iron nitride with a content of nitrogen up to 6%. For this operation a rotary furnace or a fluidized bed installation should be used at a temperature of 1100° to 1350°F (590° to 730°C). By passing through the installation, loaded with iron powder, anhydrous ammonia with at least 98% concentration under pressure of up to 8 inches (200 mm, water column), the ammonia will be dissociated in hydrogen and atomic nitrogen, which at the mentioned temperature, 1100° to 1350°F (590° to 730°C) will diffuse into the iron powder, forming iron nitride (mostly Fe_4N).

The duration of this nitriding process depends on the grading of the iron particles (about 5 minutes for 1 micron, 8.5 hours for 10 microns and more than 48 hours for 20 microns). A prior grading of the iron powder will indicate the nitriding time.

Before and after the nitriding process, the installation should be washed by passing through argon or molecular nitrogen.

Carbide Addition Agents

F. H. Perfect; U.S. Patent 3,982,924; September 28, 1976; assigned to Reading Alloys, Inc. has found that carbide addition agents comprising (a) from about 20 to 60 atomic percent of carbon, (b) about 25 to 60 atomic percent of at least one metal selected from the group consisting of vanadium and columbium, and (c) about 10 to 40 atomic percent of at least one metal selected from the group consisting of iron and manganese, are useful alloy addition agents for steel.

Since the various metals (M and M_1) which make up the carbide addition agents may, and generally do, vary in purity, the proportions thereof required to provide carbide addition agents of a given composition will vary accordingly. In addition, during the sintering operation employed in forming the carbide agent, a portion of the carbon reactant reacts with the oxygen in the oxides of the metals M and M_1. Thus, a sufficient amount of carbon must be present to obtain the reduction of the metal oxides as well as to form the desired carbides.

For these reasons, the respective amounts of M, M_1 and C utilized in preparing the carbide addition agents are expressed in terms of the composition of the desired carbide addition agent. M, M_1 and C are combined in proportions so as to provide a carbide having a composition, in terms of atomic percent from about 10 to 40 atomic percent M_1, about 25 to 60 atomic percent M, and about 20 to 60 atomic percent C. The atomic ratio of C to M in the carbide addition agents does not exceed about 2:1 (MC_2) and preferably is 1:1 (MC), for it has been found that formulations producing carbide addition agents in the form MC_2 and greater carbon contents yield, after sintering, products having a large amount of soot and excessive porosity.

In addition, formulations having a M_1 to M atomic ratio greater than 1:1 [$(M_1)_2M$] have a tendency to fuse during the sintering operation thereby limiting their usefulness as addition agents. Those carbides having a final ratio of C to M below about 1:3 are also not desirable since they have very poor strength after sintering and generally have large amounts of oxygen. Preferably, the atomic ratio of C to M is not less than 1:2 (M_2C).

In preparing the desired carbide addition agents, the carbon, oxide of the metal M_1, and oxide or combinations of oxides of the metal M are each reduced to the desired particle size, combined so as to form a substantially uniform mixture, and then compacted, as for example, with simple mechanical pressure, pelletizing or agglomeration. Pelletizing is accomplished in conventional pelletizing equipment forming pellets one-half to one inch in diameter. The compacting can be carried out in any suitable apparatus such as briquetting machine through the application of pressures on the order or 7,000 psi or more.

When pressures on the order of 7,000 psi are utilized, the resultant compacted mass generally has a density of from about 2 to about 3 g/cc; it is of course, possible to increase the density of the compacted mass by increasing the pressure applied during the compacting procedure. The compacts can be made in any desired shape and size. Preferably, from about 50 to about 500 grams of a mixture of carbon, oxide of the metal M_1, and oxide of the metal M are compacted to form a cylindrical or spherical shape having a diameter in the range of about one-quarter to about one inch. Other suitable forms include briquettes of a rectangular cross section and hockey puck shapes about one-half to one inch thick and about two to four inches in circumference.

The compacted mixture of carbon and oxides of the metals M and M_1 is then sintered in any convenient furnace, such as a brick kiln or rotary kiln furnace, at a temperature of from about 1200° to about 1400°C in a reducing atmosphere. At these temperatures, no fusion of the material occurs and as the heating progresses, the oxides are reduced with the liberation of carbon monoxide. The sintering operation preferably is carried out in a reducing atmosphere until complete, generally requiring from about 2 to about 3 hours. No special pressure conditions are required for the sintering operation.

During the sintering operation, the compacted mixture shrinks in size with a corresponding weight reduction. This shrinkage is beneficial since it enhances the density of the carbide product. After sintering, the carbide addition agents generally have a density of from about 3.5 to about 5 g/cc.

After sintering, the carbide is cooled in the furnace or auxiliary cooler, under the reducing atmosphere until the carbide addition agent reaches a temperature of about 400°C when it may be removed to cool to ambient temperatures in air.

The resultant metal carbide addition agents contain carbides of the metals vanadium or columbium, or a combination of vanadium and columbium, in sufficient quantities to be a practical and convenient source of the carbide material in the preparation of high-strength steel alloys. In preparing a high strength, low-alloy steel employing a carbide addition agent of this process, the agent is simply added to molten iron or steel where it dissolves readily and rapidly (usually in less than a minute).

The metal carbides have sufficient density so that upon their addition to molten iron and steel they immediately sink to the steel-slag interface where they readily dissolve, rapidly forming desired high-strength, low-alloy steels. Metal carbides of this process provide a very efficient means of forming high strength iron or steel alloys with vanadium or columbium, since, as a general rule, it had been found that more than 80% of the vanadium or columbium present in the metal carbide addition agent is incorporated into the final steel alloy.

Powder Containing Phosphorus

In the manufacture of articles by compacting and sintering steel powder it is desired that the articles shall neither expand nor shrink during the sintering process. *P.F. Lindskog and L.-E. Svensson; U.S. Patent 3,836,355; September 17, 1974; assigned to Hoganas AB, Sweden* have found it useful to incorporate into the steel powder an iron-phosphorus alloy powder having the approximate composition Fe_3P, with a maximum particle size of 75 μm at the most, preferably 45 μm at the most, and with a phosphorus content of between 13 and 17%, preferably 15.6%, in a quantity to give the article a phosphorus content of 0.2 to 3% by weight.

In addition, if necessary, a suitable pressing lubricant such as zinc stearate in the usual proportion, can be added. In order that the constituents of the powder described above do not segregate after the mixing operation, one way of diminishing the risk of segregation is to add 50 to 200 g thin mineral oil per metric ton of powder continuously during filling of the mixer. This brings about an attachment of the fine particles (in this case the iron-phosphorus alloy particles) to the coarser steel particles.

A further improvement in this respect is obtained if the powder mixture is exposed to a heat treatment at 650° to 900°C for 15 minutes to 2 hours in a reducing atmosphere and subsequently the lightly sintered cake formed during the heat treatment is broken up by gentle comminution. The small iron-phosphorus alloy particles which before this treatment were loosely attached to the steel particles are now solidly bonded to the latter, which effectively prevents segregation without affecting the compressibility to any appreciable degree.

Example: Three powder mixtures A, B and C, were prepared. The sponge iron powder with a maximum particle size of 147 μm was used for all these mixtures. The mixtures consisted of the following components:

Mixture A:
- 97.4% sponge iron powder
- 1.8% iron-phosphorus alloy powder with a phosphorus content of 25% and a maximum particle size of 44 μm
- 0.8% zinc stearate powder

Mixture B:
- 96.2% sponge iron powder
- 3.0% iron-phosphorus alloy powder with a phosphorus content of 15% and a maximum particle size of 44 μm
- 0.8% zinc stearate powder

Mixture C:
- 99.2% sponge iron powder
- 0.8% zinc stearate powder

The mixtures A and B thus contained 0.45% P, whereas mixture C was free of phosphorus.

Cylindrical compacts with a height of 8 mm, a diameter of 8 mm and a density of 6.75 g/cm^3 were pressed. The pressing was done in an automatic press at a rate of 35 pieces per minute. The die had been made from high-speed steel which had been hardened to a Rockwell C hardness of 63. The die wear was determined by measuring the diameter of the compacts at regular intervals. It was found that the diameter increased by 0.6 μm per 10,000 pieces when mixture A was used, whereas mixture B which had been made according to the process gave an increase of 0.3 μm per 10,000 pieces.

The pure sponge iron powder C gave an increase of 0.25 μm for the same number of pieces. Mixture B thus wore the die insignificantly more than the pure sponge iron powder (C), whereas the die wear was doubled with mixture A.

The three powder mixtures were compacted into tensile test bars in a special die for this purpose using a compacting pressure of 400 MN/m^2. The density in the green (unsintered) state was about 6.5 g/cm^3 in all cases which shows that iron-phosphorus alloy addition did not lower the compressibility of the mixtures. The tensile test bars were sintered at a temperature of 1120°C for one hour in an atmosphere of dissociated ammonia. When they had cooled to room temperature they were tested for density, dimensional change during sintering, tensile strength and elongation. The results are given in the table below.

	Mixture		
	A	B	C
Density, g/cm^3	6.60	6.60	6.55
Dimensional change, %	−0.45	−0.45	−0.20
Tensile strength, MN/m^2	280	290	180
Elongation, %	7	8	7

In another phosphorus steel process, this one for manufacturing sintered details having little tendency to brittleness ruptures, *L.-E. Svensson, J.R. Tengzelius and P.F. Lindskog; U.S. Patent 4,093,449; June 6, 1978; assigned to Hoganas AB, Sweden* mix the steel with a phosphorus powder characterized by a maximum particle size of 20 μm, preferably a maximum particle size of 10 μm. The phosphorus content of the ferrophosphorus powder should exceed 2.8% and in order to reduce the wearing of the tools the phosphorus content should be less than 17%.

If the ferrophosphorus powder is manufactured by grinding piece goods the phosphorus content should exceed 12% and preferably be between 14 and 16%. The phosphorus content of the preferred mixture is between 0.2 and 1.5%. The iron-ferrophosphorus mixture is heated with or without the addition of oil in a reducing atmosphere to a temperature of between 650° and 900°C for a period of 15 minutes to 2 hours to improve the protection against segregation.

Example: A ferrophosphorus powder having a phosphorus content of 15.8 weight percent was divided into size classes of: 0-5 μm, 5-10 μm, 10-20 μm and 20-40 μm by means of a wind-sieve device. The different powder fractions were mixed with very pure iron powder having the maximal particle size of 150 μm. The phosphorus content of the mixture was 0.6 weight percent as this content has proved to give a distinct evidence as to the brittleness tendency of sintered materials.

Seven impact strength test bars were manufactured from each mixture at 588

MPa. They were sintered in cracked ammonia at 1120°C for 60 minutes. The bars were not impaired by any indications of fracture and were treated in a Charpy apparatus at room temperature. The mean value of the impact strength (I) of these seven bars as a function of the particle size of the phosphorus was determined.

The data indicate that the material manufactured from the mixture having a ferrophosphorus powder particle size of 5 to 10 μm has the greatest toughness. The material having ferrophosphorus particles of greater size than 15 μm does however provide for brittle sintered details.

In yet another related process by *J.R. Tengzelius, P.F. Lindskog and L.-E. Svensson; U.S. Patent 4,090,868; May 23, 1978,* the iron or steel powder is mixed with a phosphorus powder containing in all less than 4% impurities which are at the sintering temperature more easily oxidized than the main components iron and phosphorus. This iron-ferrophosphorus mixture is also heated with or without the addition of oil in a reducing atmosphere to a temperature of between 650° and 900°C for a period of 15 minutes to 2 hours to improve the protection against segregation.

Oxide Coated Ferrous Steel Powder

N.L. Church; U.S. Patent 3,837,845; September 24, 1974; assigned to The International Nickel Co., Inc. describes the forging of powder alloy steels characterized at the temperature of forging by a microstructure in which the alloy steel powders are comprised essentially of the ferrite and austenite phases, these phases mutually coacting such that grain growth is retarded during recrystallization. Advantageously, at the forging temperature both the ferrite and austenite are present in a volume percentage of at least about 10% with their respective grains not exceeding an average ASTM grain size of about 10, the grain size most advantageously being not greater than about ASTM 12.

Upon bringing the steels to the desired microstructural condition and forging at a temperature which does not deleteriously disrupt this condition, there is improved die filling, lower forging loads for a given configuration, less oxidation, the likelihood of decreased die wear, breakage or distortion, lower cost, etc.

The following alloy steel compositions are considered particularly beneficial since in addition to manifesting low flow strengths at the prerequisite forging temperature they afford high strength at room temperature: about 0.05 to 0.15% carbon; about 0.8 to 2.5% silicon; up to about 2% manganese; about 0.5 to 4% nickel; about 0.2 to 2% molybdenum; up to about 2%, e.g., 0.8 to 1.75%, copper; up to 0.2%, e.g., 0.02 to 0.12%, columbium; up to 0.15 or 0.25%, e.g., from 0.01 or 0.03 to 0.12%, oxygen, the balance being essentially iron and impurities. Such compositions can be compacted and forged to virtually 100% density at about 1400° to 1500°F, e.g., 1450°F, in contrast to temperatures on the order of 1650° to 1800°F commonly employed in conjunction with current low alloy steel P/M (powder metallurgy) forging practice.

The alloy powders should not exceed about 500 microns (including oxide film), preferably being less than 275 microns, an advantageous powder mix being one containing not more than 25% of powder less than 40 microns in size, the remainder being up to 225 or 275 microns. The surface of the powder particles is substantially, if not completely, comprised of an oxide film but this is beneficial. This inhibits grain growth within each particle as well as inhibiting growth across interparticle boundaries. Though up to possibly 1% oxygen can be present in the steels, high oxygen impairs toughness, detracts from compacting and sintering prior to forging and retards interparticle welding during forging. The oxygen level should not exceed about 0.25% and the oxide film thickness should be less than about 5 or 10 microns.

To illustrate the advantages of the process, the following data are given.

Several steel compositions were prepared by atomizing prealloyed metals, the chemistry of which is given in the table below. The atomized powders were made from air induction melted heats using a magnesia crucible. Electrolytic iron and nickel were first charged and melted and then molybdenum pellets were added. The melts were heated to 2850°F and pig iron (to add carbon for carbon boil) added. Following a 5-minute boil half the silicon (ferro form) was added together with pig iron and, if any, ferromanganese, ferrocolumbium and copper. The remaining silicon was then added. The melts were tapped at 2850°F, atomized with argon at 400 psi under an argon atmosphere and the prealloyed powders were then screened into various mesh sizes. Argon was used since water atomization facilities were unavailable.

	Components, %						
Alloy	C	Si	Ni	Mo	Cu	Cb	O
1	0.069	1.22	1	0.24	1.4	0.015	0.06
2	0.11	2.38	2.6	0.93	1.0	0.071	0.04
3	0.11	2.44	2.5	0.89	NA**	NA	0.05
4	0.072	1.25	0.98	0.28	NA	NA	0.058
5	0.085	1.19	0.96	0.58	NA	0.02	0.07
6	0.084	1.28	0.98	0.50	1.6	NA	0.05
7	0.12	2.35	2.8	1.8	NA	0.037	0.058
8	0.10	2.39	2.8	1.7	1.4	NA	0.048
9	0.10	2.47	3.1	1.8	1.5	0.02	0.05
10*	0.14	1.64	1.03	0.35	NA	NA	NA

*Contained 0.12% Ti and 0.002% B and 0.65% Mn
**NA = Not added

While all the alloys above were tested, a few specific examples will be given as illustrative.

Example 1: This example is illustrative of a low-alloy steel combining high yield strength and good ductility, it being a specific objective of the process to provide a steel with a yield strength of at least 90,000 psi together with a tensile ductility of about 10% or more.

Using −40 to +100 mesh particles, Alloy 1 was sealed in a mild steel metal can (3 inches inside diameter, 6 inches long, ¼ inch thick walls) and rolled (starting temperature of 1650°F) to fully dense (no obvious voids) ⅜ inch thick plate, the mill speed being about 40 ft/min. The can was machined off and specimens for room temperature and hot tensile tests were cut from the plates, aged about 4 hours at 1000°F and air cooled. The following tensile properties were obtained at room temperature: yield strength, 125,100 psi; ultimate tensile strength, 125,100 psi; elongation (in 0.55 inch), 11.5%; reduction in area, 30.5%; R_C hardness, 27.2.

To simulate expected powder forging behavior, the steel was tensile tested at various true strain rates. Flow strength was determined at the various strain rates at 1450°F. The data include a similarly processed and currently used powder forging steel (Alloy A) nominally containing about 0.5% C, 0.35% Mn, 0.01% P, 0.02% S, 0.45% Ni and 0.55% Mo.

True Stress (psi)		True Strain in/in/min
Alloy 1	**Alloy A**	
4,570	14,200	0.006
9,160	16,930	0.062
14,980	26,050	0.606

At the temperature of 1450°F Alloy 1 of the process had a markedly lower flow strength at all three strain rates. This should contribute to enhanced die filling and/or result in less die wear and distortion in hot forging. It should be emphasized this was achieved notwithstanding that the yield strength thereof exceeded that of Alloy A by about 36,000 psi, the elongation of the latter being only 7.5%. Actually, Alloy 1 had a lower flow stress at 1450°F than Alloy A at 1650°F. Alloy 1 upon being can-hot-rolled, heated at 1450°F for ½ hour and cooled (approximating air cooling) had a fine-grained microstructure which by visual observation consisted of about 10 to 20% ferrite, less than 2% austenite (as determined by x-ray), the balance being the decomposition products of austenite. Grain size was less than ASTM 12.

Example 2: Alloy 2 is illustrative of a moderately higher strength steel than Alloy 1 but with less ductility. It was processed in the same manner as Alloy 1 and displayed a yield point of 130,700 psi, an ultimate tensile strength of 149,000 psi, an elongation (in 0.55 inch) of 5.5%, a reduction in area of 15% and a hardness of R_C 33.5. Its flow strength at 1450°F is given below together with the commercially produced Alloy A of Example 1.

True Stress (psi)		True Strain in/in/min
Alloy 2	**Alloy A**	
5,740	14,200	0.006
10,570	16,930	0.064
18,290	26,050	0.528

Alloy 2, tested as in Example 1, had a microstructure of about 25 to 35% ferrite the remainder being similar to Alloy 1. Grain size was about ASTM 12-14.

Low Content of Oxide Inclusions

P.F. Lindskog, P.G. Arbstedt and E.G. Wastenson; U.S. Patent 3,899,319; August 12, 1975; assigned to Hoganas AB, Sweden describe a powder mixture for manufacturing alloy steel wherein the metal powder portion consists of a mixture of two powders, viz. a first atomized prealloyed steel powder and a second nonatomized alloy powder comminuted from a solidified melt. The alloying elements are distributed in such a way that elements, the oxides of which are easily reducible, preferably nickel, copper, molybdenum and/or cobalt substantially enter into the atomized prealloyed steel powder and the oxidation sensitive, especially chromium and manganese enter into the finely comminuted powder.

The powder is produced by melting iron and 0.2 to 10% of alloying elements, the oxides of which have an absolute value of the free energy of formation below 120 kcal/mol O_2 (502 kJ/mol O_2) at 1000°C, water atomizing the melt and finally annealing in a reducing atmosphere (suitably cracked ammonia). Depending upon the quality of the iron raw material the atomized prealloyed powder can contain up to 0.4% accessory elements.

The content of such accessory elements, the oxides of which have an absolute value of the free energy of formation above 150 kcal/mol O_2 (627 kJ/mol O_2) at 1000°C, should not exceed 0.1%. Particularly the contents of silicon, titanium and aluminum should not exceed 0.04%, 0.03% and 0.03%, respectively. The powder should have such a particle size distribution that more than 90% and preferably more than 97% of the powder passes a sieve having a mesh opening of 175 μm (80 Tyler mesh; method for sieve analysis of granular metal powders, MPIF Standard 5-46, Metal Powder Industries Federation, New York, U.S.A.).

Other alloying elements, the oxides of which have an absolute value of the free energy of formation exceeding 120 kcal/mol O_2 (502 kJ/mol O_2) at 1000°C, are molten possibly under addition of at most 75%, preferably at most 50%, iron and/or other metals with easily reducible oxides and carbon before casting. The ingot is crushed and comminuted to a fine alloy powder. The total content of elements, the oxides of which have an absolute value of the free energy of formation exceeding 150 kcal/mol O_2 (627 kJ/mol O_2) at 1000°C, should not be higher than 1% in the finely comminuted alloy powder.

The two components are now mixed in the proportions 80 to 99% atomized prealloyed powder and 20 to 1% finely comminuted alloy powder. A powder mixture produced in this way provides a material having small and few oxide particles which is shown by the following example.

Example: Two powders were produced, one by atomizing by means of water and subsequent annealing in cracked ammonia (A), the other by mixing finely ground ferroalloy powder and water-atomized, molybdenum-alloyed steel powder (B). The two powders (A) and (B) had the composition of 1% Mn, 1% Cr, 0.5% Mo, balance Fe. The ferroalloy powder used in powder (B) had the composition 25% Mn, 25% Cr, 7% C, balance Fe, and an average particle size of 4 μm according to Fischer (methods for determination of average particle size of metal powder by the Fischer sub-sieve sizer, MPIF Standard 32 to 60, Metal Powder Industries Federation, New York, U.S.A.). In both cases graphite was added in such an amount that the carbon content of the powders amounted to 0.5%.

Green compacts in the form of cylinders with 25 mm diameter and 30 mm length were pressed from both powders. The compacts were heated, one group at 1120°C in hydrogen gas atmosphere with a dew point of −20°C and maintained at these temperatures for 30 minutes. From the furnace the compacts were rapidly brought to a die where the cylinders at the elevated temperature were compressed to full density.

In order to estimate the content of oxide inclusions the cylinders were analyzed with regard to total oxygen content. Further, the number of oxide inclusions having a linear dimension in any direction larger than 100 μm, was counted on a cross section of each cylinder. The results are shown in the following table.

Temp. °C	Material	Oxygen Content %	Number of Oxide Inclusions 100 μm/cm^2
1120	A	0.54	116
	B	0.20	10
1200	A	0.23	47
	B	0.06	3

The above example shows the advantage of the process. When Cr and Mn, the oxides of which have a free energy of formation of −126 and −140 kcal/mol O_2 (−528 and −586 kJ/mol O_2) at 1000°C respectively, were added as a finely comminuted ferroalloy (powder B) the forged product obtained a considerably lower oxygen content and lower number of large inclusions than in the case of alloying chromium and manganese already in the atomized steel powder (powder A).

Thermal Agglomeration to Produce High Green Strength

E. Klar and E.K. Weaver; U.S. Patent 3,888,657; June 10, 1975; assigned to SCM Corporation have found that by selecting a mass of small enough atomized particulates, thermally agglomerating them into a self-supporting cake, then breaking the cake into small agglomerate particles which are substantially larger in average mesh size than that of the original particulates, e.g., up to seven times larger, advantageously about 1.5 to 7, and preferably 3 to 7 times larger, the maximum ratio of the pore volume to weight of an individual agglomerate particle can be approached in a practical way to yield a product of useful, pourable size for powder metallurgy.

The green strength of the product takes a steep rise from a low value to a desirably high one when such maximum ratio is approached according to this process. The most dramatic green strength obtained is for milled cakes having original particulate makeup of −400 mesh (with only a trace on 100 mesh) milled to a resulting agglomerate particle size of 140 to 100 mesh (with only a trace on 100 mesh). Lesser, but valuable, green strength can be obtained for milled cakes having original particulate makeup of 100 to 140 mesh milled to 80 to 100 mesh resulting agglomerate particle size.

Example 1: Seventeen pounds of copper were melted in a 20 kW Ajax induction furnace. The melting was done in a graphite crucible. Molten copper was water-atomized in accordance with the principles of the U.S. Patent 2,956,304 at a rate of 40 pounds of metal per minute under a water pressure of 2,000 psi and a water velocity of 400 feet per second. The orifice of the tundish feeding zone measured three-sixteenths of an inch. The yield was 92% finer than 100 mesh and 55% finer than 325 mesh. Thermal agglomeration was done in a laboratory tube furnace with these atomized copper particulates loosely piled in a boat to a bed depth of 1½ inches.

Temperature of thermal agglomeration was 1500°F under H_2 gas atmosphere, and total time thereof was about 10 minutes. The resulting self-supporting porous cake was hammermilled into product agglomerate particles virtually all but a trace passing 80 mesh, the size distribution being about 14% finer than 325 mesh, 20% between 200 and 325 mesh, 12% between 80 and 100 mesh, and the balance between 100 and 200 mesh. This powder had green strength of 1,380 psi. In contrast, starting atomized particulates have a green strength of 140 psi when tested the same way.

Examples 2-5: Examples 2-5 are shown in Table 1.

Table 1

Example No.	2	3	4	5
−325 mesh particulates in feed	63%	69%	72%	37%
Thermal agglomeration				
temperature, °F	1800	1325	1200	1400
Time, minutes	5	15	15	10
Atmosphere	H_2	H_2	H_2	H_2
Apparent Density g/cc. according to ASTM B212-48	2.96	2.60	—	2.90
Flow (sec/50g) according to ASTM B213-48	24.0	26.4	—	24.2
H_2 Loss according to ASTM E159-63T	—	—	0.09%	.08%
HNO_3 Insolubles according to ASTM E194-62T	—	.008%	—	—
Density of compacted samples according to ASTM B331-63T	—	6.31	6.42	6.47
Green strength of atomized starting particulates, (psi)	30	109	—	21
Green strength of agglomerate product particles, (psi)	980	917	706	544

The same steps were followed as in Example 1, except that the proportion of finer than 325 mesh particulates in the mass thereof subjected to thermal agglomeration ranged from 37 to 72%, with all but a trace passing 100 mesh. The porous cake formed was milled to virtually all but a trace passing 100 mesh, the size distribution being about 20% finer than 325, 30% between 200 and 325 mesh, and the rest between 100 and 200 mesh.

Compacts for green strength testing were made with the addition of 0.5% stearic acid in Examples 3, 4 and 5, and with no additive in Example 2.

Examples 6-9: Copper was treated as in Example 1, except that time and temperature of thermal agglomeration were varied from 1000° to 1500°F for 7 to 25 minutes as indicated in Table 2. The starting atomized particulates were essentially the same as in Example 1. The porous cake formed was milled to virtually all but a trace passing 100 mesh, the particle size distribution being about 18 to 27% finer than 325 mesh, 35 to 55% between 100 and 200 mesh and the balance between 200 and 325 mesh.

Table 2

Example No.	6	7	8	9	Conventional Powder
Thermal agglomeration					
temperature, °F	1000	1200	1200	1500	
Time, minutes	25	10	20	7	
Apparent Density g/cc	2.32	2.40	2.26	2.50	2.88
Flow, sec/50g	30.8	29.4	31.6	29.0	26.0
H_2 Loss (%)	—	.11	.06	.06	0.27
HNO_3 Insolubles	—	—	—	0.017%	0.07
Green Density at 12 tsi, (g/cc)	6.14	6.15	6.08	6.07	6.06
Green Strength at 12 tsi, (psi)	1746	1691	1844	1486	960

Section A—Samples of the product agglomerate particles of these Examples 6-9 were compacted with 0.5% stearic acid, then sintered for 15 minutes at 1500°F, then compared to compacts made in like manner from a conventional compacting grade of copper powder made by reducing copper oxide. The results appear in Table 3.

Table 3

Example No.	6	7	8	9	Conventional Powder
	Sintered Properties (Plain Copper)				
Green Density g/cc	6.32	6.38	6.24	6.33	6.30
Compacting Pressure, (tsi)	13.3	14.0	13.3	13.9	16.5
Sint. Density, g/cc	6.11	6.15	6.03	6.09	6.33
Sint. Modulus of Rupture, (psi)	17,500	18,600	16,200	14,900	9,400
Dimensional Change[1] (%)	+1.06	+1.29	+1.30	+1.52	+0.47

[1]Based on 1.2500" die size

Section B—For comparison, the same conventional compacting grade of copper powder was dosed with tin powder and treated in the manner above. Samples of the product agglomerate particles of these Examples 6-9 were compacted with 0.5% stearic acid and 10% tin powder, then sintered for 15 minutes in hydrogen at 1500°F. The tin powder used in all instances had the following characteristics: apparent density 3.90 g/cc and virtually all but a trace finer than 100 mesh. The properties of these compacts are shown in Table 4.

Table 4

Example No.	6	7	8	9	Conventional Powder
	Sintered Properties[1] (Copper-10% Tin)				
Green Density g/cc	6.32	6.30	6.31	6.29	6.32
Compacting Pressure, (tsi)	11.8	11.5	12.0	11.9	14.2
Sint. Density g/cc	6.00	5.98	5.97	5.72	5.85
Sint. Modulus of Rupture, (psi)	29,000	28,900	28,600	24,500	22,600
Dimensional Change[1] (%)	+2.21	+2.23	+2.05	+2.65	+2.59

[1]Based on 1.2500" die size.

The foregoing illustrates some of the desirable properties of this process, for example: adaptability of the metal for powder metallurgy; good green strength, high purity, high sintered strength both with and without tin, and excellent compressibility.

Atomized, Prealloyed 316L Stainless Steel Powder

O.W. Reen and D.J. McMahon; U.S. Patent 4,028,094; June 7, 1977; assigned to Allegheny Ludlum Industries, Inc. describe the production of atomized, prealloyed stainless steel powder having an apparent density of at least 2.0 g/cm^3. The steel consists essentially of, by weight, up to 0.3% carbon, up to 2.0% manganese, up to 0.05% phosphorus, up to 0.04% sulfur, up to 2.0% silicon, 17.5 to 18.0% chromium, 10.0 to 14.0% nickel, 2.5 to 3.0% molybdenum, balance iron and residuals. It has an average particle size in excess of 50 microns.

Four atomized, prealloyed powders (A through D) having an apparent density in excess of 2 g/cm^3 were prepared. The chemistry of the powders is shown in Table 1.

Table 1

	Composition (wt. %)								
Powder	C	Mn	P	S	Si	Cr	Ni	Mo	Fe
A	0.010	0.02	0.008	0.003	0.83	16.30	10.25	2.80	Bal.
B	0.024	0.13	0.010	0.004	0.79	17.69	10.15	2.76	Bal.
C	0.017	0.18	0.013	0.020	0.79	17.83	13.30	2.17	Bal.
D	0.012	0.04	0.006	0.003	0.82	17.85	13.53	2.84	Bal.

Twenty compacts of each powder were formed by: blending the respective powders with 0.5 weight percent of stearic acid, a well-known lubricant; by double-action pressing of the powder in a 1-inch diameter die; and by sintering the green compacts in hydrogen at 2200°F. The density of the sintered compacts was approximately 6.4 g/cm^3.

Each group of 20 compacts was tested by subjecting it to a 5% neutral NaCl spray in accordance with ASTM Designation B117. Frequent examinations were made to observe when rust occurred. Testing was continued until 10 compacts in a group exhibited rust or for 500 hours, whichever occurred first. The results of the testing appear below in Table 2. Table 2 notes the number of compacts exhibiting rust, the number of compacts which appeared about to rust (indicated by discoloring or surface etching), and the number of compacts with no rust, for each time of observation.

Table 2

Resistance of Compacts to 5% Neutral Salt Spray (Hours Exposed)
Rating: "0" - No Rust; "1" - Potential Rust; "2" - Rust

Group	24	48	72	96	168	192	216	230	254	326	350	374	398	422	500
A	14–0	12–0	12–0	12–0		8–0									
	1–1	1–1	—	—		1–1									
	5–2	7–2	8–2	8–2		11–2									
B	20–0	20–0	20–0	20–0		12–0	11–0	10–0		9–0	9–0	9–0	7–0		
						4–1	1–1	2–1		2–1	2–1	2–1	3–1		
						4–2	8–2	8–2		9–2	9–2	9–2	10–2		
C	16–0	15–0	11–0	11–0	5–0										
	1–1	—	4–1	2–1	4–1										
	3–2	5–2	5–2	7–2	11–2										
D	20–0	20–0	20–0	20–0		20–0	20–0	20–0	20–0	20–0	20–0	20–0	20–0	20–0	20–0

Although Group A and B compacts have roughly the same chemistry, except for chromium, the corrosion resistance of Group B compacts is considerably superior to that of Group A compacts; and significantly, it is the Group B compacts, and not the Group A compacts, which satisfy the narrow chromium range for the process. Likewise, although Group C and D compacts have roughly the same chemistry, except for molybdenum, the corrosion resistance of the Group D compacts is considerably superior to that of the Group C compacts; and significantly, it is the Group D compacts, and not the Group C compacts, which satisfy the narrow molybdenum range for the process. Also, corrosion resistance of the Group D compacts is superior to that of the Group B compacts; and significantly, it is the Group D compacts, and not the Group B compacts which satisfy the narrow preferred nickel range.

Magnesium and Chromium for Improved Hardenability

S. Ito, Y. Morioka, Y. Kajinaga, M. Nitta and I. Sakurada; U.S. Patent 3,890,136; June 17, 1975; assigned to Kawasaki Steel Corporation, Japan have developed raw powders to be used for production of low-alloy steels having an excellent hardenability by powder metallurgy. They consist of 0.90 to 2.50% by weight of Mn, 0.35 to 1.50% by weight of Cr, 0.10 to 1.00% by weight of Mo, not more than 0.10% by weight of Si and the remaining being substantially Fe, provided that the sum of Mn and Cr is 1.70 to 3.10% by weight. When 0.10 to 1.00% by weight of Ni is additionally added thereto the hardenability is further improved.

Mn is the element for improving the hardenability and in this process, this element is added for such a purpose. In less than 0.90% of Mn, the effect for improving the hardenability is low, so that the lower limit of Mn is defined to be 0.90%. It is preferable in view of the hardenability that the Mn content is larger but the process is characterized in that the improvement of the hardenability is attained by the coexistence of the defined amounts of Mo and Cr. Therefore, the Mn content has the upper limit in view of the hardenability and when the content exceeds 2.50%, the hardenability deteriorates. Furthermore, when the Mn content exceeds 2.50%, the oxygen content abnormally increases and the powder granules themselves become hard, so that the cold compactibility becomes difficult.

Cr is the element which contributes to the improvement of the hardenability together with Mn and Mo. Thus, it is necessary to determine the Cr content in relation to the contents of Mn and Mo. The lower limit of the Cr content is 0.35% in view of the hardenability. The upper limit of the Cr content should be determined from this view but when the Cr content is large, the viscosity of the molten steel becomes high and unless the temperature is raised abnormally, the atomizing nozzle before granulation is clogged, the oxygen content increases and the cold compactibility of the powders becomes very difficult. Thus the upper limit of Cr is defined to be 1.50% by weight.

It has been shown that Mo is an element which is effective for the improvement of the hardenability, the annealing resistance, the annealing brittleness and the like, but this process is characterized in that Mo is an element which improves and promotes the hardenability owing to Mn and Cr. However, Mo is expensive and therefore it is necessary to define the range of Mo in economical and effective view. The Mo content must be 0.10 to 1.00% by weight in relation to the amounts of Mn and Cr.

The lower limit of the sum amount of Mn and Cr must be 1.70% in view of the hardenability within the range of 0.90 to 2.50% of Mn and 0.35 to 1.50% of Cr.

Even if it is preferable in view of the hardenability that the sum amount of Mn and Cr is larger, when the amount is too large the oxygen content of the powder increases and instead the hardenability deteriorates and also the mechanical properties are impaired, so that the upper limit of the sum amount of Mn and Cr is defined to be 3.10%. Moreover, when the sum amount of Mn and Cr exceeds 3.10%, the hardness of the powder granules themselves increases and the cold compactibility of the powders becomes difficult.

The abovementioned Mn, Cr and Mo are the elements which develop the effect by being present while it is preferable that Si be absent.

High Density by Simultaneous Reduction, Decarburization and Softening

Y. Kondo and M. Ohhori; U.S. Patent 3,887,402; June 3, 1975; assigned to Kozo Yoshizaki, Japan have developed a process for producing high density steel powders for powder metallurgy. The molten stream of low carbon steel is powder atomized by a high pressure water jet or inert gas jet to become powders.

The method is characterized in that low carbon steel having 0.02 to 0.38%, preferably 0.02 to 0.26% of carbon, or low carbon alloy steel, having alloying elements, such as Cu, Ni, Mo, Co, Cr, Mn, V, W, etc., and having 0.02 to 0.38%, preferably 0.02 to 0.26% of carbon is melted and the stream of the obtained molten steel is atomized by high pressure water jet or high pressure inert gas jet to become fine powders.

The low carbon steel powder or low carbon alloy steel powder, after drying includes 0.20 to 0.50%, preferably 0.20 to 0.35% oxygen content. It is represented by the percentage by weight of oxygen, decreased by reduction in the powder due to heating in H_2 gas at 1050°C for 60 minutes. This value is hereinafter described as the quantity of oxygen or the oxygen content. The carbon and the oxygen content in the powder are given to meet the equation: $C\% = K \cdot O\%$ ($K = 0.375$ to 0.75%; O% shows the quantity of oxygen in percent by weight), decreased in the reaction in case of reduction by heating in an inert gas; C% shows the necessary quantity of carbon (percent by weight) to be consumed in the reaction in case of reduction by heating in inert gas, being treated in heat at 600° to 1250°C in inert nitrogen gas or inert argon gas, whereby the powder can be reduced so as to make the oxygen content below 0.20%, preferably below 0.15% in the powder, and decarburized so as to make the carbon content below 0.02%, preferably below 0.01%, and simultaneously softened.

It is important that the oxygen content of the powder after water atomization or gas atomization should be made as small as possible between 0.20 and 0.50%, preferably between 0.20 and 0.35%, and kept constant.

After drying, the powders are heated in an inert gas such as nitrogen or argon, whereby the reduction, decarburization and softening of the powders are carried out simultaneously.

Example: With use of an electric furnace, low carbon steel scraps were melted. The quantity of carbon was so controlled as to be 0.10%.

The molten steel was continuously discharged from a tundish, having a hole of 12 mm ϕ, by gravity into an atomizing tank, in which water was present in the lower part. From a water jet nozzle, installed below the tundish, water was impinged to the molten steel stream under the pressure of 65 kg/cm^2 at the flow of 950 l/min. The molten steel was pulverized to powder.

The atomizing tank is closely sealed. The air in the tank was replaced by nitrogen gas before water atomization. The atmosphere of nitrogen gas was maintained during the atomization. The quantity of oxygen, contained in the powder after atomization, was controlled to be 0.30%.

The chemical composition of the atomized low carbon steel powder was carbon, 0.103%; silicon, 0.01%; manganese, 0.09%; phosphorus, 0.015%; sulfur, 0.013%; oxygen, 0.295%; and nitrogen, 0.001%. After coarse particles were removed, the size distribution was as follows:

Mesh	Percent
+80	1.5
−80+100	6.1
−100+150	13.3
−150+200	20.4
−200+250	24.0
−250+325	12.8
−325	21.9

100 kg of this powder were heated at 850°C for 1.5 hours in a belt conveyor type of inclined horizontal continuous muffle furnace with the nitrogen gas, having the purity above 99.9% and the dew point of −30°C, being flowed in the quantity of 10 m^3/hr.

The powder particles appeared slightly welded to each other after heating, but were fragile and easily restored to the original size by a light smashing.

After heating, the powder contained 0.002% of carbon, 0.036% of oxygen and 0.001% of nitrogen. The apparent density was 3.10 g/cc and the flow rate, 24.9 sec/50 g. The hardness of the powder was 87 on micro-Vickers. The oxygen content and the carbon content were sufficiently small without any increase of nitrogen. The hardness was sufficiently decreased and the powder was softened.

The optical microstructure of the powder after treatment in heat shows that the density is high without any pores and the recrystallization proceeds fully due to the heating.

About 6 g of the obtained powders were taken up and filled in a metal mold. With zinc stearate being used for die lubrication, the compaction forming was carried out at 5 t/cm^2. The powders were formed in a cylindrical shape of 10 mm in diameter and about 10 mm in height.

The green density of the compact was a high density of 7.08 g/cc. The edge stayability (Ratra value) showed a sufficient value of 0.46%.

Thus, the powder, obtained by this process, had the sufficiently small contents of oxygen and carbon, and the green density of the compact was so high that the powder was possessed of the characteristics and the high density, excellent for powder metallurgy.

In the case where powder produced by a conventional method was heated at 800° to 900°C in the presence of dissociated ammonia gas containing 75% hydrogen, the density of the compact formed under a pressure of 5 t/cm², was 6.75 to 7.00 g/cc.

High and Low Density Water-Atomized Powders

H.-C. Chao, J.H. Gross, R.R. Judd and R.L. Rueckl; U.S. Patent 3,900,309; August 19, 1975; assigned to United States Steel Corporation have developed a method for producing water-atomized steel powders with an apparent density greater than 3.2 g/cc, and preferably greater than 3.4 g/cc.

The method is applicable to water-atomized steel powders from virtually any source. Water-atomized steel powders generally contain impurities, primarily in the form of oxides, that must be removed before the powder has commercial value for the production of powder-metallurgical parts. In order to produce steel powders with maximum compressibility, it is also desirable that the final powders have a carbon content below about 0.10%, and preferably below 0.01%. However, it is generally impractical to provide an initial steel melt with such a low carbon content. Therefore, such steel melts may contain up to 0.8% C, but preferably less than about 0.15% C, and the carbon content of the atomized powders is thereafter lowered by annealing in a decarburizing atmosphere.

Atomization by high-pressure water jets results in rapid quenching of the liquid metal droplets during the early stage of the atomization process. Therefore, even if a relatively low carbon steel were employed (i.e., eliminating the need for decarburization) in the atomization process, it would still be necessary to anneal the powders to effect both softening and lowering the oxygen content thereof (to a value below about 0.2%). The initial oxygen contents of the as-water-atomized particles is generally far in excess of 0.2%, generally about 1.0%.

As a result of this high surface oxygen content and the particle configuration thereof, the as-atomized particles will pack to a high apparent density, i.e., well in excess of 3.2 g/cc. However, after the requisite annealing and reduction of the oxygen content thereby, the apparent density will be within the range of 2.8 to 3.2 g/cc. Annealing is conducted at temperatures of 1400° to 2100°F, in a reducing atmosphere such as hydrogen or dissociated ammonia for a time sufficient to effect the desired softening and reduction of impurities. This annealing treatment not only purifies the steel powder, but causes the particles to stick together in the form of a sintered cake, thereby necessitating a breaking-up of the cake to return it to powder consistency.

This process performs a true grinding operation in a disk mill employing a shear mechanism for comminution. It has now been found that by regulation of such a grinding operation, the final apparent density can be tailored to specific requirements, depending on the size distribution of the original as-atomized particles.

The size distribution of the as-atomized powder may be determined by conventional screen analysis. This screen analysis is then employed to develop the particle size characteristic (PSC) of the powder. It has been found that unduly coarse, as-atomized particles cannot be ground to achieve the desired objects of this process. Thus, to achieve the requisite grinding, it is necessary that at least 80%, and preferably greater than 95%, of the as-atomized particles be finer than 80 mesh (U.S. Series). While a number of different methods are available for defining PSC value, for purposes of this process, this value is determined in the following manner.

A cumulative weight percentage is first determined of the particles that are retained on U.S. Standard 100-, 140-, 200-, 230- and 325-mesh screen and the pan fraction. Thereafter, the so-determined cumulative percentages are totalled and divided by 100. Thus, utilizing this definition, an increase in PSC reflects a coarser particle size distribution and a low PSC is indicative of a fine particle size distribution.

It has been found that the apparent density increases as:

a. The PSC value of the as-atomized particles is increased,
b. The disk speed of the mill is increased, and
c. The mill gap is decreased.

It was found, within the specified temperature range of 1400° to 2100°F, that apparent density could be slightly increased by decreasing the annealing temperature. However, the use of lower temperatures would necessitate the employment of longer annealing periods. As a practical compromise of these competing effects, i.e, the achievement of high apparent density within a reasonably short annealing time, it is therefore preferable to anneal within the range of about 1700° to 1900°F.

Using a similar technique, *H.-C. Chao, J.H. Gross, R.R. Judd and R.L. Rueckl; U.S. Patent 3,954,461; May 4, 1976; assigned to United States Steel Corporation* have found that the apparent density of molding-grade, water-atomized steel powder can be significantly decreased to less than 2.8 g/cc by (a) employing powders with a finer particle size distribution, (b) decreasing the rotational speed of the disks and (c) increasing the mill gap.

Plasma Burner with Finely Ground Ores

H.R. Von Waclawiczek and K. Ermisch; U.S. Patent 3,862,834; January 28, 1975; assigned to Fried Krupp GmbH, Germany provide a method for making steel from finely ground, rich iron ores using a plasma burner and operating continuously from ore to finished steel, which method reduces heat losses and makes possible a plant of low cost and small size.

The method for reducing iron ore dust, comprises forming the dust into a cloud, transporting the cloud in reducing gas, reacting the cloud with the gas for enriching the dust, and introducing the transported cloud into the plasma stream of a plasma burner whose plasma stream is penetrating into the bath of a melting furnace.

The process provides the advantage that the continuously resulting, finished steel can be continuously submitted to further working, especially in a continuous casting mold and rolling mill trains following such a mold.

The iron oxide recovered from the exhaust gases of the refining units and the melting hearth, so-called LD-dust, can be fed back into the process and thus need not represent a loss.

The reduction of the ore begun in the heated transport device is continued in the plasma stream of the plasma burner and finished in the melting hearth or following refining unit.

In one embodiment, a cracked gas is used as the reducing gas. This cracked gas is produced by means of nuclear reactor heat loss in a heated cracking tube from coal, oil, or natural gas. In this way, it is possible to combine the basic process extremely economically with a nuclear power plant.

In another embodiment, the reducing gas is a cracked gas produced from natural gas and furnace exhaust gas in a cracking tube heated by the heat fed to the heated transport device. This gives a further improvement in the heat utilization.

Thin Flake

B.G.S. Lundgren; U.S. Patent 4,063,942; December 20, 1977; assigned to SKF Nova AB, Sweden provides a metal product which is well suited for the manufacture of substantially dendrite-free metal powder intended for powder metallurgical purposes, which has an amorphous to compact-grained structure and a chemically unaffected surface, and is free from dissolved and/or trapped gas.

It is a metal flake product consisting of a plurality of relatively thin, brittle and easily crushed, substantially dendrite-free metal flakes of amorphous to compact-grained structure.

The brittleness of the flakes varies with their hardness. With flakes of hardened steel the hardness should be at least about HRC = 60 to make them brittle and easy to crush.

According to the process, the metal flake product can further be manufactured by causing molten metal, which is of such a composition that rapid cooling of thin layers of the melt results in relatively brittle, crushable films, to impinge in a vacuum or protective gas upon at least one relatively cold metal surface of great cooling capacity, moving rapidly and generally across the direction of delivery of the molten metal, to form at least one discrete, relatively thin, flake-shaped layer of molten metal on the metal surface, and causing the relatively thin layer to solidify extremely rapidly on the cold metal surface of great cooling capacity, to form a relatively thin, brittle and easily crushed, substantially dendrite-free metal flake of amorphous to compact-grained structure. The metal powder can then be produced simply by crushing or grinding the metal flakes, e.g., in a ball mill.

For ordinary tool steels, the temperature of the moving metal surface should always be a minimum of 200°C lower than the temperature at which solidification is completed.

Because the molten metal is made to form a thin layer or film on the cold metal surface of great cooling capacity, considerably faster solidification can be achieved than by conventional methods of producing spherical metal powder from a melt. Because of the extremely rapid solidification, an essentially completely dendrite-free, very fine-grained to amorphous structure is obtained, and only negligible quantities of protective gas have had time to dissolve in the melt. It is also possible to carry out solidification so rapidly that completely dendrite-free flakes are obtained, and by working in a vacuum the risk of absorbing protective gas into the melt can be eliminated. When using the process, the cooling rate must be at least 10^4 °C/sec, preferably at least about 10^5 °C/sec, and expediently at least about 10^6 °C/sec, at least in the temperature range for solidification.

To facilitate the breaking up of the metal flakes into powder of the required particle size, the parameters which determine the flakes' dimensions during manufacture should be so mutually adjusted that the thickness of the flakes is at the most about 0.5 mm, preferably at most about 0.1 mm. Expediently, the parameters are also so mutually adjusted that the ratio of the flakes' length to thickness is at least 100, and the ratio of the flakes' width to thickness is at least about 20, and the ratio of the flakes' length to width is at most about 5.

Dendrite-Free Film

In related work *B.G.S. Lundgren; U.S. Patent 4,069,045; January 17, 1978; assigned to SKF Nova AB, Sweden* has developed a steel powder suited for powder metallurgical purposes consisting of an amorphous to compact-grained, essentially dendrite-free material with irregularly cornered particle shape. Such a steel powder may be produced by causing molten steel to form at least one discrete, relatively thin film on a relatively cold metal surface of great cooling capacity, causing the thin film to solidify extremely rapidly on the metal surface to form a brittle amorphous to compact-grained, in principle completely dendrite-free steel film, and crushing or grinding the brittle film into a powder of an irregularly cornered particle shape.

So that the metal films or flakes can be easily broken up into powder or the required particle size, the parameters which during manufacture determine the dimensions of the films or flakes should be so mutually adjusted that the thickness of the films or flakes is at most about 0.5 mm, preferably at most about 0.1 mm. Expediently, the parameters are also so mutually adjusted that the ratio of the foils' or flakes' length to thickness is at least 100, the ratio of their width to thickness is at least about 20, and the ratio of their length to width is at the most about 5.

Copper Addition to Improve Strength

Y.-T. Chen; U.S. Patent 4,170,474; October 9, 1979; assigned to Pitney-Bowes has found that adding 0.5 to 1.5% copper to known mixtures of 1.0 to 2.5% Ni, 0.15 to 0.30% Mn, 0.3 to 0.7% Mo, 0.3 to 0.7% C and 0.5 to 1.0% zinc stearate, the balance being iron, has resulted in a metal powder which, when compacted, sintered and heat treated, results in a metal part having unusually good properties. This is particularly true of the metal part that results from the sintering stage. As is known, it is advantageous to have a sintered piece with high strength as it may be subjected to stresses during the heat treatment stage.

In adding the copper to the known composition, it was found that the tensile strength increased after heat treating, the fracture toughness increased after heat treating and the machineability was substantially improved. In regard to machineability, it was found that drill bits used to machine the heat-treated products made in accordance with this invention lasted from 50 to 100% longer.

Example: A composition was prepared having the following ingredients:

Material	Amount
Copper (150 RXM, Glidden Metals Corp.)	0.82%
Graphite (Grade 1651, Southwestern Graphite Co.)	0.55%
Zinc Stearate PM (Penick Corp.)	0.75%
Prealloyed Powder (Ancorsteel 4600 v, Hoeganaes Corp.)	
Ni	1.8%
Mo	0.6%
Mn	0.25%
Fe	Balance

A 1.25" x 0.5" x 0.25" transverse rupture bar was compacted at 50 psi and sintered at 2050°F for 15 to 30 minutes, with a dew point of 35° to 55°F and under endothermic atmosphere. There was only 0.0006" shrinkage in length. After carbonitriding at 1550°F for 30 minutes, the base was oil quenched and tempered at 350°F for one hour. There was only 0.0008" expansion. In addition to maintaining stable dimensions, high strength and toughness was also achieved.

A number of samples of the above dimension from both the known and the process compositions were made in similar processes. In one series of tests the percentages of components as stated were kept constant and the amount of copper was varied from 0.77 to 1.22%. In another series of tests the percentages of components were kept constant as stated in the example and the carbon content was varied from 0.35 to 0.55%. All such samples were found to give superior results.

The samples resulting from the example were found to have a transverse rupture strength of approximately 160,000 psi after sintering and a transverse rupture strength of approximately 200,000 psi after heat treating for the disclosed composition. This compares with a transverse rupture strength of approximately 141,000 psi for the known composition in the sintered condition and approximately 196,000 psi in the heat treated condition. The disclosed composition was found to have a fractured toughness as sintered of approximately 21,000 psi/in^2 and 23,000 psi/in^2 in the heat-treated condition. This compares with the known composition having an as-sintered fractured toughness of approximately 21,000 psi/in^2 both in the sintered and heat-treated condition.

With respect to machineability, a test was run wherein a drill bit with a load of 24 pounds was applied to the above samples and rotated at a speed of 1,000 rpm. These loads were applied to samples which had a thickness of approximately ¼". For the disclosed material it was found that approximately 11 seconds were required to drill through a sample and for the standard material approximately 15 seconds were required. Even more significant was that the drill bit showed considerably more wear after drilling through the standard material than it did after drilling through the disclosed material.

The tensile strength of samples made from the disclosed composition was measured at 81,000 psi sintered and 125,000 psi heat treated whereas samples made from the known composition were found to have 75,000 psi and 110,000 psi strengths respectively.

As a consequence, it has been found that substantially better results are achieved in making parts using the known composition when 0.5 to 1.5% of copper is added.

Cobalt and Nickel Alloys

COBALT

Reduction of Cobalt Oxide Derived from Cobalt Ammine Carbonate

B.N. Doyle, W.H. Seibt, K.M. Sarkar and M.R. Benz; U.S. Patent 4,093,450; June 6, 1978; assigned to Sherritt Gordon Mines Limited, Canada have developed a process for producing ultrafine cobalt powder by using an aqueous solution of cobalt ammine carbonate, with the concentration of cobalt ions being in the range of from about 1 to 20 g/l. The solution is heated to drive off ammonia and carbon dioxide and precipitate ultrafine cobalt oxide. The cobalt oxide precipitate is then separated from the solution and heated in a reducing atmosphere to reduce the cobalt oxide to ultrafine cobalt powder.

Example: 77 kg of a commercial grade cobalt powder with an average particle size of about 50 microns were leached in an ammoniacal ammonium carbonate solution containing 180 g/l NH_3 and 65 g/l CO_2. The leach was carried out for 3 hours at a temperature of 80°C under a total pressure of 550 kPa using air as an oxidant, the partial pressure of oxygen gas being 110 kPa. The final volume of the solution was 980 liters and the concentration of cobalt ions was 78 g/l, indicating that over 99% of the cobalt had dissolved. There were about 7 mols of total NH_3 present, compared to approximately 1.3 mols of CO and 1.5 mols of CO_2.

After undissolved solids had been filtered off, the leaching solution was diluted with about 10 times its own volume of water to reduce the concentration of cobalt ions to 6 to 7 g/l. The solution was then boiled with steam at 240 kPa for 3 hours to drive off NH_3 and CO_2 and precipitate cobalt oxide. The Fisher number of the cobalt oxide precipitate was 1.06.

Slurry from the precipitation step was passed to a settling tank and allowed to settle for 1 hour, after which the supernatant liquor was decanted. The remaining slurry was agitated, passed through a 100 mesh screen, and filtered over a pan filter, with the resultant filter cake then being washed.

The cobalt oxide cake was then fed at a controlled rate onto a moving belt passing through a reduction furnace containing a hydrogen atmosphere. The entrance and exit of the furnace were purged with nitrogen and the cobalt material was maintained at a temperature of 630°C. The speed of the moving belt was such that the cobalt material remained in the furnace for approximately 6 hours.

The cobalt powder product was discharged from the furnace into a container purged with argon, then pulverized in an enclosed disc pulverizer purged with argon, and packaged in air-tight polyethylene bags, which were then sealed in steel drums.

After pulverizing, the final cobalt powder product had a Fisher number of 1.35 and the oxygen content was 0.56%.

Hydrogen Reduction of Cobaltous Carbonate

W.J. Huppmann and W. Kunda; U.S. Patent 3,994,716; November 30, 1976; assigned to Sherritt Gordon Mines Limited, Canada have developed a process for producing cobalt powder having a particle size of about 2 microns or smaller. The process involves two basic operations. Firstly, finely divided cobaltous carbonate is prepared by precipitation from a cobaltous ammine-ammonium sulfate solution by reacting the solution with carbon dioxide. Secondly, the cobalt carbonate precipitate is separated from the solution and is dry-reduced with hydrogen under controlled conditions of time and temperature to produce finely divided elemental cobalt powder.

The aqueous cobaltous ammine-ammonium sulfate solution has a free ammonia to cobalt molar ratio of at least about 2.0. The solution is heated in a closed reaction vessel to a temperature within the range of about 50° to 120°C. It is actively agitated and reacted with the carbon dioxide under a partial pressure of carbon dioxide within the range of from about 20 to 300 psi to precipitate cobalt from the solution in finely divided cobaltous carbonate form. Secondly, the cobaltous carbonate precipitate is separated from the solution and heated under a hydrogen atmosphere at a temperature in the range of about 400° to 700°C for a time sufficient to reduce the precipitate to metallic cobalt powders containing less than about 2% oxygen. The metallic cobalt powder is cooled under nonoxidizing conditions before exposing it to the atmosphere.

In one modification of the basic process, ultrafine cobalt particles, about 1 micron in size and smaller, having desirable magnetic properties, are produced by depositing a small quantity of a refractory oxide forming compound, such as magnesium hydroxide or yttrium hydroxide, on the particles of cobaltous carbonate precipitate prior to the reduction with hydrogen. During the reduction step, this compound functions to prevent migration of cobalt atoms during heating thus inhibiting the growth of the cobalt particles. Upon exposure to air after cooling, the refractory metal compound is converted to submicron refractory oxide particles. These particles, which are substantially inert and present only in a very small amount, do not adversely affect the magnetic properties of the powder in any way while they serve to stabilize the very fine cobalt particles by greatly inhibiting their pyrophoricity.

A powder product obtainable by this procedure consists of nonpyrophoric magnetic particles of cobalt of a size predominantly no larger than about 1 micron and preferably no larger than a single magnetic domain (0.8 micron). The cobalt particles have a minor amount of refractory oxide particles fixed in the surfaces thereof which serve to stabilize the powder, permitting handling and substantially preventing spontaneous oxidation of the particles on exposure to air such that the oxygen content of the powder (excluding that associated with the refractory oxide particles) remains substantially unchanged.

Another modification of the process is to grind the $CoCO_3$ precipitate, such as by wet ball milling, prior to the reduction step as a means of further decreasing the particle size of the precipitate.

Hydrometallurgical Process from Mixed Metal Sulfides

H.F. Wallace and D.R. Weir; U.S. Patent 4,108,640; August 22, 1978; assigned to Sherritt Gordon Mines Limited, Canada have developed a process for recovering metallic cobalt from an aqueous ammoniacal solution containing dissolved cobalt and other metals as ammine sulfates with substantially all the dissolved cobalt being in cobaltic form. The process includes contacting the solution with a water immiscible liquid ion exchange reagent dissolved in an inert organic diluent to selectively substantially completely extract the other metals from the solution and produce an organic extractant loaded with the other metals and an aqueous cobalt bearing raffinate substantially free of the other metals.

The cobalt bearing raffinate is separated from the loaded organic extractant and the loaded organic extractant is washed to remove substantially all ammonia therefrom. The other metals are stripped from the washed organic extractant which is then recycled to treat further solution. Cobalt is recovered from the raffinate.

Example 1: A sample of damp nickel-cobalt-zinc mixed sulfides filter cake had the following composition (%): Ni, 15.7; Co, 27.3; Cu, 0.006; Fe, 0.038; Zn, 14.8; Mn, 0.001; S_T, 32.7; $S(SO_4)$, 1.78; S^O, 0.01; insolubles, 3.15; and H_2O, where S_T is the total analysis of all forms of sulfur, $S(SO_4)$ is the analysis for hexavalent sulfur combined to form the sulfate radical, and S^O is elemental sulfur. For this mixed sulfide, the molar ratio of sulfur to the total of Ni + Co + Zn was 1.07:1. A sample of this mixed sulfide containing 445 g of solids on a dry basis was charged along with 250 g of ammonium sulfate and 290 g of ammonia added as reagent grade ammonium hydroxide, and sufficient water to form a total slurry volume of 2.5 liters into a 1 gallon capacity low pressure autoclave.

The autoclave was equipped with a mechanical agitator, outside gas heating and a water cooling system. The agitator had a marine three blade type impeller at the bottom and a four blade paddle type impeller located 4 inches from the bottom. The speed of agitation was 800 rpm. The autoclave charge was leached for 3 hours at 80°C under an oxygen partial pressure of 1.76 kg/cm^2 and resulted in a total autoclave gauge pressure of 5 kg/cm^2. The resulting leach solution had the following analysis (g/l): Ni, 26.5; Co_T, 45.2; Co^{++}, 0.1; $Co_{(hex)}$, 17.4; Zn, 24.8; S, 82.8; and Mn, 0.001 where $Co_{(hex)}$ was the analysis of the cobalt present as cobalt hexammine sulfate. Extractions of 99.2, 99.0 and 95.3% were observed for nickel, cobalt and zinc, respectively.

Example 2: A sample of unwashed nickel-cobalt-copper mixed metal sulfide filter cake had the following analysis:

	Ni	Co	Cu	Fe	S	NH_{3F}	NH_{3T}	CO_2
Solids (dry %)	22.0	11.6	6.12	4.05	28.2			
Liquor (g/l)	14.1	0.153	0.001			68.0	73.7	67.5

where NH_{3F} was titratable ammonia and NH_{3T} was the total ammonia as determined by the Kjeldahl method. Samples of this filter cake weighing 1550 g and containing 21.7% solids were leached in a one gallon titanium lined autoclave for 2.5 hours at 104°C. In addition to the damp mixed metal sulfide filter cake, the charge also consisted of 312.5 g of ammonia added as reagent grade ammonium hydroxide, variable quantities of ammonium added as reagent grade ammonium hydroxide, variable quantities of ammonium sulfate as shown in Table 1 and sufficient water to make a total slurry volume of 2.5 liters. The oxygen partial pressure during the leach was variable and is also shown in Table 1 along with the metal extractions obtained.

Example 3: A portion of the damp nickel-cobalt-copper mixed metal sulfide described in Example 2 was tray dried overnight at 90°C. Several 200 g samples of this dried mixed metal sulfide were leached in an autoclave under 1.0 kg/cm^2 oxygen partial pressure for 2.5 hours. Ammonia and ammonium sulfate additions as well as the leach temperature were variable and are given in Table 2. The total slurry volume was 2.5 liters. The metal extractions are also shown in Table 2.

Example 4: A 36.8 kg sample of the unwashed mixed metal sulfide filter cake described in Example 2, was charged into a 25 gallon stainless steel autoclave, equipped with a 7 inch diameter impeller which rotated at 900 rpm. To the mixed metal sulfide in the autoclave was also added 32.7 liters of ammonium hydroxide solution containing 220 g/l ammonia and 5.75 kg of ammonium sulfate. The charge was leached in this autoclave at 104°C and

Table 1

NH_3-$(NH_4)_2SO_4$ Leach of Unwashed Mixed Sulphide Slurry
Feed material - unwashed mixed sulphide slurry #4.
Equipment - 1 gallon titanium lined autoclave

Leach Time, Hours	Temp. °F	O_2 Pressure, psig	Charge, gm	NH_3 Addition, gpl	$(NH_4)_2SO_4$ Addition, gpl	Final Residue Analysis, % Ni	Co	Cu	Fe	Extraction, %[1] Ni	Co	Cu	Final Residue Wt., gm
2.5	220	50	2000	125	100	4.51	25.9	0.700	17.5	96.5	50.9	97.5	87.0
2.5	220	50	1550	125	100	1.16	0.483	0.184	49.5	99.7[2]	99.7[2]	99.7[2]	24.8
2.5	220	25	1550	125	0	2.21	1.54	0.384	44.4	99.1	98.8	99.4	29.1
2.5	220	25	1550	125	50	1.66	0.914	0.323	47.0	99.3	99.3	99.5	27.8
2.5	220	25	1550	125	170	1.84	0.766	0.445	48.2	99.3	99.4	99.4	28.5
2.5	220	25	1550	125	225	1.09	0.671	0.257	44.4	99.5	99.5	99.6	29.1
2.5	220	15	1550	125	100	1.47	0.754	0.271	48.4	99.4	99.5	99.6	26.1
2.5	220	15	1550	125	50	1.43	0.923	0.256	46.3	99.4	99.3	99.6	30.2

[1]Based on residue analysis unless otherwise noted.
[2]Based on product analyses.

Table 2

NH_3-$(NH_4)_2SO_4$ Leaching of Dried Mixed Sulphides
Charge: 200 gms dried unwashed mixed sulphide #4, NH_3 - variable, $(NH_4)_2SO_4$ - variable, H_2O to 2.5 liters.
Conditions: 15 psig O_2, 220° F unless otherwise noted, 2.5 hour leach time

NH_3 Addition gpl	$(NH_4)_2SO_4$ Addition gpl	Final Residue Analysis, gpl Ni	Co	Cu	Fe	Extraction, %* Ni	Co	Cu	Final Residue Wt., gm
100	100	8.83	3.25	0.611	37.0	95.6	96.9	98.9	39.5
125	100	4.96	2.03	0.520	38.7	97.6	98.1	99.1	36.3
150	100	4.77	1.21	0.526	43.3	98.0	99.0	99.2	32.1
125	100	5.38	1.78	0.540	42.9	97.7	98.5	99.2	32.2
125	0	14.9	4.00	0.506	25.8	89.3	94.5	98.7	57.2
125	50	8.82	2.46	0.528	37.4	95.6	97.7	99.0	37.9
125	75	8.58	1.99	0.490	39.0	95.9	98.2	99.2	34.5
125	125	4.38	1.33	0.870	44.6	98.2	99.0	98.7	31.0
125	100	3.26	1.32	0.558	46.4	98.7	99.0	99.2	29.6
125	100	4.49	2.30	0.795	41.4	98.0	97.7	98.4	37.3

*Based on residue analysis.
**Leach temperature = 150° F.
***Extreme drying conditions employed.

3.3 kg/cm^2 oxygen partial pressure for 2.5 hours. Extractions of 99.7%, 99.3% and 99.9% were obtained for nickel, cobalt and copper respectively.

Example 5: A sample of leach solution obtained by the ammoniacal leaching of a nickel-cobalt-copper mixed metal sulfide had the following concentration (g/l): Ni, 39.8; Co, 27.4; Cu, 16.6; Fe, 0.004; Zn, 0.025; S, 80.4; Co^{++}, 1.19; $Co_{(hex)}$, 9.90; NH_{3F}, 79.0 and NH_{3T}, 135.2. The contacting equipment used in this example consisted of ten laboratory size mixer settler units with a mixing chamber volume of 200 ml and a settler volume of 400 ml. The mixers were equipped with 1 inch pumping impellers which provided interstage pumping in addition to mixing of the two phases.

Copper and nickel were extracted from the aqueous feed solution by contacting it countercurrently with 4 stages of extraction in 25 vol % LIX64N (a commercially available mixture of 2-hydroxy-5-nonyl-benzophenoxime with a small quantity of 5,8-diethyl-7-hydroxy-6-dodecanone-oxime available from General Mills Chemical Corp.) and 75 vol % commercially available kerosene. The loaded organic was stripped by contacting it countercurrently in two stages using a feed of 250 g/l sulfuric acid. The loaded and stripped organic streams were each washed countercurrently in two stages of water wash.

Distilled water was pumped countercurrently in two stripped organic wash cells to remove entrained acid from the stripped organic and was then fed countercurrently to the two loaded wash cells to remove dissolved and entrained ammonia from the loaded organic. Most of the ammonia present in the loaded organic was washed from the organic in the first stage of contact with the wash solution while the acid washed from the stripped organic wash circuit served to neutralize traces of ammonia which were not removed in the first stage of the loaded organic wash.

The feed rates were 100 ml/min organic, 12 ml/min aqueous feed, 18.6 ml/min acid strip feed, 18 ml/min water wash. Because of the high organic to aqueous phase ratios, it was necessary to incorporate an aqueous recycle in each mixer settler where a portion of the settled aqueous phase from the settler was recycled back to the mixer feed and in this manner a mixer phase ratio of approximately 1:1 was maintained.

The cell by cell aqueous and organic analysis are given in Table 3. The raffinate contained approximately 0.05% of the feed nickel and less than 0.006% of the feed copper. The ratio of Ni + Cu to Co in the extract was 120:1.

Example 6: A sample of ammoniacal solution obtained by acid leaching and ammonia oxidation of a nickel-cobalt-zinc mixed sulfide had the following analysis: Ni, 24.5; Co, 22.8; Co^{++}, 0.041; $Co_{(hex)}$, 9.47; Cu, 0.026; Mn, 0.136; S_T, 63.2; NH_{3F}, 70.0 and NH_{3T}, 127.0. The organic-aqueous contactors were laboratory mixer settler units with a 600 ml mixer chamber and an 800 ml settler. Each mixer was equipped with a 1.5 inch pumping impeller. The extractant was 25 vol % LIX64N in "Napoleum 470" (Kerr McGee Corp.). The cell arrangement was similar to that described in Example 5. The nickel was stripped from the loaded organic with 125 g/l sulfuric acid and in order to ensure better phase disengagement in the wash circuit, the wash feed was made from 20 g/l sulfuric acid neutralized to a pH of 7 with ammonia. The feed rates were 105 ml/min organic, 15.4 ml/min aqueous feed, 10 ml/min strip feed solution (125 g/l sulfuric acid) and 14.3 ml/min wash solution feed. The results for a 6 hour period of continuous operation are shown in Table 4.

Example 7: The same feed was used as in Example 6. The cell arrangement was modified to incorporate four stages of extraction, two stages of loaded organic wash, three strip stages and one stage of stripped organic wash. In order to ensure that ammonia did not transfer to the strip circuit in the loaded organic phase, the pH of the final stage of loaded organic wash was maintained in the range of 5.5 to 7.0 by the controlled addition of 125 g/l sulfuric acid to the mixer. Feed rates were 106 ml/min organic, 16 ml/min aqueous, 8.8 ml/min strip feed solution (125 g/l H_2SO_4) and 13.6 ml/min wash solution feed. The results for 22 hours of operation are shown in Table 5.

Table 3

	Concentration (gpl)					Co/Ni or Ni/Co	Remarks
	Ni	Co	Cu	NH_3F	H_2SO_{4F}		
Aqueous							
Feed	39.8	27.4	16.6			.7	
Extr. 1	16.3	27.6	.08	47.2		2	
2	4.28	27.1	<.01	32.6		6	
3	.28	27.2	<.01	22.4		97	
4	.06	26.2	<.01	17.4		437	
Wash 1	.598	5.02	.0036	7.3		8	loaded organic
2	1.16	.59	.0056		NIL	2	wash.
Wash 1	17.2	.19	5.74		178	90	
2	3.17	.004	.88		233	792	
Wash 1	1.13	.002	.0084		3.6	565	stripped organic
2	.100	.0012	<.001		NIL	83	
wash.							
Raffinate	.020	23.8	<.001	10.5		1190	
Organic							
Feed	.008	.002	.052				
Extr. 1	4.04	.027	1.44				
2	3.36	.027	.084				
3	1.78	.030	.062				
4	1.00	.031	.058				
Wash 1	5.25	.036	1.43				Washed loaded
2	4.50	.028	1.41				organic.
Strip 1	.027	.020	.224				
2	.011	.016	.023				
Wash 1	.0010	.022	.041				Washed stripped
2	.008	.024	.041				organic.

	S_T	NH_{3T}	Co^{++}	Co^{HEX}
Additional **Raffinate analyses.**	79.1	97.5	.328	16.9

Table 4

		Aqueous						Organic					
	Hrs	1	2	3	4	5	6	1	2	3	4	5	6
E4	Ni	0.0030	0.0030	0.0030	0.0030	0.0030	0.0045						
E3	(gpl)												
E2													
E1								5.51	4.68	4.45	4.25	4.20	4.10
LOW1		0.0045	0.0030	0.0035	0.0040	0.0050	0.0055	5.72	4.57	4.45	4.25	4.20	4.10
LOW2								6.08	5.04	4.85	4.85	4.68	4.64
S1		34.6	42.4	43.6	40.4	41.2	43.6	1.43	1.33	1.31	1.00	0.888	0.917
S2								0.08	0.144	0.238	0.191	0.182	0.195
SOW1								0.036	0.054	0.102	0.150	0.166	0.191
SOW2								0.037	0.049	0.084	0.125	0.130	0.152
LOW1	Co	0.034	0.100	0.151	0.142	0.191	0.180						
S1	(gpl)	0.138	0.164	0.191	0.204	0.238	0.281						
SOW2								0.023	0.056	0.111	0.185	0.215	0.298
E4 Co/Ni		7566	7566	7566	7566	7566	5044						
S1 Ni/Co		250	258	228	198	173	155						

Raffinate Comp. = 22.7 gpl Co

Table 5

(Part 1)

AQUEOUS

Process Stage	Material	Time (hrs.) 2	4	6	8	10	12	14	16	18	20	22
E4	Ni	0.0040	0.0034	0.0048	0.0032	0.0056	0.0042	0.0038	0.0040	0.0038	0.0052	0.0040
E3	(gpl)	0.070	0.046	0.074	0.056	0.052	0.046	0.050	—	0.061	0.046	0.099
E2												0.975
E1												9.21
LOW1		0.018	0.021	0.038	0.013	0.027	0.045	0.033	0.035	0.029	0.026	0.037
LOW2												1.26
S1		55.0	53.9	52.4	51.6	51.7	51.8	50.4	52.3	51.6	49.3	45.6
S2												25.6
S3												4.26
SOW												2.83

(continued)

Table 5: (continued)

(Part 1)

AQUEOUS

Process Stage	Material	Time (hrs.) 2	4	6	8	10	12	14	16	18	20	22
LOW1	Co	0.373	0.438	0.382	0.378	0.500	0.380	0.329	0.363	0.362	0.306	0.316
LOW2	(gpl)											0.245
S1		0.335					0.379					0.363
S2												0.128
S3												0.030
SOW												0.019
LOW	Zn	.258					0.294					0.228
S1	(gpl)	0.695					0.519					0.559
E4	NH_{3F}	32.5					32.0					33.5
E3	(gpl)	N.S.S.					44.5					43.3
E2												47.0
E1												56.0
LOW1		7.5					6.0					7.5
S1	Free	17.7					24.2					33.7
S2	H_2SO_4											79.0
S3	(gpl)											119.0
SOW												2.4
E4	Co/Ni	5500	6470	4583	6875	3928	5238	5790	5500	5790	4230	5500
S1	Ni/Co	164					137					126

(Part 2)

ORGANIC

Process Stage	Material	2	4	6	8	10	12	14	16	18	20	22
E4	Ni											0.176
E3	(gpl)											0.362
E2												1.65
E1												4.65
LOW1												4.80
LOW2												5.00
S1												1.75
S2												0.180
S3												0.089
SOW		0.078					0.060					0.027
SOW	Co	0.560	0.570	0.610	0.600	0.600	0.616	0.620	0.630	0.620	0.630	0.636
	Cu	0.0025					0.0025					0.0015
	Zn	0.032					0.029					0.020
	Mn	0.0014					0.0002					0.0004

Example 8: Approximately 20 liters of a composite sample of raffinate from Examples 6 and 7 and other experiments was boiled to drive off most of the volatile free ammonia. The solution was then pumped through a column of 12 x 40 mesh activated carbon to remove the last trace of entrained organic material. The pH of the solution was then adjusted to 7.0 with sulfuric acid and the solution heated to 150°F and 340 g of cobalt powder was added to convert the cobaltic species to the cobaltous species. Sulfuric acid was added to maintain a pH of 7.0 during the reaction. After 60 minutes, the solution was filtered to remove excess cobalt powder.

The filtrate had the following analysis: Ni, 0.200; Co, 30.6; Cu, 0.0007; Fe, 0.0016; Mn, 0.018; Zn, 8.00; Co^{++}, 30.6 and NH_{3F}, 17.8. The solution was treated with hydrogen under pressure to produce metallic cobalt powder. A charge of 2,250 ml reduction feed solution was charged in a high pressure 1 gallon autoclave along with sufficient ammonium hydroxide to yield a free ammonia to cobalt molar ratio of 2:3, and 10.3 ml of 2.5% "Acrysol A-3" solution. Initially 200 g of wet fine cobalt powder (cobalt nucleation powder) was added.

The content of the autoclave was heated to 121°C and hydrogen was introduced to a partial pressure of 24.6 kg/cm². After 25 to 35 minutes when all hydrogen consumption had ceased, the autoclave was vented, discharged and the cobalt powder separated from the reduction end solution. The reduction end solution had a final pH of 7.35 to 7.50. A series of reductions or "densifications" was carried out wherein the cobalt powder from the previous densification was added back to the autoclave along with a fresh charge of reduction feed solution, ammonia and Acrysol A-3. After eight such densifications the bulk powder was washed and dried. Chemical analysis and physical characteristics of the cobalt powder are given in Table 6.

Table 6

Chemical Analyses of Cobalt Powders

	Co Plant Powder	D1	D2	D3	D4	D5	D6	D7	D8	D8 Reduced End Solution
Ni+Co+Cu									99.5	
Ni	not	0.315	0.289	0.337	0.338	0.421	0.457	0.495	0.520	0.017
Co	available									4.36
Cu									0.065	
Fe									0.0028	
S_T		0.0208	—	0.0376	0.0320	0.0294	0.0270	0.0254	0.0276	98.0
C		0.146	0.175	0.171	0.171	0.179	0.171	0.168	0.179	
Mn									0.0001	
Mg									0.007	
Zn									0.084	7.74
O_2									0.319	

Physical Characteristics of Cobalt Powders

	D8 Co Powder
Apparent Density	3.07
Fisher No.	22.0
+100	2.1
−100 +150	9.3
−150 +200	37.9
−200 +250	10.7
−250 +325	30.4
−325	9.6

Rare Earth Alloys by Gas Transport Reactions

C. Schueler; U.S. Patent 3,855,017; December 17, 1974; assigned to BBC Brown, Boveri & Company, Ltd., Switzerland has developed a process wherein powderized compounds of cobalt and one or more rare earth metals are produced in a gas transport reaction.

The gas transport reaction may take place in such a manner that the reaction components are heated in a closed reaction system or space into which the reaction components have been introduced in their solid state form in the presence of a transport substance, especially iodine. The reaction components and the transport substance form compounds which are volatile at the reaction temperature whereby the equilibrium constant of the reaction system is such that a reaction between the volatile cobalt compound and a rare earth metal is possible so that cobalt is released and a further reaction is possible between the volatile rare earth metal compound and cobalt whereby the rare earth metal is released.

The gas transport reaction may also be accomplished according to a modified embodiment wherein the cobalt in powder form and the rare earth metal or mixtures of rare earth metals in any suitable form, for example in coarsely comminuted form, are heated in a common reaction chamber to a temperature at which the rare earth metal or the rare earth metal mixtures evaporate to a substantial extent and at which temperature the cobalt vaporizes insignificantly.

In both types of reaction the compound which is thermodynamically stable under the selected stoichiometry and the selected reaction temperature is obtained after a suitably long reaction time has expired and after intermediate phases have been formed.

Example: *Cobalt Samarium*—11.76 g of cobalt and 7.20 g of samarium in the form of lumps, shavings or powder were weighed and placed in a tantalum ampoule having a length of 100 mm and a diameter of 20 mm equivalent to a volume of 31.5 cc; 0.22 g of iodine was added as a gas transport agent. The excess of samarium above the stoichiometric ratio of 5:1 of the end product Co_5Sm was about 20% in order to bind the iodides to be precipitated at the end of the reaction and in order to compensate for entrained oxygen-containing impurities in the starting materials and for reactions with the vessel wall.

Grinding weights of molybdenum, tungsten or tantalum were also introduced, for example 90 g of molybdenum pins of a diameter of 1 mm and a length of 5 mm.

The tantalum ampoule was then vacuum sealed and fused under vacuum into a quartz ampoule. The latter was rotated at a temperature of between 750° and 1150°C, e.g., 850°C, and at 10 to 500 rpm, e.g., 100 rpm. The reaction to CO_5Sm by isothermal gas transport takes place under these conditions, partly by way of thermodynamically unstable intermediate phases. The lower the selected reaction temperature is the more time the reaction will require. This fact and the material with which the reaction components, especially the iodine, come into contact (vessel walls, grinding weights) determine the abovementioned lower temperature limit. Nevertheless, such lower limit is not critical. The reaction is carried out for a period ranging from a few hours to several days, for example 24 hours, and then the ampoule is cooled and opened in ethanol.

After washing out the iodides (mainly SmI_3), Co_5Sm powder is obtained with a coercive force of $H_C > 20,000$ Oe. This high coercive force is caused by the grinding weights. Without the latter, the resulting product has a coercive force of only $H_C \approx 5,000$ Oe. There is yet no explanation for this surprising effect of the grinding weights. It would appear that the fact that the powder grains are kept small, approximately 1 to 5 μ diameter as compared to 20 to 100 μ diameter without grinding weights, due to the avoidance of primary sintering plays some part. Probably, however, suitable lattice defects are introduced by plastic deformations which serve as anchoring sites for the Bloch walls whereby magnetic reversal is made more difficult.

In order to vary the pressure reaction parameter, an inert gas may be provided in the reaction chamber which is small in order to reduce the transport distances. The temperature reaction parameter can be varied locally and/or as a function of time within the thermodynamic limits determined by the end product.

Cobalt-Base Reduced Powder

Y. Saito, M. Iijima and O. Mayama; U.S. Patent 4,131,450; December 26, 1978; assigned to Mitsubishi Kinzoku KK, Japan have developed a process for producing a cobalt-base reduced powder. They use a main raw material powder comprising cobalt oxide powder, chromium oxide powder and carbon powder with at least one metal oxide powder selected from the group consisting of manganese oxide powder, vanadium oxide powder, titanium oxide powder, tantalum oxide powder, niobium oxide powder, boron oxide powder, hafnium oxide powder, tungsten oxide powder, molybdenum oxide powder, iron oxide powder, nickel oxide powder and copper oxide powder. These powders are mixed and pulverized to prepare a mixed powder; the mixed powder is then reduced by heating under vacuum or at a temperature of 700° to 1150°C in a reducing atmosphere of a dew point of −20°C to obtain a reduced sponge-like mass. The sponge-like mass is finely pulverized, thereby manufacturing a cobalt-base reduced powder.

A cobalt-base sintered alloy produced by using the cobalt-base reduced powder thus manufactured has a relatively wide range of diffusion sintering temperatures applicable at the time of sintering thereof and has furthermore a high sintered density.

Example: A mixed powder consisting of 700 g of cobalt oxide (II) (CoO) powder, 227 g of chromium oxide (III) (Cr_2O_3) powder, 152 g of tungsten oxide (VI) (WO_3) powder and 86 g of carbon (C) powder, having a particle size of up to 100 mesh, was pulverized by a ball mill for 24 hours to prepare a fine mixed powder having an average particle size of 1.5 μm.

Then the fine mixed powder was reduced by heating to 1150°C and holding at this temperature for 2 hours in a gaseous hydrogen flow of a dew point of −40°C to obtain a reduced sponge-like mass consisting essentially of, in weight percentage:

Chromium	31.0%
Tungsten	12.0%
Carbon	2.5%
Cobalt and incidental impurities	balance

The reduced sponge-like mass could easily be pulverized by a hammer mill to a size of up to 100 mesh, thus permitting manufacture of a cobalt-base reduced powder of the process.

The cobalt-base reduced powder thus obtained had a sponge-like structure and an irregular exterior shape, and showed substantially the same chemical composition as that of the reduced sponge-like mass. Properties of the cobalt-base reduced powder are indicated in Table 1. For comparison purposes, properties of another cobalt-base alloy powder, having substantially the same chemical composition as that of the cobalt-base reduced powder, manufactured by the conventional atomizing process shown in Table 1.

Table 1

	Apparent density (g/cm^3)	Apparent particle size distribution (wt. %)			Primary powder particle size (μm)
		100 to over 250 mesh	250 to over 325 mesh	up to 325 mesh	
Co-base reduced powder manufactured by the process	3.78	24	11	65	5.6
Co-base alloy powder manufactured by the conventional atomizing process	4.03	56	19	26	at least 50

As shown in Table 1, the particle size of the cobalt-base reduced powder manufactured by this process is very fine as compared with that of the alloy powder manufactured by the conventional atomizing process.

Then, with the cobalt-base reduced powder manufactured by the process and that manufactured by the conventional atomizing process as the material powders, respectively, two kinds of green compact were formed from the material powders under a pressure of 6 tons/cm^2. This was followed by the sintering of the green compacts at a room temperature of 1230°C for 1 hour under vacuum, to manufacture two kinds of cobalt-base sintered alloy.

The green densities and the sintered densities of these two cobalt-base sintered alloys thus manufactured are indicated in Table 2.

Table 2

	Green density (g/cm^3)	Sintered density (g/cm^3)
Co-base sintered alloy produced from Co-base reduced powder manufactured by the process	5.6	8.2
Co-base sintered alloy produced from Co-base alloy powder manufactured by the conventional atomizing process	5.8	6.0

As shown in Table 2, while the green density of the green compact produced from the cobalt-base reduced powder manufactured by this process is slightly lower than that of the green compact produced from the powder manufactured by the conventional atomizing process, the sintered density of the former is far higher than that of the latter.

High Hafnium Carbide-Containing Alloys

N.J. Grant and R.K. Robinson; U.S. Patent 4,084,964; April 18, 1978; assigned to Massachusetts Institute of Technology have developed a class of cobalt-based alloys and the like containing hafnium carbide (HfC) of particle size finer than about 3 microns and containing HfC in the range 2 to 15% by volume. The class has good strength characteristics at both very high and very low temperatures, it exhibits very good stability, and it has excellent oxidation characteristics even at temperatures at 2000°F and above. In addition, the alloy has ductility and low-temperature toughness characteristics that make it attractive for wrought processes.

The cobalt-based alloy contains 10 to 30 weight percent chromium, 4 to 12 weight percent of tungsten and/or molybdenum, 3 to 15 weight percent hafnium, sufficient carbon to produce hafnium carbide in the alloy, and the balance cobalt. It is understood that up to 30% of the cobalt can be substituted by nickel or nickel plus iron. The process includes the steps of raising the temperature of the metal composition until it is in the liquid state and then rapidly cooling the liquid melt at a rate greater than the order of 10^2 and as high as 10^7 °C per second, thereby to form a solid having secondary dendritic arm spacings of 10 microns or finer.

The necessary rapid cooling can be effected by atomization of the liquid melt into fine droplets which are rapidly quenched in an atomizing fluid or are quenched against a metallic substrate, for example, copper rolls. Fluids that can be used include inert gases, air, steam, water, nitrogen, and hydrogen. The use of high quenching rates sharply optimizes the formation of HfC (i.e., preferentially forms HfC) to the general exclusion of the less desirable carbides which would form at slower solidification and cooling rates, carbides such as the chromium and molybdenum or tungsten carbides.

Example 1 below describes the preparation of a single alloy using this process; Example 2 is more general and the process outlined includes the process of Example 1.

Example 1: To illustrate the concepts described herein, one powder metallurgy (P/M) Co-HfC alloy is described: a Co-21, Cr-10, Ni-, 9HF, 5Mo, 0.6 C alloy was melted, hydrogen-vacuum atomized, consolidated by hot isostatic pressing (HIP), hot extruded and evaluated. A stable, fine and uniform dispersion of HfC (nominally 3 atom percent) stabilized the fine, equiaxed, as-extruded grain structure for short term exposures at temperatures in excess of 2400°F (1316°C). [Regular Co base alloys are in fact molten (in part) at 2350°F and higher]. Stored energy due to thermomechanical working (TMT) of the as-extruded rod (cold work plus intermediate anneals) was retained up to 4 hours exposure at 2300°F (1260°C) as indicated by tensile property data in Table 1 below:

Table 1

Alloy	YS (0.2%) ksi	UTS ksi	Elong %	R.A. %
As HIP and extruded	132	187	4	4
30% reduction in area with intermediate anneals	200	250	13	14
30% reduction in area with intermediate anneals + 4 hours 2300°F	174	235	13	14

Thermomechanical working was found effective to increase room temperature ultimate tensile strength and yield strength from 187 and 132 ksi to 250 and 200 ksi, respectively. Such working increased the stress rupture strength for 100 hour life at 1800°F from 5 to 9.3 ksi; ductilities of this alloy are comparable to typical P/M alloys exceeding the ductilities of precision-cast cobalt alloys by factors of 2 to 20. Oxidation resistance in static air at 2000°F (1093°C) is comparable to or exceeds that of the best commercially available cobalt-based super alloy. Grain structure, even after swage-anneal treatments, was found to be very stable to annealing treatment up to 100 hours at 2300°F.

Example 2: Cobalt-based hafnium-carbon alloys, with varying amounts of nickel, chromium and molybdenum in solution were dispersion hardened with up to 12 weight percent HfC, using the rapid quenching powder metallurgy approach. Alloys containing the desired amount of Hf + C and solid solution additives were obtained initially in vacuum-melted ingot form using conventional ingot technology, or alloying additives were put into the melt prior to atomization, or combinations of each method were employed. Stoichiometric, substoichiometric, and superstoichiometric additions of carbon were made so as to be able to form exactly the compound HfC, HfC with excess Hf and HfC with excess carbon contents, respectively.

When Hf was added to the melt just prior to atomization, it was stirred into the melt after all other alloying additions had been made. After stirring in the Hf addition, the melt was held at about 2800°F (1500°C) for about one-half hour prior to atomizing. Using this method, complete homogenization of the melt was effected without undue oxidation-reaction of the hafnium with its environment, e.g., atmosphere and/or crucible. The molten alloys were atomized using steam, rotating electrode, and hydrogen-vacuum techniques to form rapidly quenched alloy particles having average sizes of 2,000, 350 and 100 microns, respectively. The only powder requiring subsequent cleaning was the steam atomized powder which was subjected to repeated cycles of one hour each in hot aqueous, near alkaline potassium permanganate, followed by tumbling in inhibited HCl (hydrochloric acid) repeated up to 4 to 10 cycles.

After atomization (and cleaning for steam atomized powder), the powders were loaded into mild steel cans which were subsequently sealed except for a vacuum outgas tube through which the loaded can was evacuated while heating to about 700°F (371°C), at which temperature the evacuation tube was crimped and welded shut. The evacuated and sealed cans were then cooled and placed in a hot isostatic pressing (HIP) autoclave which was subsequently pressurized from 15,000 to 30,000 psi for 1 to 2 hours at temperatures from 2000°F (1093°C) to 2325°F (1274°C) using argon gas as the pressurizing fluid.

After hot isostatic pressing or otherwise consolidating the powders into a fully dense form, the now consolidated powders were either further worked using hot or warm rolling or hot extrusion or the consolidated powders were tested and evaluated in the hot isostatically pressed form. The majority of the alloys were both HIP and extruded; therefore, operating parameters for both of these operations were found and are given in the thesis. Standard extrusion methods were employed with extrusion preheat temperatures ranging from 2000°F (1093°C) to 2050°F (1121°C) and extrusion ratios varying from 11:1 to 22:1.

Consolidated alloys were examined in detail using various mechanical, chemical, physical and metallographic techniques. The testing procedures employed are known to workers in this art and, further, are discussed in great detail in the Robinson thesis. It should be noted that the discussion is applicable regardless whether the consolidated-powder, metallurgy product was hot isostatically pressed only, or whether it was subsequently hot or warm worked by extrusion and/or rolling methods.

In addition to examining the produced properties of these alloys, additional metallurgical treatments were conducted to enhance the strength, structure, and stability of the alloys. High temperature solutionizing, aging, grain coarsening and oxidation treatments were employed. All treatments at temperatures up to and including 2000°F (1093°C) were conducted in a static furnace-air atmosphere. For temperatures exceeding 2000°F (1093°C), test specimens and samples were either vacuum sealed in Vycor or quartz tubing, or they were packed in alumina boats which were subsequently placed in ceramic tubes continuously purged with flowing argon gas. Specimens used for the oxidation studies were placed in loosely-fitting, high purity alumina tubes to collect any loose scale. In all cases thermocouples were plced in the immediate vicinity of the specimens. All cooling was air cooling unless water quenching is otherwise specified.

Thermochemical treatments were conducted using the standard swage-anneal techniques. All HIP plus extruded P/M alloys were given swage-anneal treatments consisting of cold swaging up to a total of 40% reduction in area (RA) with intermediate anneals after about 10% RA for 1 hour at 1500°F (816°C), 1800°F (982°C), and 2300°F (1260°C) for selected alloys. The majority of the swaging was done at M.I.T. on a Fen 3F swager. The mild steel cladding was removed from the extruded rods by dissolving in a 1:1 nitric acid solution prior to swaging.

NICKEL AND COBALT

Catalytic Hydrogen Reduction from Aqueous Solutions

E.C. Chou, R.P. Crnojevich and H. Koehler; U.S. Patent 3,989,509; November 2, 1976; assigned to Amax Inc. describe a method for efficiently precipitating metal powder selected from the group consisting of nickel, cobalt and copper from aqueous solutions using sulfur-free hydrogen at elevated pressure and temperature. The improvement comprises adding to the solution a small but effective amount of a catalyst as an antiplating agent. The catalyst is an alkali metal or ammonium salt of a copolymer of maleic acid anhydride and an unsaturated branched chain aliphatic olefinic hydrocarbon containing 4 to 10 carbon atoms.

Example 1: *Additive Testing*—Hydrogen reduction batch tests were carried out in a two-liter stainless autoclave with a glass liner. The conditions under which the tests were conducted except for the particular catalyst used were as follows:

Temperature	190°C (375°F)
Hydrogen partial pressure	450 psig
Agitation (2 impellers)	1,200 rpm
Nucleation time	30 min
Volume of solution	600 ml

Nickel ammonium sulfate feed solution	Ni - 41 gpl
$(NH_4)_2SO_4$	125 gpl
NH_3:Ni mol ratio	2:1

After sealing, the autoclave is purged with nitrogen first and then with hydrogen. The nucleating catalyst (for seeding tests only) and antiplating agents are introduced into the autoclave with required hydrogen pressure when the temperature reaches 150°C (302°F). The temperature then reaches 190°C (375°F) during the hydrogen injection and the temperature maintained during the test.

As soon as the nucleation is finished, the hydrogen supply, agitators and heat source are shut off. The autoclave is cooled, disassembled, emptied and cleaned with mild nitric acid between runs to insure that no nickel is left to catalyze the subsequent reduction runs. The so-called antiplating agents which showed considerable plating (at least 3% or higher) or poor reduction efficiency, were eliminated. The agents which appeared to show promise were again used through three densifications using the nickel powder as nuclei for the seed run.

The nickel powders were evaluated by using a micromesh screen (5 to 53 microns), the size distribution being determined by wet screening, drying and weighing. Apparent density was determined by weighing 1 cm^3 of the metal powder.

Tests conducted using 0.3 gpl of the sodium salt of maleic acid anhydride-diisobutylene copolymer alone resulted in a percent reduction of 98% and percent plating loss of only 0.3%.

However, in running seeding tests ferrous ions were employed as a nucleating catalyst in an amount corresponding to 0.3 gpl.

Using the conditions stated above, except for the inclusion of ferrous ions as a nucleation catalyst (0.3 gpl), tests were conducted comparing materials known by certain trade names with the catalyst of the process. The materials tested are given as follows:

Table 1

Dispersants	
Acrysol A-3	Polyacrylic acid $(CH_2CHCOOH)_n$
Tamol 850	Polyacrylic acid $(CH_2CHCOOH)_n$
Tamol N	Sodium salt of sulfonated formaldehyde and naphthalene
Flotation promoters	
Aerofloat 242	Aryl dithiophosphoric acid
Sodium Aerofloat	Sodium diethyldithiophosphate
Aero Promoter 845	Anionic sulfonate promoter

(continued)

Table 1: (continued)

Surfactant	
Aerosol OT	Anionic (2-ethyl sodium sulfosuccinate)
Flocculants	
Aerofloc 550	Anionic flocculant
Primafloc C-3	Cationic polyelectrolyte
Flotation reagent	
Aero 301 Xanthate	Sodium secondary butyl xanthate
Mineric B	Copper floater
Catalyst used in the process	Sodium salt of maleic acid anhydride, diisobutylene copolymer

0.3 gpl of each of the foregoing was used together with 0.3 gpl ferrous ions in the form of ferrous sulfate. The percent reduction was determined for the nickel solution stated before together with the amount of plating on autoclave parts. A percent plating of at least about 3% is deemed to be undesirable based on the total nickel reduced. The following results were obtained under the conditions given.

Table 2

Additives	Percent Plating	Percent Reduction
Acrysol A-3	3	95
Tamol 850	0.5	84
Tamol N	5	97
Aerofloat 242	3	25
Sodium Aerofloat	10	25
Aero Promoter 845	25	99
Aerosol OT	50	25
Aerofloc 550	50	25
Primafloc C-3	5	98
Aero 301 Xanthate	5	90
Minerec B	*	75
Catalyst used in the process**	0.3	99

*Severe plastering.

**Sodium salt of maleic acid anhydride-diisobutylene copolymer.

As will be clearly apparent, the catalyst employed in carrying out the process provides markedly improved combination of results, that is, high reduction efficiency (99%) coupled with low plating losses (0.3% plating).

On the other hand, the remaining additives tested either have too high a plating loss and/or too low a reduction efficiency. For example, Acrysol A-3 and Tamol 850 (polyacrylic acids) do not provide the desirable combination of results, the former exhibiting a relatively high plating loss of 3% coupled with an overall reduction efficiency of 95%, while the latter provides a low plating loss coupled with a relatively low overall reduction efficiency.

Example 2: *Catalyst Comparison*—Tests were also conducted comparing the additive Tamol 850 (polyacrylic acid) with the catalyst of the process (sodium salt of maleic acid anhydride-diisobutylene copolymer) in which an average of three densification runs was employed, each of the foregoing additives being used in an amount of about 0.3 gpl in the initial nucleation run and about 0.2 gpl in densification run. The results obtained are given as follows:

Table 3

. . . . Conditions	
Temperature	190°C (375°F)
Hydrogen partial pressure	450 psig
Agitation (2 impellers)	1,200 rpm
Nucleation time	30 min
Densification time	20 min
Nucleating catalyst	0.3 gpl Fe^{++}
Volume/test	600 ml
Nickel ammonium sulfate feed solution	
Ni	41 gpl
$(NH_4)_2SO_4$	125 gpl
NH_3/Ni mol ratio	2:1

	Tamol 850	Process Catalyst
Percent reduction		
Nucleation	85	98
Densification, average of 3	91	99
Apparent density, g/cc		
Nucleation	2.07	0.62
Third densification	2.26	3.00
Screen analysis, distribution, %		
Nucleation		
+53 μ	0.14	34.32
+38 to 53 μ	0.60	19.66
+20 to 38 μ	10.14	32.36
+15 to 20 μ	9.20	5.12
+10 to 15 μ	44.12	4.87
+5 to 10 μ	30.88	2.64
-5 μ	4.92	1.02
Third densification		
%53 μ	28.36	25.29
+38 to 53 μ	43.68	18.51
+20 to 38 μ	23.46	36.12
+15 to 20 μ	0.84	11.09
+10 to 15 μ	1.54	7.53
+5 to 10 μ	0.70	0.59
-5 μ	1.24	0.14

(continued)

Table 3: (continued)

	Tamol 850	Process Catalyst
Nickel powder–assay after 3 densifications, %		
Ni	99.0	99.0
Fe	0.0025	0.0025
S	0.002	0.002

The nucleation runs, that is, the runs before densification, showed the process catalyst to be superior to Tamol 850, in addition to being superior following the densification runs. Following the third densification run, the apparent density of the nickel powder produced in accordance with the process was higher, this being attributed to the fact that a better size distribution is obtained with the process. It will be noted that with either additive, a high purity nickel product is obtained, the process one being more efficient, however.

Example 3: *Nucleating Agent Testing*—Some tests were conducted in which no additives were employed and in which Tamol 850 was compared to the catalyst of the process with and without nucleating agents. Where the nucleation agent employed was ferrous ions, the amount was 0.3 gpl, and where the nucleation agent was NaCN plus Na_2S, the amount was 1.5 gpl NaCN combined with 0.08 gpl Na_2S.

The following autoclave conditions were employed:

Conditions	
Temperature	190°C (375°F)
Hydrogen partial pressure	450 psig
Agitation, 2 impellers	1,200 rpm
Nucleation time	30 min
Volume of test	600 ml
Nickel ammonium sulfate solution	
Ni	41 gpl
$(NH_4)_2SO_4$	125 gpl
NH_3/Ni mol ratio	2:1

The results are given in Table 4 as follows:

Table 4

Additives	Percent Reduction	Percent Plating*
None	5	75**
Tamol 850	44	45
Tamol 850 plus Fe^{++}	84	0.5

(continued)

Table 4: (continued)

Additives	Percent Reduction	Percent Plating*
Sodium salt of maleic acid anhydride, diisobutylene copolymer (A)	98	0.3
(A) plus Fe^{++}	99	0.3
(A) plus NaCN + Na_2S	99	0.2

*Percent nickel plated to total nickel reduced.
**Includes a precipitate that was sticky.

As will be noted, Tamol 850 used alone does not provide the results of catalyst (A) used alone. Even when Tamol 850 is used with ferrous ions as a nucleating agent, only 84% of the nickel in solution is reduced, while, by contrast, catalyst (A) employed in the process provides markedly improved results.

Thus, it is preferred that the catlayst be employed with nucleating agents, such as those selected from the group consisting of ferrous ions in an amount ranging from 0.05 to 0.5 gpl and the combined nucleating agents NaCN plus Na_2S in a total amount ranging from about 0.1 to 2 gpl, each being present in an amount of at least 0.05 gpl.

Caustic Washing to Remove Silicon

R. Crnojevich, E.I. Wiewiorowski and L.R. Tinnin; U.S. Patent 4,149,875; April 17, 1979; assigned to Amax Inc. describe a method for decreasing the silicon content of relatively pure nickel and/or cobalt powder, wherein silicon is present as an impurity substantially in the oxidic state. The method comprises mixing the metal powder in an aqueous solution of an alkali metal hydroxide, e.g., sodium hydroxide, and maintaining the powder in contact with the solution until the silicon content of the powder has been reduced to the desired level.

The caustic wash treatment can be carried out over a preferred range of NaOH concentrations of, for example about 5 to 20% NaOH by weight. The method may be carried out at atmospheric pressure at temperatures below the boiling point, for example, in the range of 120° to 160°F (49° to 71°C). The time of treatment may depend on the silicon content and on the physical characteristics of the powder. Generally, the time does not exceed about five hours and is usually from about 0.5 to 2 hours. It is essential that the metal powder be intimately mixed (suspended) with NaOH solution during the treatment. The metal slurry pulp density can advantageously be about or above 50% by weight. The pulp density can range broadly from about 10 to 70%. In tests of the process the reduction in silicon content of various samples was up to about 70 to 80%, e.g., from a level of about < 200 ppm to a level of < 40 ppm, and generally to a level in the range of about 10 to 30 ppm Si.

The process is applicable to the reduction of silicon from relatively pure metals containing up to about 500 ppm and usually up to about 200 ppm silicon. The caustic treatment is generally applied to nickel and/or cobalt metal powders containing amounts of silicon in excess of 50 ppm. However, it also applies to powders containing less than about 50 ppm silicon.

In addition to silicon, other impurities soluble in caustic may be reduced, such as Fe, Al, Cr and Zn, among others.

Example: Each of a number of nickel powder samples produced by hydrogen reduction obtained from refinery production runs was treated with 5 and 10% NaOH solutions, respectively, for 1 hour at 150°F (65°C) at a pulp density of 50% solids with the following results.

	Untreated Powder	5% NaOH Treatment		10% NaOH Treatment	
Samples	Si Content ppm	Final Si Content ppm	Silicon Reduced %	Final Si Content ppm	Silicon Reduced %
(1)	56	22	61	10	82
(2)	56	30	46	<10	>84
(3)	117	<10	>92	<10	>92
(4)	250	68	73	54	78
(5)	92	57	38	31	65
(6)	73	14	81	14	81
(7)	162	10	94	<10	>94
(8)	40	22	45	<10	75
(9)	163	14	91	16	90
(10)	46	<10	88	14	70
(11)	61	<10	>84	<10	>84

While Samples (4) and (5) at 5% NaOH concentration showed a substantial reduction in silicon content in one hour of treatment, Sample (4) being reduced to 68 ppm at 73% reduction and Sample (5) to 57 ppm at 38% reduction, it will be appreciated that the residual silicon content can be further reduced by merely increasing the time of treatment to substantially above one hour and/or by increasing the NaOH concentration. Thus, at 10% NaOH concentration, Sample (4) was reduced to 54 ppm and Sample (5) to 31 ppm in one hour. Further treatment of Sample (4) in excess of one hour will reduce the residual silicon content to less than 50 ppm.

Thus, it is clearly apparent from the table that the treatment of the powders with alkali metal hydroxide (e.g., NaOH, KOH, etc.) is very effective in reducing the silicon content to below 50 ppm. The solution concentration may range from about 5 to 30% by weight of alkali metal hydroxide. Where the powder being treated contains relatively high amounts of silicon, e.g., in excess of 100 ppm, the concentration of the solution is preferably on the high side, such as from about 10 to 20% by weight or higher to reduce the silicon to a level not exceeding 50 ppm. In any event, the time of treatment at a particular alkali metal solution concentration should be sufficient to provide a final residual silicon of less than 50 ppm, particularly where the initial silicon content is substantially in excess of 50 ppm.

Oxide Strengthening

S. Baranow; U.S. Patent 4,010,024; March 1, 1977; assigned to Special Metals Corporation has developed a process for preparing metal having a substantially uniform dispersion of hard filler particles, by admixing metal powder from the group consisting of nickel, cobalt and alloys thereof and oxide particles having a negative free energy of formation at 1000°C of at least as great as that of aluminum oxide. The mixture is milled for a period of time sufficient to effect a substantially uniform dispersion of the oxide particles in the metallic powder, the milling being in an oxygen containing atmosphere containing sufficient oxygen to substantially preclude welding of particles of the metallic powder to other particles of the metallic powder. The oxygen is present in an amount greater than that encountered when powders are milled in the air often present in a mill, and is supplied from an external gaseous source or from an oxygen-bearing compound which is admixed with the metal powder and oxide particles. The dispersion strengthened powder has a Fisher No. of less than 15, preferably less than 10. It is heat treated to remove excess oxygen therefrom and subsequently blended with powder which is not readily reduceable by hydrogen.

The dispersion strengthened metal powder produced according to the process is suitable for consolidation by any number of methods, e.g., extrusion, rolling, swaging and forging. The following examples are illustrative of several aspects of the process.

Example 1: A charge of 30 pounds of carbonyl nickel, 0.48 pounds of yttrium oxide and 1.5 pounds of ammonium bicarbonate were placed in an attritor and milled for eight hours under a blanket of argon. During milling, the bicarbonate was continuously decomposed as evidenced by the smell of ammonia emanating from the attritor. The powder was discharged from the attritor, a fresh charge was added, and the entire process was repeated. Both lots of powder had a Fisher No. of approximately 5.5. The two lots of powder were subsequently combined and passed through a hydrogen push-pull furnace at the rate of 30 pounds of powder per hour. The peak furnace temperature was 1750°F.

A small portion of this powder was packed into a mild steel container of 2⅝" o.d., evacuated, sealed, and extruded at 2050°F, at an extrusion rate of 12:1. The extruded material was cold swaged 70%. After 25% working, the 2000°F ultimate tensile strength was 8.0 ksi whereas after 70% working, the ultimate tensile strength was 13.1 ksi. Specimens were tested after a one hour anneal at 2000°F. These properties are typical for material of this class; pure nickel similarly worked has a tensile strength of about 3 ksi at 2000°F.

Example 2: About 20 pounds of a 150 mesh master alloy powder consisting of 76% chromium and 24% aluminum was milled for two hours under argon in an attritor, and discharged subsequent to cooling for about one-half hour. Most of the powder product of Example 1 was combined with this master alloy powder and tumbled in a twin shell blender for about an hour. The blended powder was milled in two batches for one-half hour each, and combined. The composition of the combined powder was 16% chromium, 5% aluminum, 1.2% yttrium oxide, balance nickel. About 70 pounds of the powder was pressed to 50% of its theoretical density in a mild steel can of 7¾" o.d.

The can was evacuated, sealed and extruded at 2050°F at a reduction in area of 15:1. Subsequent to recrystallization at 2450°F, it was found that a medium grain size structure had been developed with a desirable cube-on-edge texture; such a texture is necessary for high thermal fatigue resistance. Transverse mechanical properties at 2000°F showed a stress rupture life in excess of 20 hours at 5.5 ksi and an ultimate tensile strength of 10.3 ksi.

Example 3: Two batches of powder each containing 19.7 pounds of nickel and 0.3 pounds of yttrium oxide were milled for four hours in an attritor, in an atmosphere of flowing air. The milled powder had a Fisher No. of 4.0. After milling, the batches of powder were combined and hydrogen reduced in a push-pull furnace operating at a temperature of 1700°F. The reduced powder was then pulverized and blended with 21% of a master alloy consisting of 76% chromium and 24% aluminum. The new blend was then milled for one hour in two batches under a blanket of argon. The milled powder was subsequently pressed in a mild steel can to a density of 65% of theoretical density.

The canned billet was then evacuated, sealed and extruded at 1950°F, at an extrusion ratio of 13:1 through a rectangular die. After subjecting the extruding slab to a 5% reduction by rolling at 1850°F and heat treating at 2450°F to effect recrystallization, tests were performed at 2000°F. Transverse mechanical properties exhibited a stress rupture life in excess of 20 hours at 6 ksi. Longitudinal mechanical properties exhibited a stress rupture life in excess of 20 hours at 8 ksi.

In further work by *S. Baranow; U.S. Patent 4,156,053; May 22, 1979; assigned to Special Metals Corporation,* the milling time is reduced to less than 2 hours.

In addition to the obvious benefit of increased efficiency of production, shortened milling periods are additionally accompanied by other benefits which include less oxygen in the final product, a higher yield and easier cleanup of the milling media. Moreover, long milling times disadvantageously lead to welding between the powder and the milling media, and to the production of highly cold-worked particles which cannot be cold consolidated.

The following example is illustrative of several aspects of this process.

Example: Two hundred and sixty grams of an 80 Cr-20 Al alloy were crushed to an average particle size of 4 microns and subsequently mixed with 1,024 g of carbonyl nickel and 16 g of Y_2O_3. The carbonyl nickel had an average particle size of less than 10 microns, and that for the Y_2O_3 was 150 angstroms. The proportions of the components were chosen to produce a batch of powder consisting essentially of 16 Cr, 4 Al, 1.2 Y_2O_3 balance Ni. The admixed powder was milled in an attritor, under argon, for one-half hour and subsequently discharged.

The powder was then packed into a mild steel container of 2⅝" o.d. which was evacuated, sealed and extruded through a 0.4 x 1.2 inch die at 2050°F. A specimen was then recrystallized at 2450°F and tested. It was found to have a cube-on-edge texture and a 2000°F life of 13 hours at a stress level of 9 ksi. In addition, at failure its elongation was 18.4% and its reduction in area was 21.0%. Material such as this is of sufficient strength, and inherent corrosion resistance to be utilized as an uncoated turbine vane in most jet engine applications.

Reduction to Produce Agglomerated Coating

C.E. O'Neill, R. Sridhar and M.C.E. Bell; U.S. Patent 3,914,124; October 21, 1975; assigned to The International Nickel Company, Inc. describe a process for reducing nickel oxide using a moving bed of nickel oxide containing volatile impurities and having a controlled sulfur content. It contains at least one additive selected from the group consisting of magnesia, calcium oxide or compounds heat decomposable thereto in an amount sufficient to react with the nickel oxide during reduction to provide an antiagglomerating coating on a substantial portion of the nickel oxide granules and it is contacted with a reducing gas at temperatures high enough to reduce the nickel oxide and to refine it.

The addition agent is advantageously added to the nickel oxide as an aqueous solution of a water soluble compound of magnesium or calcium, e.g., magnesium sulfate or chloride or nitrate. The process is used in conjunction with high temperature roasting of nickel sulfide, with roasting possibly followed by chlorination to control the copper content, to provide a combination of high temperature treatments that is effective in providing a refined metal product.

Reduction can be conducted in any commercially available reactor as long as the reactor provides good gas-solid contact to promote the reducing reactions.

Nickel oxide can be reduced at temperatures as low as 500°C but volatile impurity elimination becomes more effective with increasing temperatures. Improved throughput rates and more effective impurity elimination are achieved at reduction temperatures of about 600°C or higher, e.g., 750°C. Nickel oxide is advantageously reduced at temperatures of about 800°C, most advantageously between about 850° and 1100°C. Reduction at temperatures between about 900° and 1000°C is particularly advantageous in that throughput rates and impurity elimination, particularly lead and/or bismuth, are maximized while problems associated with sticking are minimal.

Lead and/or bismuth are most effectively eliminated during reduction at temperatures above about 850°C. At 1000°C, a nickel product containing less than about 0.01% lead and less than about 0.002% bismuth can be produced whereas the lead and bismuth contents of nickel obtained by reducing the oxide at less than 850°C temperatures would be about 0.05 and 0.005%, respectively. The extent of refining is dependent upon the initial impurity content, the reduction temperature, and the residence time at reduction temperatures, and the extent of actual refining obtained in practice is a matter of balancing product specifications with economics to optimize the various parameters. The use of reduction temperatures in excess of about 850°C to obtain impurity removal is possible only with use of the basic oxide to prevent agglomeration.

Example 1: Samples of granular nickel oxide, produced by roasting nickel sulfide above its softening temperature (i.e., nickel sulfide was fluid bed roasted at temperatures between 1100° and 1200°C), having a size distribution as follows:

Distribution, Weight % of Total	Size, Mesh TSS
1.6	+20
19	−20 +35
23.1	−35 +48
47.7	−48 +65
8.6	−65

and analyzing in weight percent as follows:

Ni	Cu	Fe	S	Ca	Mg	SiO_2	Al
77.1	0.7	0.18	0.017	0.095	0.024	0.57	0.17

were mixed with varying quantities of different additives, and the mixtures were fed to an externally heated rotary tube furnace. The additives, which have a size distribution of 100% minus 200 mesh TSS, and the amounts of additives employed are shown in Table 1. The term "trace" employed in Table 1 is approximately 200 ppm.

The mixtures of nickel oxide and additive were heated to different reducing temperatures for about three hours. During these tests a gaseous atmosphere containing, by volume, 7.8% carbon monoxide, 5.3% carbon dioxide, 12.1% hydrogen, 13.5% water vapor and 6.13% nitrogen, which atmosphere simulated atmospheres obtained by partially combusting natural gas, was passed through the rotating tube furnace at uniform rate such that upon completion of the test the amount of reductant passing through the furnace was equivalent to two to three times the stoichiometric amount required to reduce nickel oxide.

The degree of reduction obtained in terms of the weight percent of oxygen in the final product is reported in Table 1 for each of the tests. In all instances, the nickel to sulfur ratio in the reduced product was greater than about 9,500:1, confirming that the product was refined at least with respect to the sulfur content.

During reduction at higher temperatures (e.g., 1000°C) the average particle size and the distribution thereof should shift downward due to shrinkage induced by sintering. Thus, if no agglomeration occurs, one would expect a decrease of the fraction of the larger particles like the −20 +35 mesh TSS fraction. The presence of agglomeration during reduction is indicated by a positive increase in the fraction of the larger particle sizes. The change of the −20 +35 mesh TSS fraction of each of the tests is reported in Table 1.

Table 1

Additive	Amount of Additive (wt %)	Temperature (°C)	Increase in -20 + 35 Mesh Fraction (wt %)	Oxygen in Reduced Product (wt %)
Dolomite	trace	750	9.2	7.9
Dolomite	3	750	4.0	8.9
Dolomite	trace	1000	67.0	2.3
Silica	3	1000	56.8	1.2
Dolomite	3	1000	-15.0	0.7
Dolomite	2	1000	-15.9	0.6
Magnesia	3	1000	-4.5	2.0
Lime	3	1000	6.6	1.4
Limestone	3	1000	25.2	1.5
Magnesium sulfate*	0.75	1000	-7.5	<0.1

*Sprayed onto the nickel oxide at 920°C. Other additives well mixed with the nickel oxide before reduction.

A number of significant conclusions can be made from the results reported in Table 1. First, it is quite clear from the oxygen content of the final products reduced at different temperatures that greatly improved reduction kinetics are realized at the higher reduction temperatures. Second, even at reducing temperatures of only about 750°C, the solid additive must be added in amounts of at least about 0.2% in order to minimize sticking and agglomeration. Third, at reducing temperatures of about 1000°C a 3% silica addition is only a little more effective in controlling agglomeration than is only a trace addition of dolomite. Fourth, and highly significant, dolomite becomes more effective at higher reduction temperatures, possibly because at such higher temperatures thermal decomposition of the dolomite is insured. Fifth, lime is more effective in minimizing agglomeration than limestone, possibly because at the reduction temperatures employed limestone is not completely thermally decomposed and the additives must be in the oxide form to be effective. Sixth, magnesia-containing additives are more effective than calcium-containing additives. Seventh, lower amounts of additive are required when added to the nickel oxide as an aqueous solution which insures uniform distribution.

Example 2: The effects of sulfur in increasing the reduction kinetics are shown in this Example. Nickel oxide samples having the composition and particle size distribution given in Example 1 were mixed with 3% dolomite and the mixtures were then reduced at different temperatures in an atmosphere containing 20% hydrogen and 80% nitrogen (which simulated a partially combusted atmosphere containing hydrogen and carbon monoxide. The times to reduce the oxide by 75 and 95% (i.e., the time required to eliminate 75 and 95% of the oxygen from the oxide) were ascertained. The times and temperatures are reported in Table 2.

Table 2

	600°C		700°C		800°C		900°C		1000°C	
	75%	95%	75%	95%	75%	95%	75%	95%	75%	95%
Activation	Time (min) for Various Percent Reductions									
None	>400	–	–	–	200	–	–	–	52	>200
H_2SO_4	52	160	33	120	32	70	31	60	29	50

For comparative purposes, nickel oxide discharged from the fluid bed roasters into fluid bed coolers was activated with sulfur by adding sulfuric acid to the fluid bed coolers in amounts to provide the nickel oxide with a sulfur content of about 0.3%. Samples of the sulfur activated nickel oxide were reduced in the identical manner described above at various temperatures to ascertain the time required for 75 and 95% reductions. These results are also reported in Table 2.

Comparison of the results for nonactivated and sulfur-activated nickel oxide confirms that sulfur activation increases the reduction kinetics at all temperatures. The reduced products, after elutriation, contained less than about 0.5% sulfur. Thus, the sulfur activation increased the reduction kinetics while the dolomite addition not only minimized sticking but also aided sulfur removal after the sulfur has served its catalytic function.

Example 3: This Example confirms that reduction of nickel oxide by the process in accordance with the present invention is effective in producing an improved metal product. Samples of granular nickel oxide having the following analysis:

Ni	Cu	Co	Fe	S	Ca	Mg	SiO_2	Al
(wt %)								
75.0	0.64	0.76	0.29	0.05	0.2	0.021	0.035	0.15

O	As	Pb	Bi	Sb	Se	Te	Sn	Zn
(wt %)			(ppm)					
21.0	0.12	0.035	33	74	15	4	17	<5

were mixed with 3% −200 mesh TSS dolomite and were reduced at different temperatures for varying lengths of time in the same manner as described in Example 1. The conditions and results are reported in Table 3.

Table 3

. . Reduction Conditions . .		. . Metal Product . .	
Temperature (°C)	Time (hr)	O_2 . . . . (wt %)	Pb . . . . .
Initial oxide	–	21	0.035
750	6	0.2	0.034
800	6	0.1	0.023
900	6	0.1	0.013
950	4	0.1	0.008
950	6	<0.1	0.005
1000	3	<0.1	0.006
1000	6	<0.1	0.008

The results in Table 3 confirm that at temperatures of 750°C the nickel to lead ratio is increased to about 2,800:1 from about 2,100:1 in the oxide, i.e., more than about 25% of the lead is eliminated during the reduction refining operation. However, at temperatures about 800°C, particularly above 900°C, lead elimination becomes materially more effective. The improved reduction kinetics at the high reduction temperatures are also apparent from the results presented in Table 3.

Example 4: This Example confirms that high reduction temperatures are effective in eliminating impurities, such as antimony, arsenic, bismuth, selenium, tellurium and tin. Nickel oxide having the composition given in Example 3 was mixed with 3% −200 mesh TSS dolomite and reduced at 900° and 1000°C for six hours in the manner described in Example 1. The results are reported in Table 4. The results in Table 4 confirm that high reduction temperatures, particularly reduction temperatures above 900°C, are highly effective in eliminating impurities.

Table 4

	Reduction Temp., °C.	Bi	Sb	Analysis (ppm) Se	Te	Sn	As
Calculated Metal		42	94	19	5	22	1646
Reduced Metal	900	49	17	15	4	17	750
Reduced Metal	1000	19	11	13	4	16	730

NICKEL

Thermal Decomposition of Nickel Carbonyl Using Nitrogen Oxides

D.M. Llewelyn; U.S. Patent 3,820,977; June 28, 1974; assigned to The Interna-

tional Nickel Company, Inc. has found that nickel carbonyl can be thermally decomposed to nickel powder or pellets in the presence of nitric oxide, nitrogen trioxide or nitrogen peroxide in concentrations of 1 to 5,000 ppm or more. By suitable choice of gas concentrations and temperatures in the range 230° to 350°C the powder can be produced with substantially spherical particles or with low carbon content or both.

Below 230°C so small a proportion of the carbonyl is decomposed to powder that the process is not practicable on an industrial scale. Above 350°C a high proportion of filamentary aggregates are formed. A very suitable temperature is 290°C.

The amount of oxide of nitrogen required varies with the temperature, decreasing as the temperature decreases. Considering nitric oxide, and assuming that low carbon content is required, some reduction in the carbon content is obtained with very small amounts of nitric oxide, i.e., as little as 1 ppm particularly if the vessel walls have previously been nitrided and the process is operated continuously. When a steel reactor is employed, the walls of the reactor can be initially nitrided by introducing ammonia into the reactor and heating the reactor to nitriding temperatures, e.g, 500°C, for at least 1 hour, e.g., 3 hours. It is found that as the concentration of nitric oxide increases, the carbon content falls and then rises again. Typically the nickel carbonyl is introduced into the decomposer, as a gas mixture of carbon monoxides containing 8% nickel carbonyl, and at this concentration and at 290°C the concentration of nitric oxide should be from 50 ppm to 200 ppm. Above 250 ppm the carbon content of the powder can actually be higher than if no nitric oxide is added.

Converted into percentage of the carbonyl by volume, at 290°C the nitric oxide should be from about 0.06% to 0.25% of the carbonyl, whatever the concentration of the carbonyl.

In controlling the carbon content of the carbonyl nickel powder, at lower temperatures the nitrogen oxide concentration in the gaseous mixture is correspondingly lowered, being from 25 to 100 ppm, i.e., 0.03 to 0.12%, at 230°C, and at higher temperatures it is correspondingly increased, being from 100 to 300 ppm, i.e., 0.12 to 0.4% at 320°C.

When the object is to produce spherical powder, the concentration of nitric oxide at 290°C should be at least 0.09% of that of the carbonyl. If the temperature is lower, this minimum can be correspondingly reduced, but at 230°C should be at least 0.012%. Likewise at higher temperatures the minimum concentration of nitric oxide must be higher, being at least 0.2% at 320°C.

If low carbon content is not important, the concentration of nitric oxide can be considerably higher, but so far as the production of spherical powder is concerned there is no advantage in exceeding 1.25% or even 0.625% of the carbonyl concentration.

Decomposition of Hydrocarbonyl

L. Bakker; U.S. Patent 3,966,886; June 29, 1976 has developed a method for the preparation of metal hydrocarbonyls of metals of Groups I-B, II-B, III-B, IV-B, V-B, VI-B, VII-B and VIII of the Periodic Table of the Elements and the attendant method for isolation of pure metals by the separation of a single metal as a metal hydrocarbonyl from any mixture and the subsequent decomposition of the metal hydrocarbonyl to metal powder.

It comprises contacting a solid material having a maximum particle dimension of 1 mm and containing, for example, nickel present as any of free metal, metal oxides, metal sulfides, metal salts, metal esters and mixtures thereof with a mixture of hydrogen and carbon monoxide in a molar ratio of from 3:1 to 1:3 at a temperature of 185°±15°C and a pressure of from 1 to about 250 atm, preferably 15 to 150 atm, for a time sufficient to convert substantially all of the nickel in the solid material to nickel hydrocarbonyl, and separating the nickel hydrocarbonyl from the reacting system while maintaining the system at the temperature and pressure.

After separation from the system, the products can be isolated by reducing their temperature while under pressure to a point below the freezing or crystallization point. Alternatively, the products can be immediately reduced to metal powder by reducing or increasing the temperature. As a still further alternative, if the product comprises a mixture of metals as metal carbonyls and/or metal hydrocarbonyls, the various components can be separated by conventional methods such as distillation, crystallization, extraction and the like. This is true even if there is a mixture of isotopes of a single metal in the starting materials.

Example 1: A molybdenum ore concentrate containing 5 to 10% molybdenum sulfide as well as other materials was charged to a reaction vessel capable of withstanding high pressures and elevated temperatures. It is best the ore is finely divided, i.e., no larger than about 325 mesh U.S. Sieve Series. The vessel was charged with synthesis gas (an equimolar mixture of hydrogen and carbon monoxide) and heated to a temperature of 200°C and a pressure of 10,000 psi. In this way, the hydrocarbonyl and small quantities of carbonyl were formed. The reaction may be carried out batchwise or continuously.

The carbonyl-hydrocarbonyl mixture was vented while keeping the pressure constant. Liquid carbonyls were separated from the gaseous hydrocarbonyls. Both the carbonyl and the hydrocarbonyl were decomposed by lowering the pressure to yield pure molybdenum metal powder. The molybdenum metal powder can be used as such or oxidized to MoO_3. The carbon monoxide and hydrogen gases were returned to the system with fresh synthesis gas.

An alternate route is to convert the molybdenum sulfide into the chloride and to proceed from there in much the same way. Still another route is to roast the sulfide to the oxide and to either convert the oxide into the molybdenum hydrocarbonyl or to reduce the metal oxide to the metal which is then extracted from the ore as a hydrocarbonyl and decomposed to a powder by reducing the pressure and temperature.

Example 2: In accordance with another example of the process, a cobalt ore containing cobalt sulfide was employed. The first step was to divide the ore finely. Next, the cobalt sulfide was roasted to cobalt oxide. The cobalt oxide-containing ore was then charged to a reaction vessel capable of withstanding high pressure and elevated temperatures. The vessel was charged with synthesis gas (an equimolar mixture of hydrogen and carbon monoxide) and heated to a temperature of 200°C and a pressure of 5,000 psi. The molar ratio of hydrogen to carbon monoxide may vary from 1:3 to 3:1. The foregoing process resulted in the formation of cobalt hydrocarbonyl. This hydrocarbonyl was reduced to powdered cobalt metal form by reducing the temperature to −26°C and the pressure to atmospheric.

Example 3: Ferro nickel is ground to a 300 mesh powder (U.S. Sieve Series) and reacted with synthesis gas for 24 hours at 1,500 psi and 195°C. After equilibrium is reached, the reactor is slowly vented at a rate of 3 ft^3/min of gas/100 g of ferro nickel, keeping the temperature and pressure constant in the reactor with the addition of fresh systhesis gas, into a receptacle at room temperature and atmospheric pressure where about 80% of the cobalt in the ferro nickel is recovered as metal powder.

Example 4: Nickel linoleate was reacted with synthesis gas for 12 hours at 1,500 psi and 195°C. After 12 hours when equilibrium had been reached, the reactor was slowly vented at 3 ft^3/min of gas/100 g of nickel linoleate, keeping the temperature and pressure constant in the reaction vessel with the addition of fresh synthesis gas, into a receptacle at room temperature and pressure where all the nickel was recovered as a 325 mesh (U.S. Sieve Series) powder.

Heat Treatment of Nickel Oxalate

D.B. Pall; U.S. Patent 3,923,496; December 2, 1975; assigned to The United States Energy Research and Development Administration has developed a method of producing nickel powder in the range of 0.1 to 40 μ which comprises reacting oxalic acid with aqueous nickel nitrate to produce nickel oxalate. The nickel oxalate is heated in an atmosphere of carbon dioxide at a temperature in the range of 450° to 850°C in the presence of a small amount of a particle grain growth inhibitor such as magnesium nitrate to produce a substantially pure nickel powder of uniform particle size. Thereafter the powder is heated in a mildly oxidizing atmosphere at an elevated temperature of less than 100°C to render the powder nonpyrophoric.

Example 1: *Preparation of a 0.4 Micron Nickel Powder*—77.1 pounds of a special reagent grade nickel nitrate [$Ni(NO_3)_2 \cdot 6H_2O$] are dissolved; the solution is clarified through a small filter press, pumped to a 75 gallon glass-lined tank, and made up to a 2 molar solution by the addition of H_2O and steam. Then 0.035% magnesium (figured on nickel basis) as 4.1 g MgO which has been completely dissolved in HNO_3, is added. 36.8 pounds of technical oxalic acid ($H_2C_2O_4 \cdot 2H_2O$) are dissolved in hot water, and made up to a 2 molar solution in a 20 gallon stoneware vessel.

The nickel nitrate solution is then heated to 97°C, and the oxalic acid solution is heated to 85°C. The oxalic acid solution is added to the nickel solution uniformly over a 10 minute period; the strike is maintained at 97°C for 20 minutes. The precipitate is decanted with interim washing three times, and the solid nickel oxalate is collected in a wash press, washed and dried at 110°C for approximately 18 hours.

All the nickel oxalate ($NiC_2O_4 \cdot 2H_2O$), approximately 44 pounds, is placed in a large rotary calciner externally heated by gas. The powder is then heated at a uniform rate in a CO_2 atmosphere, until the temperature reaches 450°C at the rate of approximately 100°C per hour. The calciner is rotated at 2 to 3 rpm and rocked forward and backward mixing while the temperature rises from 390° to 450°C. The powder is then cooled to less than 100°C, and aerated overnight with a slow stream of air to render it nonpyrophoric. Finally, it is removed from the calciner, and passed through a mechanical disintegrator.

This powder has a mean crystal diameter averaged by volume of 0.4 micron, as determined by the tinting strength method. An estimate of the size by means of the electron microscope gave the same result. An analysis for magnesium showed only 0.008% magnesium (figured on nickel basis) which shows that only approximately one-fourth of the magnesium was carried down with the nickel oxalate precipitate.

Example 2: When the previous example is repeated, carrying the temperature to 500°C, the diameter of the powder is 0.5 micron, by the tinting strength method.

Example 3: *Preparation of a 0.1 Micron Nickel Powder*—154.2 pounds of a special reagent grade nickel nitrate [$Ni(NO_3)_2 \cdot 6H_2O$] are dissolved in water and filtered through a clarification press. The solution is pumped to a stainless steel tank, and made up to a 2 molar solution by the addition of water and steam. Then 0.65% magnesium (figured on nickel basis) or 149.2 g MgO which was completely dissolved in HNO_3 are added.

73.6 pounds of technical oxalic acid ($H_2C_2O_4 \cdot 2H_2O$) are dissolved in hot water and made up to a 2 molar solution in a 40 gallon crock. The nickel nitrate solution is then heated to 97°C, and the oxalic acid solution is heated to 85°C. The oxalic acid is added uniformly over an 18 minute period, and the strike is digested at 97°C for 20 minutes. Washing by decantation is repeated three times; the nickel oxalate is filtered in a wash press, washed and dried at 110°C for approximately 18 hours.

The nickel oxalate ($NiC_2O_4 \cdot 2H_2O$), approximately 50 pounds, is then placed in a large rotary calciner, heated by gas. The powder is then heated at a uniform rate of 100°C per hour in a CO_2 atmosphere, until the temperature of the powder reaches 460°C (the furnace is rotated at 2 to 3 rpm and rocked from 390° to 460°C to get end-to-end mixing). The powder is cooled to less than 100°C, then aerated overnight with a slow stream of air to render it nonpyrophoric, removed from the calciner, and micropulverized.

This powder has a mean crystal diameter averaged by volume of 0.1 micron, as determined by the electron microscope. An analysis for magnesium showed only 0.11% magnesium (figured on nickel basis) which shows that only approximately one-sixth of the magnesium was precipitated.

The particle size of fine nickel powders increases very rapidly if they are heated above about 400°C. The increase in particle size is affected by both the time and the temperature, maximum size being reached after about 15 to 30 minutes. Because of this fact, the preparation of nickel powders of selected size would be technically precluded except for the discovery of particle size inhibitors of the class consisting of magnesium, the alkali metals and the alkaline earth metals. When a member of this group is present during heating, even in very minute quantities, particle growth is markedly restrained.

Substantially complete reduction of nickel oxalate to nickel powder is accomplished by heating pure nickel oxalate in a nonoxidizing chamber at a temperature around 450°C. If no inhibitor is present the particle size of the product will be on the order of 1 micron, and as the heating is continued or as the temperature is raised, the particle size will increase with some rapidity. However, if a small amount of magnesium is present, a very much smaller particle size is obtained. The relationship of the various factors such as the percentage of inhibitor and the temperature are explained in further detail in drawings contained in the patent.

Controlled Geometry by Hydrazine Precipitation

J.D. Grundy; U.S. Patent 4,089,676; May 16, 1978; assigned to Williams Gold Refining Company, Inc. has developed a method for producing nickel powder of controlled geometry.

An appropriate nickel compound is dissolved in water in a suitable vessel under ambient pressure conditions and a reducing agent or precipitation agent in the form of hydrazine is added. The preferred agent is hydrazine hydrate, although hydrazine hydrochloride, hydrazine sulfate and other hydrazine salts can be employed. Upon addition to the hydrazine, a nickel complex is formed immediately as evidenced by a pink to blue precipitate. An antifoam agent, for example Union Carbide #7600, is added, and a measured amount of an alkali metal base, preferably sodium hydroxide, dissolved in water is added. While sodium hydroxide is preferred, potassium hydroxide and the other alkali metal hydroxides could be used.

The measured amount of alkali metal base is determined according to the desired geometry of the nickel powder end product. The resulting mixture or slurry is heated to a temperature in the range from about 88° to about 92°C with stirring or similar agitation. Upon heating, the color of the original nickel complex changes to a purple-pink color. A measured amount of alkali metal base, i.e., sodium hydroxide, is added and the temperature is raised to a range from about 94° to about 96°C. The temperature is maintained until it is observed that nickel powder is precipitated.

The entire method is carried out under ambient pressure conditions. In carrying out the foregoing method, the alkali metal base need not be added in two parts. For example, it is advisable to add the base in two steps when making relatively large surface area powders to control the reaction rate.

The precipitated nickel powder is removed from the bottom of the vessel and filtered, preferably by means of a suction or vacuum type filter. The nickel powder then is washed, using deionized water, and then dried. The resulting nickel powder end product of controlled geometry then can be screened according to desired size. The nickel powder end product advantageously is observed to be nonpyrotechnic and nonpyrophoric. Thus, the finely divided particles of the nickel powder prepared according to the method are observed to be noncombustible.

The method is illustrated in further detail by the following examples.

Example 1: The method was carried out in the manner described above, and nickel salts from which nickel metal powder can be made according to this process include nickel acetate, nickel carbonate, nickel chloride hexahydrate, and nickel sulfate hexahydrate, the latter being known in the plating industry as single nickel salt.

Example 2: The method was carried out in the manner described above, and it was determined that nickel ammonium sulfate, nickel nitrate, nickel oxide, and nickel sulfate heptahydrate (known to the plating industry as double nickel salt) would not provide nickel metal powder from the method described above.

Examples 3 Through 7: These examples are summarized in Table 1 as follows:

Table 1

Ex. No.	Ni Salt (g)	Ni (g)	N_2H_4 (ml)	NaOH (g)	NaOH/Ni (g/g)	Powder Surface Area (m^2/g)	Particle Size (μm)	Particle Shape
3	550	122.82	300	10	0.08	0.98	2.74	spherical
4	550	122.82	300	30.5	0.25	6.88	0.78	spherical-irregular
5	716	159.88	400	40	0.25	7.10	0.57	spherical-irregular
6	250	55.83	125	60	1.07	10.5	0.32	irregular
7	22,171	4,950.78	11,550	5,331.5	1.08	11.69	0.14	irregular

The data presented in Table 1 was obtained by carrying out the method as described above for each of the Examples 3-7 with the indicated amounts of nickel salt in grams, hydrazine in milliliters, and sodium hydroxide in grams. Nickel sulfate hexahydrate was the nickel salt. The powder surface area in square meters per gram was obtained by the instrumental BET method. The particle size in micro meters is an average particle size obtained by the Fisher Sub Sieve Size procedure.

The examples presented in Table 1 indicate that increasing the caustic, i.e, sodium hydroxide, relative to the nickel of the nickel salt results in a nickel powder end product of greater surface area. Similarly, decreasing the proportion of sodium hydroxide to nickel in the salt gives a nickel powder end product of lesser surface area.

Similarly, increasing the proportion of sodium hydroxide to nickel in the salt results in a nickel powder end product of decreasing particle size, and decreasing the proportion of base to nickel results in increasing particle size. In addition, the resulting nickel powder end products of lower powder surface area were observed to have particles of spherical shape, those of medium powder surface area were observed to be spherical with surface irregularities, and those of greater powder surface area were observed to vary from spherical to relatively irregular particle shapes.

Recovery from Aqueous Ammoniacal Ammonium Sulfate Solution

V.N. Mackiw, D.R. Weir, D.J.I. Evans and D.G. Maschmeyer; U.S. Patent 3,816,098; June 11, 1974; assigned to Sherritt Gordon Mines Ltd., Canada have developed a process whereby, substantially pure, particulate elemental nickel is recovered from nickel compounds and particularly from basic nickel carbonate. The compounds are soluble in aqueous ammoniacal ammonium sulfate solution and contain impurities including particularly zinc, manganese and/or magnesium. First substantially all the nickel content is extracted from the impure nickel compound by dissolving it in an aqueous ammoniacal ammonium sulfate solution containing ammonium sulfate in excess of the stoichiometric amount required to react with the nickel to form nickel diammine sulfate. The resulting nickel diammine sulfate solution is separated from undissolved residue and is reacted with hydrogen at elevated temperature and pressure to reduce dissolved nickel to elemental form and to precipitate it in powder form.

The nickel-depleted solution, referred to as "reduction end solution," containing zinc, manganese and/or magnesium as well as small amounts of other impurities as dissolved sulfates is recycled to the nickel compound dissolution step and the quantity of such solution recycled is controlled such that the impurities level in the reduction end solution is maintained at a predetermined level above the level that would result if no solution were recycled but below that level at which zinc, manganese and/or magnesium report in the nickel powder product in sufficient quantity to contaminate the product. The portion of the reduction end solution which is not recycled is bled from the process circuit.

Example: A pilot scale test run using the process was carried out as follows: nickel bearing ammoniacal ammonium carbonate leach solution derived from leaching reduction roasted laterite ore was reacted with ammonium sulfate to remove sufficient cobalt to produce a nickel to cobalt molar ratio in the solution of about 700:1. This solution was then boiled to distill off ammonia and carbon dioxide to effect precipitation of substantially all dissolved nickel as basic nickel carbonate.

Approximately 1,500 pounds of wet basic nickel carbonate precipitate was produced in this manner. The wet precipitate analyzed as follows: (weight percent) Ni, 55.0; free H_2O, 56.3; S_{total}, 2.51; Mg, 0.170; Mn, 0.172; Zn, 0.108; Co, 0.072; Fe, 0.032; Al, less than 0.001; $NH_{3\ total}$, 0.25 and CO_2, 8.7.

This wet basic nickel carbonate precipitate was dissolved in aqueous ammoniacal ammonium sulfate solution in 27 350 gallon batches utilizing the following operating conditions: target nickel content (gpl), 50.0±0.5; total S, 125±5; free NH_3, 35; temperature, 180°F; time, 40 minutes; percent dissolution, at least 99%. The make-up solution for dissolving the first batch of basic nickel carbonate precipitate was prepared with approximately 11,000 pounds of commercial grade ammonium sulfate obtained from an extraneous source.

To simulate the effect of a commercial operation in which a constant level of impurities is maintained after an initial period of impurities buildup, this solution was spiked with zinc, manganese and magnesium to raise the levels of these impurities in the solution to about 2 gpl for zinc and manganese and 2.5 gpl for magnesium. Subsequent batches of basic nickel carbonate precipitate were dissolved in partially evaporated reduction end solution obtained as hereinafter described with reference to the reduction step of the process.

The solution and the undissolved residue from the redissolution operation were transferred to batch settling tanks where 5 hours retention time was allowed to effect separation of the solids from the solution. The undissolved residue separated from the redissolution solution in this manner had the following average chemical analyses: (weight percent dry weight basis) Ni, 33.2; S_T, 4.71; Mg, 0.788, Mn, 3.25; Zn, 3.50; Co, 0.285 and Fe, 2.25. The weight percent of nickel, sulfur and metal impurities contained in the basic nickel carbonate fed to the process circuit which was removed from the circuit via the settling tank underflow solids was as follows: Ni, 1.16; S_T, 5.55; Mg, 7.3; Mn, 40.0; Zn, 58.3; Co, 9.5 and Fe, 95.5.

The solids free solution from the redissolution operation was contacted with air at 400 psi and at a temperature between about 410° and 425°F for about 35 minutes to oxidize substantially all sulfur values in the solution to sulfate form. Composite analysis of the solution from the sulfur oxidation step was: (gpl) Ni, 44.2; S_T, 121.0; Mg, 2.40; Mn, 1.984; Zn, 1.910; Co, 0.180; Fe, 0.0166 and Al, less than 0.001.

Nickel was recovered from the oxidized solution by direct reduction with hydrogen using the following procedure: a reduction cycle consisting of an initial nucleation reduction and 49 subsequent "densification" reductions was commenced by pumping 120 pounds (dry weight) of fine (1 micron) pure nickel seed powder obtained from an extraneous source into a high pressure agitator equipped autoclave with 145 imperial gallons of water. Discharge was followed with a 15 minute nitrogen purge, the contents of the autoclave were then heated to 230°F followed by a water vapor purge. The temperature was raised to 350°F, the nickel powder was allowed to settle and the water was finally discharged.

145 gallons of oxidized feed solution was then charged into the autoclave, heated to 350°F and hydrogen was applied to give a total pressure of 550 psig. During heating, sufficient 200 gpl aqua ammonia was added to maintain the reduction end solution pH at about 7.0 to 8.0. Overall reduction operating conditions were: number of reductions or "densifications," 49; feed volume (imperial gallons) per densification, 145; reduction temperature (°F), 350; total pressure (psig), 550; volume 2.5% Acrysol A 3 per reduction (ml), 280 (0.42 ml/l); average reduction end solution pH, 7.5; average aqua ammonia addition, 2 imperial gallons per reduction; average reduction time, 20 minutes.

When each reduction was completed, the autoclave agitators were turned off, the nickel powder was allowed to settle and finally the barren end solution was discharged to a flash tank. Upon completion of the cycle of 49 reductions, the autoclave was discharged with agitators running. The nickel powder and end solution were separated in the flash tank. Composite analyses of the reduction end solution was: (gpl) Ni, 0.538; S_T, 126.8; Mg, 2.46; Mn, 2.06; Zn, 2.04; Co, 0.19; Fe, 0.014; Al, less than 0.001; Cu, 0.0003 and $(NH_4)_2SO_4$, 515.

Approximately 2% of the total solution entering the process circuit was bled from the circuit (primarily in the form of analytical samples) to maintain levels of zinc, manganese and magnesium in the circulating solution substantially constant while at the same time maintaining the sulfur balance in the system. The reduction end solution was heated to evaporate sufficient water to maintain the water balance in the system while, at the same time, concentrating the ammonium sulfate to about 600 to 650 gpl. This solution was then recirculated in the circuit to repulp batches to basic nickel carbonate precipitate as previously described.

The nickel powder product was water washed to remove entrained ammonium sulfate, filtered and dried. Analyses of the final powder product was as follows: (weight percent) Ni, 99.75; Co, 0.052; Cu, 0.0005; Fe, 0.013; Mg, 0.0011; Mn, 0.0016; Zn, 0.0027; Al, less than 0.001; Cr, less than 0.001, Ca, 0.0020, Si, 0.0037; S, 00.20; C, 0.019 and O_2, 0.137.

A representative sample of the powder was briquetted and sintered under hydrogen atmosphere at 1750°F for 80 minutes. The S, C and O_2 analyses of the final product were: S, 0.004; C, 0.006 and O_2, 0.036. It can be noted that the sintering step lowered the sulfur, carbon and oxygen impurities by a factor of 4 or greater resulting in a final nickel analysis of 99.88% in the sintered briquettes.

In a related process by *W. Seibt and D.R. Weir; U.S. Patent 4,148,632; April 10, 1979; assigned to Sherritt Gordon Mines Limited, Canada,* wherein an impure nickel compound contains a major proportion of nickel values and a minor proportion of other metal values relatively soluble in ammoniacal ammonium sulfate solution, and a minor proportion of nickel values and other metal values relatively insoluble in the solution, the impure nickel compound is treated with ammoniacal ammonium sulfate solution so that the relatively soluble nickel and other metal values are dissolved in the solution with the relatively soluble nickel values being dissolved substantially as nickel diammine sulfate; thus the relatively insoluble nickel and other metal values form an undissolved residue.

The residue is separated from most of the ammoniacal ammonium sulfate solution so as to provide a substantially residue-free solution of nickel diammine sulfate and relatively soluble other metal values on the one hand and a slurry of the residue of undissolved nickel and other metal values and the solution on the other hand. The substantially residue-free solution is subjected to a reducing operation by hydrogen gas to produce precipitation of elemental nickel powder from the residue-free solution, and the elemental nickel power is separated from the residue-free solution to produce a first nickel powder product. The slurry is subjected to a reducing operation by hydrogen gas to produce precipitation of elemental nickel powder from the slurry, and the elemental nickel powder is separated from the slurry to produce a second nickel powder product.

Recovery from Aqueous Ammoniacal Ammonium Carbonate Liquor

In another process for recovering elemental nickel *D.R. Weir; U.S. Patent 3,928,020; December 23, 1975; assigned to Sherritt Gordon Mines Ltd., Canada* uses an impure aqueous ammoniacal ammonium carbonate liquor containing, in addition to nickel, one or more impurities such as sulfur, chlorine, cobalt, magnesium, manganese, iron, zinc and copper. According to the process, the composition of the liquor is first adjusted to contain at least 1.5 mols sulfur in the form of dissolved ammonium sulfate per mol of contained nickel. The resulting liquor is then heated to drive off ammonia and carbon dioxide to produce a nickel diammine sulfate solution from which elemental nickel is recovered. At least one mol of ammonium sulfate is formed in the solution per mol of recovered nickel and the sulfate containing solution is recycled, at least in part, to make up the sulfur requirements of the first step.

Nickel Chloride Reacted with Ammonia

L. Gandon and S. Solar; U.S. Patent 3,846,117; November 5, 1974; assigned to Le Nickel, France have developed a method for producing high-purity nickel powder with predetermined physical characteristics, wherein gaseous ammonia or an aqueous solution of ammonia is reacted at a temperature between 20° and 40°C with an aqueous solution of nickel chloride containing at least 50 g/l of nickel. The resulting precipitate of hexammine complex of nickel chloride is separated by filtration and submitted to centrifugal drying at a temperature higher than 100°C in order to convert it, at least partly, into a diammine complex of nickel chloride with release of ammonia, and the diammine complex is reduced by hydrogen at a temperature between 450° and 1000°C.

The precipitation of the hexammine complex is nearly quantitative, with yields higher than 90%, and it is to be noted that an advantage of the method is that it can be carried out with very simple equipment. Furthermore, the only reagents employed are ammonia and hydrogen which can be produced with high purity, so that the risks of contamination are minimized.

The ammonia released during the drying of the hexammine complex is preferably recycled for producing fresh amounts of this hexammine complex. The diammine complex of nickel chloride obtained during the drying step is then reduced, as mentioned hereinabove, which produces, on the one hand, ammonium chloride which sublimates and, on the other hand, very pure nickel powder. The ammonium chloride is itself pure, and can be sold as produced or used for generating ammonia by means of a double-decomposition reaction with lime.

Furthermore, the production of nickel powder according to the invention makes it possible, starting from a nickel chloride solution, to obtain powders with controlled physical characteristics by controlling the precipitation of the hexammine complex of nickel chloride and the temperature at which the final reduction treatment is carried out.

The method can lead directly to the production of billets, pills, briquettes, laminates, and any other form of nickel half-product, provided that the initial ammine complexes are performed.

Example 1: *Precipitation of Hexammine Nickel Chloride by Ammonia*—1,000 ml of an aqueous nickel chloride solution with the following composition:

	Grams per Liter
Ni	207
Ca	0.47
Na	0.27
$SO_4^{=}$	0.50

are poured into a double-walled reactor with a capacity of 3 liters, and cold water is circulated through the wall.

2,800 ml of a 28%-ammoniacal solution are added to this nickel chloride solution over a period of 45 minutes. The addition takes place with stirring and keeping the temperature lower than 30°C. The obtained product is filtered, by means of a Buchner-filter, in order to eliminate as much water as possible.

The obtained product is then dried under vacuum at room temperature in the presence of phosphoric anhydride, for 24 hours.

740 g of a product with the formula $Ni(NH_3)_6Cl_2$ are obtained, which corresponds to a precipitation yield of 90% with respect to the initial amount of nickel.

The precipitated complex presents the following composition:

	Percent	
	Experimental	Theoretical
Ni	25.24	25.3
NH_3	43.37	44.0
Cl	31.50	30.6

Example 2: *Precipitation by Ammonia of the Hexammine Complex of Nickel Chloride*—1,000 ml of a nickel chloride solution with the same composition as in Example 1 are introduced in a 3 liter-reactor provided with a double wall which is water-cooled. A flow of gaseous ammonia, obtained by heating 2,800 ml of a commercial solution of ammonia at 28%, is bubbled into this solution over a period of 4 hours. The temperature is kept near 25°C and the solution is stirred during the reaction. The obtained product is filtrated and dried. 905 g of raw crystals are recovered and dried under vacuum as in Example 1. The dry product corresponds to the formula $Ni(NH_3)_6Cl_2$ and weighs 801 g, which corresponds to a yield of 98% with respect to the initial amount of nickel.

Example 3: *Preparation of Nickel Powders*—The raw hexammine complex is dried at 120°C until its conversion into a diammine complex $Ni(NH_3)_2Cl_2$.

Then, for each following test, 60 g of this diammine complex are introduced into a tube furnace through which hydrogen flows at the rate of 70 liters per hour.

A first set of tests is carried out at reduction temperatures of 450°, 550° and 750°C during one hour for the first temperature and one-half hour for the higher temperatures.

After reduction, the nickel powders are allowed to cool to room temperature in an atmosphere of hydrogen and are then subjected to analysis.

The composition of the impurities is given in Table 1 below.

Table 1

Test No.	.... Reduction .... Temperature (°C)	Duration (min)	PR*	.... Elements (%) C	S	Cl	Na	Ca
1	450	60	0.60	0.020	0.004	**	**	**
2	550	30	0.21	0.016	0.003	0.007	0.015	0.010
3	750	30	0.18	0.030	0.004	<0.004	0.015	0.010
4	750	30	0.18	0.019	0.003	<0.004	0.015	0.0082

*PR is loss on heating.
**Not determined.

On the other hand, the physical parameters of these nickel powders are given in Table 2 below.

Table 2

Test No.	DM* (μ)	Porosity** (%)	Apparent DNT***
1	2.5	75.5	0.70
2	5.2	70	0.90
3	6.2	76	1.02
4	6	74	1.06

*Average diameter measured by Fisher Sub Sieve Sizer.
**The porosity is also measured with the Fisher apparatus.
***Apparent density of the unrammed powder (g/cc).

The figures in Table 1 show that the purity of the nickel powders produced by this method is quite remarkable. On the other hand, Table 2 shows that the average diameter of the powder grains increases with the reduction temperature, which provides a first means for controlling their physical characteristics, as will be explained in greater detail below.

A second set of tests is then carried out, also producing a hexammine complex of nickel chloride by bubbling ammonia into a nickel chloride solution at 30°C, by varying the duration of the reaction. For each test, 400 ml of a nickel chloride solution are used and the same total quantity of gaseous ammonia is bubbled into it, equal to two times the stoichiometric amount, but the flow-rate of the gas is different for each test. It will readily be understood that the duration of the bubbling is different in each case and it is easy to calculate this duration.

The other steps of the treatment are as follows: transformation of the hexammine complex into a diammine complex by drying it at 120°C, and hydrogen reduction at 550°C over a period of 45 minutes.

Table 3 below shows the physical parameters of the nickel powders obtained:

Table 3

Test No.	Flow Rate NH_3 (l/min)	DM (μ)	Porosity (%)	Apparent DNT (g/cc)
5	4	2.9	77.5	0.48
6	6	2.1	80	0.44
7	8	1.5	80	0.38
9	12	1	80	0.29

It will be apparent that the average diameter of the powder grains decreases when the flow-rate of ammonia increases, which provides a second means for controlling the physical parameters of the nickel powder produced according to the invention.

Thus, if a nickel powder with, for instance, a predetermined average grain diameter is to be produced, a limited number of tests will produce the required result. If a first test shows, for instance, that the average grain diameter is too high, the required result will be obtained by lowering the reduction temperature and/or increasing the flow-rate of ammonia which precipitates the hexammine complex, in proportion to the figures of Tables 2 and 3.

B-Type Carbonyl Nickel Powder

D.M. Llewelyn; U.S. Patents 3,918,955; November 11, 1975; and 4,018,596; April 19, 1977; both assigned to The International Nickel Co., Inc., describes a process for the production of B-type nickel powder by the thermal decomposition of nickel carbonyl in a decomposer with or without the presence of a diluent gas. The rate of input of nickel carbonyl, the decomposition temperature, and the amount of any diluent gas introduced are correlated to provide that fluffy nickel powder is not deposited on the heated walls of the decomposer, and fine nickel powder suspended in gases leaving the decomposer is recovered separately, e.g., by filter, from the B-type nickel powder settling at the bottom of the decomposer.

Operating in this way not only enables B nickel powder to be obtained with a larger particle size for a given bulk density, but also enables powders having unusually low bulk density, i.e., less than 1 g/ml, e.g., 0.8 g/ml, or less than 0.8 g/ml, preferably less than 0.5 g/ml and even as low as 0.3 g/ml or less, to be obtained on a practical industrial scale. It is also found that at any given bulk density the powder produced exhibits substantially less shrinkage on sintering than the powder hitherto produced. Generally speaking, the fine particles recovered from the gases have a Fisher particle size of less than 2 μm, (microns) e.g., from 1.5 to 1.9 μm, and amount to from 10 to 17%, e.g., about 15%, by weight of the total amount of powder produced.

The nickel carbonyl vapor fed to the decomposer is advantageously diluted, preferably with carbon monoxide or other inert diluent gas, so that carbonyl concentration is preferably in the range of 10 to 80% by volume of the total gas entering the decomposer. In addition it is advantageous to introduce as a diluent a small amount of an agent for promoting solid nuclei formation for the powder, preferably oxygen, for the purpose of reducing the shrinkage on sintering without substantially changing the bulk density, as described in U.S. Patent 2,844,456. Preferably, oxygen is introduced in an amount of from 0.01 to 0.06% by volume of the total gas entering the decomposer, but larger amounts, up to not more than 0.1%, can be used, though these tend to lower the Fisher value of the powder.

Example: B nickel powder was produced by the decomposition of nickel carbonyl vapor mixed with carbon monoxide in a decomposer consisting of an upright cylindrical vessel 2 m in diameter and 10 m high, the walls of which were externally heated electrically. The internal temperatures were measured by means of thermocouples mounted on a vertical axis 20.3 cm from the walls at various levels, and the wall temperature was measured at various levels by thermocouples mounted in the walls, three thermocouples being placed at each level 120° apart around the circumference of the wall.

Two series of tests were carried out. In the first series the fine powder separated from the exit gases was returned to the bottom of the decomposer. In the second series, which are examples according to the process, the fine powder was recovered separately from the powder settling at the bottom of the decomposer. The internal and wall temperatures measured as described above were as set forth in Table 1, the internal temperatures being the same in the two series of the tests but the wall temperatures in the upper part of the decomposer being lower in the second series.

Table 1

Depth* (cm)	Temp. Point	Internal Temp. (°C)	Temp. Point	. . Wall Temp. (°C) . . Series I	Series II	Depth* (cm)
132	T_1	260-270	T_7	480-490	430-440	107
183	T_2	240-250	T_8	480-490	460-470	320
343	T_3	250-260	T_9	450-460	450-460	533
488	T_4	260-270	T_{10}	420-430	420-430	747
554	T_5	270-280				
793	T_6	260-270				

*Below top of decomposer.

The operating conditions in each series of tests are set forth in Table 2 below, together with the bulk density and Fisher particle size of the powder produced, which in each case was B powder.

Table 2

Series I				Series II			
Total Gas Flow (m^3/hr)	$Ni(CO)_4$ Concentration (% vol)	Bulk Density (g/ml)	Fisher Size (μm)	Total Gas Flow (m^3/hr)	$Ni(CO)_4$ Concentration (% vol)	Bulk Density (g/ml)	Fisher Size (μm)
620	14-18	0.52	2.33	550	14-18	0.53	2.87
620	14-18	0.64	2.44	550	14-18	0.62	2.88
620	14-18	0.76	2.35	550	14-18	0.73	2.91
620	14-18	0.87	2.65	560	14-18	0.90	3.07

In both series the input gases in each test included 0.06% by volume of oxygen.

In the tests of Series I, nickel powder accumulated on the heated walls of the decomposer as a fluffy deposit and this powder appeared in the main product, whereas with the examples of Series II there was no evidence of fluff in the product and, by inference, not on the wall of the decomposer.

Preferably, in this process the rate of nickel carbonyl input is in the range of from 70 to 117 m^3/h (cubic meters per hour), a total gas input in the range of from 500 to 650 m^3/h with the concentration of nickel carbonyl in the range of from 14 to 18% by volume of the total gas entering the decomposer, and the temperature of the inner wall of the decomposer is in the range of from 400° to 500°C and advantageously not greater than 470°C.

The diminished tendency of the powder to shrink on sintering when the fine particles are excluded is shown by the following results, in Table 3, of comparative tests for typical batches of nickel powder made without separation of fines (contra to the process) and with separation of fines (examples according to the process).

Table 3

Fines not separated			Fines separated		
Bulk density (g/ml)	Fisher size (μm)	Linear shrinkage(%)	Bulk density (g/ml)	Fisher size(μm)	Linear shrinkage(%)
0.50	2.30	10	0.5	2.60	5

The linear shrinkage on sintering was measured as the percentage reduction in length of a body of powder loosely filled into a covered graphite-coated mold, flushed with nitrogen, and heated for 20 minutes at 900°C under flowing hydrogen.

Thermoplastic Prealloyed Powder

T.E. Volin, J.S. Benjamin, J.M. Larson and R.L. Cairns; U.S. Patents 3,865,575;

February 11, 1975; and 3,930,841; January 6, 1976; both assigned to The International Nickel Co., Inc. have found that if high strain energy is induced into powder prealloyed to desired composition, e.g., virtually unworkable IN-100, the nature of the powder is altered to such an extent that the powder takes on the physical characteristic of thermoplasticity, exhibiting significantly less resistance to flow.

As a direct consequence of this phenomenon, (a) lower pressing temperatures and/or forming stresses can be used than otherwise would be the case, (b) the capacity of available equipment would be capable of producing huge shapes such as 4 to 5 foot diameter turbine discs, (c) processed powder can be put into a preform closely corresponding to the configuration of shape of the final product, (d) the preform is capable of retaining its relative ease of deformability properties through subsequent processing, and (e) limitations introduced by being otherwise restricted in the type of apparatus that can be used, e.g., the extrusion press, are greatly minimized if not obviated.

The amount of strain energy required is influenced by the hardness and strength of a given alloy at a given elevated temperature. And this to a considerable extent is largely a reflection of composition. To indicate generally when sufficient strain energy has been imparted to prealloyed powder to confer "thermoplasticity" as described herein, reference is made to Figure 3.1.

Figure 3.1: Thermoplastic Prealloyed Powder

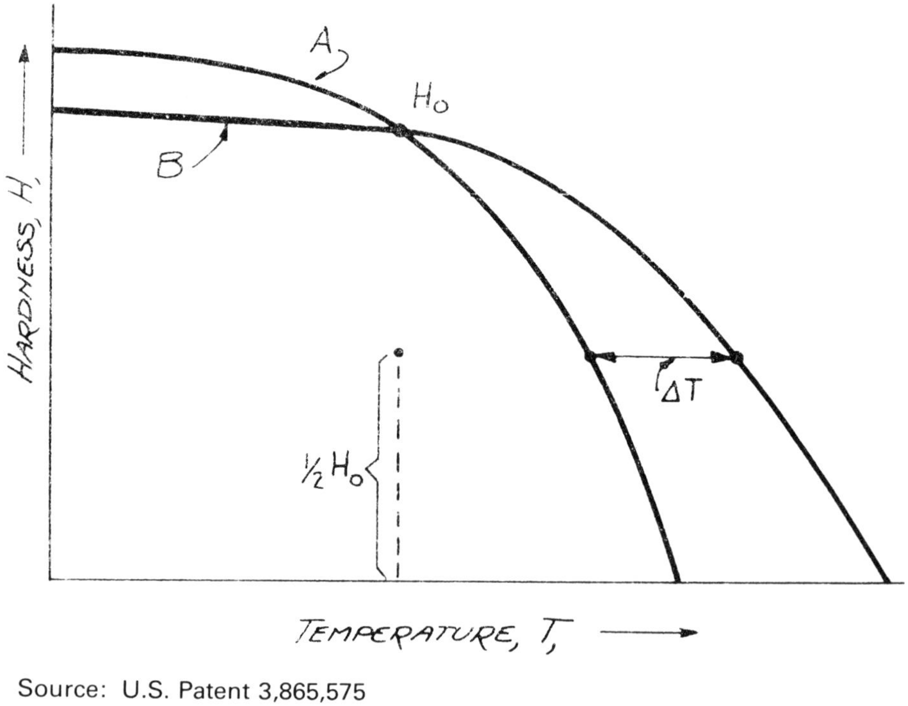

Source: U.S. Patent 3,865,575

Curve **A** represents prealloyed powder which has been subjected to strain energy, Curve **B** represents prealloyed powder of the same composition but which has not been so processed. Point $\mathbf{H_0}$ represents a common hardness value for each of the prealloyed powders at a given temperature, the respective powders having been consolidated to a density of at least 99% of theoretical.

In accordance thereof if an amount of strain energy has been imparted to prealloyed powder such that at the point ½ $H_0\Delta T/TM$ (the temperature differential, ΔT, between the respective hardness curves divided by the absolute melting temperature of the alloy, TM) is at least 1%, the prealloyed powder is deemed thermoplastic. However, this thermoplastic condition, referred to as TPC-1 (thermoplastic physical characteristic), is considered to be minimal. Preferably this ratio (ΔT/TM) should be at least 2% (TPC-2) and most advantageously at least about 5% (TPC-3). This contributes greatly to minimum flow stress and lower pressing temperatures which in turn reduce the otherwise required load on a press (or equivalent functioning equipment).

The process is not restricted to IN-100, but is primarily directed to difficultly workable superalloys and is applicable to alloys in general. Among the superalloys are those containing up to 60%, e.g., 1 to 25%, chromium; up to 30%, e.g, 5 to 25%, cobalt; up to 10%, e.g., 1 to 9%, aluminum; up to 8%, e.g., 1 to 7%, titanium, and particularly those alloys containing 4 or 5% or more of aluminum plus titanium; up to 30%, e.g., 1 to 8%, molybdenum; up to 25%, e.g., 2 to 20%, tungsten; up to 10% columbium; up to 10% tantalum; up to 7% zirconium; up to 0.5% boron; up to 5% hafnium; up to 2% vanadium; up to 6% copper; up to 5% manganese; up to 70% iron; up to 4% silicon, and the balance essentially nickel.

Cobalt-base alloys of similar composition can be treated. Among the specific superalloys might be listed IN-738 and 792, Rene alloys 41 and 95, Alloy 718, Waspaloy, Astrology, Mar-M alloys 200 and 246, Alloy 713, Alloys 500 and 700, A-286, etc. Some of these alloys are more workable than others. Other base alloys such as titanium can be processed as well as refractory alloys such as SU-16, TZM, Zircaloy, etc. The prealloys can contain up to 10% or more by volume of a dispersoid such as Y_2O_3, TLO_2, La_2O_3, etc.

Example 1: Atomized IN-100 powder containing (nominally) about 15% Co, 10% Cr, 5.5% Al, 4.75% Ti, 3% Mo, 0.02% C, 0.15% B, 0.06% Zr, balance essentially Ni, was screened to −325 mesh. It was then subjected to high energy impact in one gallon attritor for a period of about 10 hours, the impeller speed being maintained at about 350 rpm, 37 pounds of through hardened 52100 steel balls (five-sixteenths of an inch diameter) being used as the impacting elements. The ball to powder (volume) ratio was about 17:1 and a flowing nitrogen atmosphere was used.

The as-attrited powder was then canned in mild steel, evacuated, heated to 600°F under vacuum and sealed. Compaction was accomplished by heating the sealed can to 1900°F and then upsetting the same against a blank die in an extrusion press. This was followed by vermiculite cooling. Hot hardness was determined on the as-compacted material as shown in Table 1. Included for purposes of comparison are the corresponding hardness values of the IN-100 powder in the as-atomized and compacted state.

Table 1

	.... Rockwell Hardness, R_A	
Temperature	As Atomized and Compacted	Attrited and Compacted
RT*	74.5	78.5
1000	73.2	76.2
1200	72.4	74.1
1400	71.0	69.8
1600	63.2	14.4
1800	37.2	**

*Room temperature.
**Too low to determine.

The foregoing reflects the tremendous drop in hardness as the temperature exceeded about 1400°F. At 1600°F, hardness was already at the dramatically low level of 14.4. At 1800°F, the hardness of the strain energy induced (attrited) powder was at such a low level that it could not be determined on the Rockwell A scale. In marked contrast, it will be observed that the hardness of IN-100 in the as-atomized and compacted condition was, comparatively speaking, still quite high at the corresponding temperatures. At the temperature 1600°F, for example, the attrited powder in comparison with its as-atomized counterpart had a TPC-3.

Example 2: Another batch of −325 mesh atomized IN-100 powder was processed as in Example 1. In this instance, compacting in the extrusion press was conducted at 1950°F. Tensile specimens were prepared from the processed billet. One specimen was tested in the attrited and as-compacted state at 1800°F. Flow stresses were determined and were found to range from 4,870 to 11,170 psi over strain rates of from 0.0025 to 0.625 min^{-1}. At fracture, the elongation was 144% and the reduction in area was 99%. In marked contrast the flow stress for as-atomized IN-100 upset compacted at 1950°F was from 8,670 to 16,000 psi over the same strain rates, the elongation being 10% and reduction in area being 4.5%.

The rather striking differences in flow stress and/or ductility between the attrited and as-atomized powders are deemed attributable, at least in part, to the much finer grain size or dislocation structure of the attrited powder. Moreover, it is considered that there is a finer dispersion of embrittling phases as a by-product of the attriting. Compaction of the as-atomized powder is rendered most difficult at best by reason of phases on the powder surface and by the higher deforming stresses required. This contributes to subsequent problems concerning porosity and embrittlement.

For purposes of determining stress-rupture life, another tensile specimen from the attrited compacted powder was heat treated at 2100°F for 4 hours, oil quenched, heated at 1200°F for 24 hours, air cooled, and aged at 1400°F for 8 hours followed by air cooling. At a temperature of 1350°F and under a stress of 100,000 psi the specimen had a rupture life of 22.5 hours. This compares with but 2 hours for the as-atomized compacted powder treated in the same fashion. In the latter case, strength could not be improved by heat treatment.

Example 3: Batches of atomized IN-100 powder of a composition corresponding to that given in Example 1, screened to −60 +325 mesh were processed in a 5½" diameter "ball mill." The ball mill was run at about 80 rpm using ⅜" diameter, through hardened 52100 steel balls, a 10:1 ball to powder ratio and a sealed air atmosphere, the difference being that in one case the process was conducted for 5 hours as opposed to 50 hours in the second instance. Two other batches of −325 mesh atomized IN-100 powder were "attrited" as in Example 1 for periods of 1 and 3 hours, respectively.

Each of the 4 batches was compacted as in Example 2 and hardness determinations were made at various temperatures as given in Table 2. Also set forth, for convenience, are the hardness values for as-atomized and compacted (no strain induced energy) powder and the IN-100 atomized and attrited powder of Example 1 (attrited 10 hours). Finally, hardness data taken from that published for conventional as-cast IN-100 is also included.

Table 2

	Rockwell Hardness, R_A						
Temp.	As	Ball Milled		Attrited			As
(°F)	Atomized	5 Hours	50 Hours	1 Hour	3 Hours	10 Hours	Cast
RT*	74.5	74.8	75.8	76.4	77.2	78.5	69.5
1000	73.2	73.9	73.8	74.4	76.0	76.2	68.2
1200	72.4	73.8	73.2	74.1	74.0	74.1	68.2
1400	71.0	71.3	71.5	70.0	69.4	69.8	67.8
1600	63.2	61.7	55.7	47.2	33.0	14.4	65.0
1800	37.2	27.4	**	**	**	**	58.5

*Room temperature.
**Too low to measure.

The above results show that the attriting for only 1 hour rendered the IN-100 powder vastly more thermoplastic than dry ball milling for 5 hours and somewhat more deformable than ball milling for the less practical period of 50 hours, this being a direct effect of the amount of strain energy induced in the respective powders. As a practical matter, the as-cast hardness remained rather unchanged over the complete temperature range of test.

Rolling Mill

In additional work on producing thermoplastic prealloyed powders, *J.M. Larson; U.S. Patent 3,976,482; August 24, 1976; assigned to The International Nickel Co., Inc.* has developed a refined technique of achieving a continuous type process, a process which lends to minimizing contaminant pick-up that can arise with other cold working processes that might be used to induce the strain energy.

Consolidated prealloyed powders produced in accordance with this process have a highly desirable coarse grain structure, less than ASTM 5, upon solution treatment. Higher mechanical properties at elevated temperature can be expected. Turbine blade production as well as disc production should be facilitated. Moreover, the process is economical and its simplicity is decidedly attractive from the commercial viewpoint.

In this process prealloyed, highly alloyed superalloy powders of the nickel- and/or cobalt- and/or iron-base types, alloys normally difficult to hot work and fabricate using conventional technologies, are subjected to the compressive forces generated by properly spaced rolls of a rolling mill, whereby the powder particles take on the "thermoplastic" condition. Heating the so-processed powder to consolidation temperature and compacting results in considerable grain refinement in comparison with the unprocessed prealloyed powder, the thermoplastic powder manifests a markedly lower flow stress.

Subsequent forming operations can be conducted at lower temperatures and/or stresses than otherwise would be the case with conventional means, including powder metallurgy (P/M) processing. Because of this state of thermoplasticity, tremendous flexibility in operation is afforded. On the one hand, it is considered that large diameter discs, say 4 or 5′ in diameter, can be hot isostatically pressed for aircraft or industrial gas turbine use, while on the other hand intricate, complex shapes can be formed as by, for example extrusion.

Example: To illustrate the difference between the consolidating of thermoplastic powder produced in accordance with the invention and consolidating as-prealloyed powder, IN-792 powder, virtually all of a mesh size −40 +325 was divided into two equal batches. One sample was placed within a disc-shaped container (container A) formed of a superplastic alloy nominally of about 66% Fe, 26% Cr, 6.5% Ni, 0.5% Mn, 0.5% Si, 0.2% Ti, 0.05% with low P and S. The other batch of powder was passed through carbide rolls, the rolls being about 2¼″ in diameter and approximately 0.002″ dynamic gap. One roll pass was used, the rolls being rotated at about 35 rpm. This thermoplastically processed powder was then placed in a similar disc-shaped container (container B).

The disc-shaped containers were then hot isostatically pressed (HIP) at 15,000 psi for 1 hour. Container A with the conventionally processed powder was pressed at 2155°F whereas Container B was pressed at 1960°F, nearly 200°F below the former.

Upon evaluation it was found that the compact formed of powder processed in accordance herewith reached a density just about that of possible theoretical, porosity being well less than 0.07%. This was in marked contrast with the conventional product which exhibited a high degree of porosity, to wit, 1.8%, despite the 195°F higher compacting temperature. A consolidation temperature of 2250° to 2300°F might have afforded a comparable density level as achieved through the process, but it is deemed that the as-compacted grain size would have been on the order of that of the original powder particles.

Specimens of the respective compacts were also tested to determine flow stress characteristics. A temperature of 1900°F was used and the compact of the process displayed a low flow stress of 5,200 psi (0.01 min^{-1} strain rate) vs 8,900 psi (70% higher) for the conventionally processed material. It might be mentioned had the container A powder been compacted at the 1950°F temperature, it would not have even been consolidated enough to give a flow stress value.

Upon solution heating at 2225°F for 1 hour, a highly desired coarse grain, ASTM 2-3, was obtained for the thermoplastic processed product as against ASTM 5-6 for the conventional material.

Fine Particles Having Spheroidal Form

F. Montino and L. Colombo; U.S. Patent 3,850,612; November 26, 1974; assigned to Montecatini Edison SpA, Italy describe a process for preparing very finely divided, homogeneous metallic nickel powder of spheroidal shape, with a granulometric range between 0.03 and 0.7 micron and with a very narrow granulometric spectrum. It occurs by reduction with H_2 of a hydroalcoholic suspension of nickel hydroxide at temperatures between room temperature and 200°C, under partial pressures of H_2 between 20 and 100 atm, in the presence of bisacrylonitrilenickel in quantities between 15 and 0.5 g/mol of $Ni(OH)_2$, and optionally in the presence of pyridine in a quantity between 1 and 20 g/l of suspension.

The reduction reaction is conducted in the presence of a sulfur compound selected from the class consisting of hydrogen sulfide and alkali metal and alkaline earth metal sulfides in quantities corresponding to between 2 and 50 mg sulfur/mol of nickel hydroxide. Sodium and potassium sulfides and hydrogen sulfide are especially suitable.

The addition of NH_3 in quantities between 0.1 and 1 g/l of suspension is preferable, but not necessary.

According to a preferred embodiment, nickel hydroxide is suspended in a hydroalcoholic mixture in which the sulfur compound has been previously dissolved. To this suspension is then added bisacrylonitrilenickel in a quantity determined by the granulometry that one wishes to obtain in the end product, and pyridine.

The hydrogen reduction of the hydroalcoholic suspension of the nickel hydroxide, containing the sulfur compound and the additives, is preferably carried out in a stainless steel autoclave fitted with a blade stirrer and with a resistance heater.

Example 1: 178 g of $Ni(OH)_2$ still wet with water (obtained from $NiCl_2$ by precipitation with NaOH, at 90°C, and subsequent filtering and washing of the precipitate until the wash water shows the disappearance of NaCl) and corresponding to 46.3 g of dry $Ni(OH)_2$ (0.5 mol) and 132 g of H_2O, were suspended in a solution of 348 cc of water and 240 cc of monoethylene glycol so as to have on the whole 720 cc of a solution with a water/glycol ratio of 2:1. To this hydroalcoholic solution there had previously been added 4 mg of sulfur (as Na_2S) as an aqueous solution of known titre.

To the hydroalcoholic $Ni(OH)_2$ suspension, maintained under N_2, were added 5 g of bisacrylonitrilenickel, 8 cc of pyridine and 2 cc of NH_4OH (a 30% NH_3 solution). The percentage of Ni added as bisacrylonitrilenickel with respect to the total Ni amounted to 5.7%.

This suspension was then transferred to a 2 liter autoclave of AISI 316 steel, fitted with a blade stirrer driven by a magnetic coupling at 400 rpm, and with a conventional electrical resistance heater.

After eliminating the air by means of N_2, the autoclave was charged with an H_2 pressure of 40 atm, the stirrer was started, and the heating begun. Then the inside temperature was brought up to 150°C over a period of 50 minutes and maintained at that value for 15 minutes. Thereupon the autoclave was cooled down and the product discharged.

The powder was separated magnetically, washed with water and then with methanol, dried in a current of nitrogen at 80°C and then under vacuum at the same temperature.

In this way there were obtained 31 g of powder which may have as much as 2 to 5% of solvent (methanol and/or water). In the mother liquor the Ni was practically absent.

Under the electron microscope the nickel powder thus obtained showed up as spheroidal homogenous granules with a low degree of aggregation, and with an average diameter of 0.04 micron.

The granulometric count showed that 92% of the individual granules had a diameter between 0.030 and 0.057 micron.

Example 2: The following example shows that the addition of a sulfur compound enables one to obtain submicronic nickel powders with homogeneous spheroidal granules having a low degree of aggregation. However, the presence of a sulfur compound in the hydroalcoholic suspension was not a sufficient condition for obtaining powders with a granular diameter below 0.06 micron if the reduction was conducted by feeding H_2 into the autoclave which was previously heated.

A hydroalcoholic suspension (prepared according to Example 1 above) was subjected to a reduction test similar to that described above in Example 1, except for the H_2 which was here introduced with a partial pressure of 50 atm into the autoclave heated beforehand to 170°C. The temperature was maintained constant until the manometer attached to the autoclave indicated a pressure drop (35 minutes).

In this way, there was obtained a powder consisting of homogeneous spheroidal granules having a low degree of aggregation. The average diameter of the granules in this instance, however, was between 0.06 and 0.07 micron.

Nickel Halide as Interdispersion Cold-Bonding Agent

In producing mechanically alloyed composite powder particles, *G.L. Fisher and C.R. Cupp; U.S. Patent 3,865,572; February 11, 1975; assigned to The International Nickel Company, Inc.* have developed a process which comprises conducting the mechanical alloying operation in the presence of an inorganic interdispersion cold bonding agent (ICBCA) formed of a halide. The halide is in contact with the powder particles to be mechanically alloyed and present in an amount effective to provide the necessary control balance during the course of the operation, with the upper limit being about 3% by weight of the powder charge.

The control balance is such that an intimate interdispersion of constituent powders is achieved without either the formation of an appreciable amount of detrimentally large composite particles or the deleterious adherence of powders to the milling elements or mill interior surfaces and without incurring serious impairment of the mechanical properties of the composition to be produced.

As little as 0.05% of certain ICBCAs (based upon the weight of the powder charge) has been found satisfactory.

It is most advantageous that the ICBCA be sublimable or decomposable, the sublimation or decomposition temperature exceeding the operating temperature of the attritor mill (or equivalent functioning high energy machine such as the planetary or vibratory ball mills) but preferably not exceeding about 2000°F. Halides, including the metal halides, are deemed quite advantageous, small amounts of, for example, nickelous chloride and bromide having given excellent results.

By virtue of being sublimable or decomposable the interdispersion cold bonding agent can be appreciably removed during subsequent heat treatment of the mechanically alloyed composite product powder particles. As is known, the composite product powder particles produced by "MA" are hot consolidated, temperatures on the order of 1600° to 2000°F being commonly used. Thus, an ICBCA with a sublimation or decomposition temperature below about 2000°F, and advantageously below 1600°F, lends itself to this treatment.

The amount of an ICBCA employed will be largely influenced, by inter alia, the alloy composition to be produced, milling time, ball-to-powder ration, and the nature of the attriting elements, particularly the latter. Hardened steels, stainless steels, tungsten carbide, nickel and other metals as well as cermets may be used as the attriting media; however, considerably less ICBCA will usually be necessary in conjunction with the harder attriting elements such as 52100 steel as opposed to, say, nickel.

The longer milling periods may require a slightly higher percentage of ICBCA than otherwise. Generally, not more than 2 or 3% of an effective ICBCA component need be employed, a range of 0.1 to 0.5% and up to 2% being deemed quite satisfactory. Excessive amounts can be detrimental. In terms of a metal halide, for example, a range of about 0.05 to 0.5% of halide as metal halide is of advantage, the halide serving as the ICBCA.

Example: $NiCl_2$ and $NiBr_2$ were used as ICBCA to determine particle size results and ICBCA retention in processing commercial nickel 123 powder. 8 kg batches of the nickel powder were processed in a 4-gallon attritor for 5 hours at an impeller speed of 250 rpm using ⅜" diameter carbonyl nickel balls at a ball-to-powder ratio of about 11.4:1. Oxygen, about 0.05% of the charge weight, was added as air during processing. Nitrogen was also added and served to maintain the oxygen partial pressure at about 1% of the weight of gas in the attritor.

A very fine particle size distribution could be attained with both halides, approximately 90% of the processed powder being finer than about 105 μ in diameter at a halide content of 0.5%. The data indicate that a "control balance" could be achieved.

The attrited powders were then placed in metal steel cans (3½" diameter) and evacuated at 750°F to 20 hours. The cans were sealed, heated for 1 hour at 1800°F and extruded thereat to 1" diameter rods.

Chemical analyses were made to determine retained chlorine and bromine content of the extruded rods, the results appearing below.

Chlorine		Bromine	
Added %	Retained %	Added %	Retained %
0.02	0.012	0.02	0.021
0.05	0.018	0.08	0.031
0.10	0.039	0.41	0.022
0.50	0.37		

If the extrusion treatment had been carried out at a higher temperature, e.g., 200°F, it is considered that a greater amount of halide would have been removed. It is to advantage that the sublimation (or decomposition) temperature be at least 200° to 300°F below the temperature of hot consolidation.

Stearic Acid as Interdispersion Cold-Bonding Agent

In mechanically alloying powders wherein the initial constituent powders are repeatedly fragmented and cold bonded by the continuous impacting action of attriting elements, usually metal balls, for a period such that composite product powder particles of substantial saturation hardness are formed, *T.E. Volin; U.S. Patent 3,877,930; April 15, 1975; assigned to The International Nickel Company, Inc.* has found that the desired "control balance" can be attained, through incorporating an effective percentage of at least one interdispersion ICBCA.

An organic agent, including organometallic compounds, capable of reacting with or of being adsorbed on a plurality of the powder (particle) surfaces of an initial powder charge and which is capable of residing thereof in reacted or adsorbed form during at least a significant part of the "mechanically alloying" process, is effective in achieving the control balance. (Such materials include organic compounds which decompose such as to form a constituent capable of reacting with or of being adsorbed on the powder surfaces.) Only a small quantity of the organic ICBCA is required, though this is dependent upon the given conditions used in producing the desired mechanically alloyed composite product powder particles as is further described herein.

Possible ICBCAs could be: saturated hydrocarbons such as methane, ethane, propane and hexane; unsaturated hydrocarbons such as ethylene, allene, hexene and acetylene; alcohols, including methanol, ethanol, and ethylene glycol; aldehydes, e.g., formaldehyde, trioxane, acetytaldehyde, and benzaldehyde; ethers, for example, methyl or vinyl ether; ketones, such as acetone; polymers of the foregoing; organohalogen compounds, including the carbon tetrahalides such as CF_4, and polymers thereof such as polytetrafluoroethylene; organometallic compounds, for example, nickel or cobalt stearate, triethylaluminum, dicyclopentadiene-nickel, nickel tetracarbonyl; fats and fatty acids such as stearic acid, stearin, and the formic, oleic, and oxalic acids; etc.

The process is particularly applicable to the production of nondispersion hardening superalloys including those containing up to 65%, e.g., from 2% up to 25 or 35%, chromium; up to 30%, e.g., 5 to 25%, cobalt; up to 10%, e.g., 1 to 9%, aluminum; and up to 8%, e.g., 1 to 7%, titanium; including those alloys containing 4 or 5% or more of aluminum plus titanium; up to 30%, e.g., 1 to 8% molybdenum; up to 25%, e.g., 2 to 20%, tungsten; up to 10% columbium; up to 10% tantalum; up to 7% zirconium, up to 0.5% boron; up to 5% hafnium, up to 2% vanadium; up to 6% copper; up to 5% manganese; up to 70% iron; up to 4% silicon; and the balance essentially nickel.

Cobalt-base alloys of similar composition can be treated. Among the specific superalloys might be listed IN-738, Rene alloys 41 and 95, Alloys 500, 700, 713 and 718, Waspaloy, Astroloy, Mar-M alloys 200 and 246, A-286, B-1900, etc. However, such superalloys and other contemplated alloys can also contain up to 10% by volume of a refractory dispersoid material including the oxides, carbides, nitrides and borides. Such refractory dispersoids can be of various elements including yttrium, lanthanum, thorium, zirconium, hafnium, titanium, silicon, aluminum, cerium, uranium, magnesium, calcium, beryllium and the like. As a practical matter, only a very small amount of such dispersoids need be employed, e.g., up to 2% by volume. Other base alloys such as titanium and copper can be processed as well as refractory alloys such as SU-16, TZM, Zircaloy, etc.

Example: Composite product powder particles of the alloy IN-792 were produced using various ICBCAs and either nickel or steel attriting elements. Powder blends were used, each being comprised of carbonyl nickel powder (99.1% nickel approximately 4 μ in size), 0.15% Asbury flake graphite to raise the carbon content to nominal, and a low oxygen omnibus master alloy (−100 mesh of 30% Ni, Cr, Mo, Co, Al, Ti, W, Ta, B, Zr and C) prepared by vacuum induction melting and grinding in cold nitrogen. Each run consisted of 5 g of the blended powder plus the particular ICBCA used as given in Table 1.

A Spex mill was employed in each case, the respective runs being conducted for 30 minutes under an argon atmosphere (atmosphere was acetylene plus argon in Run No. 6). The Spex mill jar was cooled prior to opening, emptied and the balls replaced in the jar and processed for an additional 30 seconds in air to remove any loosely adherent powder. The total amount of drained powder was weighed to ascertain the percentage of powder recovered. An oxygen analysis was also made, and retained ICBCA was determined for Runs 4 and 6. Runs 1 and 2 are included for purposes of comparison.

Table 1

Run No.	. ICBCA (%) . .		Attriting Element (100 g)	Powder Recovery (%)	Oxygen Analysis (%)	ICBCA Retained
1	None	—	Nickel	32	0.30	—
2	None	—	Steel	57	0.17	—
3	Stearic acid	1	Nickel	81	0.46	
4	Stearic acid	0.1	Steel	81	0.22	0.32% C
5	PTFE	0.2	Nickel	65	0.31	—
6	Acetylene	*	Steel	78	0.21	0.29% C

*Atmosphere estimated to be about 10% acetylene.

The low recovery of Run 1 reflects the detrimental adherence of the IN-792 powder to the milling elements and the interior surfaces of the mill. This was reduced somewhat by using the harder steel impacting elements in Run 2. Runs 3 and 4 further illustrate the marked effect of the nature of the attriting elements, only one-tenth of the stearic acid ICBCA being required with steel balls to give the same recovery attained using the softer nickel pellets. It is considered that the recovery for the polytetrafluoroethylene (PTFE) of Run 5 would be improved using steel balls as the attriting elements.

NICKEL ALLOYS

Nickel-Aluminum with Improved Grindability

F.G. Larson, Jr.; U.S. Patent 3,839,011; October 1, 1974; assigned to The International Nickel Co., Inc. describes a Raney metal alloy product of about 50% by weight of nickel and about 50% by weight of aluminum comprising water shattered particles each consisting essentially of thin, convoluted and folded laminae having internal voids in communication with the particle surface. The particles are characterized by a tapped bulk density of about 0.3 to 1.2 g/cm^3 and consist of about 25% to about 65% by weight of a catalytic metal from the group consisting of iron, nickel, cobalt, copper, silver and mixtures thereof, preferably nickel. The balance, apart from impurities, incidental elements and promoting ingredients, is an alkalisoluble metal selected from the group consisting of aluminum, zinc, silicon and mixtures thereof, preferably aluminum.

In manufacturing the product the Raney metal alloy is melted and is poured at a superheat of at least about 50° in centrigrade units in a molten stream through an essentially horizontally flowing stream of water. It is essential that the water be flowing at a minimum velocity to shatter the Raney metal alloy because mere quenching from the liquid state is reported to result in a product which is very hard and difficult to break up. Assuming a metal pouring rate of about 1.5 kg/sec in a thin stream and a roughly rectangular stream 1.6 cm thick by 6.35 cm wide a minimum water flow rate of 350 liters per minute is necessary to provide the shattered, fragmented structure of the present process.

Table 1 contains melting point and pouring temperature data as to examples of Raney metal alloy amenable to being produced by this process.

Any of the exemplifying alloys of Table 1 can be modified by inclusion of catalyst promoters or modifiers. For example, small amounts of alkaline earth metals, e.g., calcium, barium or the like, thorium, vanadium, chromium, molybdenum or similar elements known in the metallic or oxidic forms to promote particular catalytic activity can be included in the Raney metal alloys useful as the product of this process. Likewise, small amounts of inert metal, e.g., copper in nickel catalysts or nickel in silver catalysts can be used either alone or in combination with promoters to appropriately modify the Raney catalyst produced by leaching the process product.

Table 1

Ex. No.	Catalytic Metal	Percent by Weight	Alkali Soluble Metal	Percent by Weight	Melting Point (°C)	Pouring Temperature (°C)
1	Ni	42	Al	58	1080	1200
2	Fe	55	Al	45	1230	1350
3	Co	50	Al	50	1400	1500
4	Ag	35	Al	65	620	750
5	Ni	50	Al	50	1310	1480
6	Ni	60	Al	40	1570	1650
7	Co	40	Si	60	1140	1300
8	Ni	50	Zn*	40	–	1300
9	Cu	55	Al	45	600	800

*Alloy includes 10% aluminum.

Tungsten-Boron Phase

H.F. Prasse and H.E. McCormick; U.S. Patent 3,809,546; May 7, 1974; assigned to Ramsey Corporation have developed a method of making an alloy containing a tungsten-boron phase. The alloy comprises a matrix such as a nickel-chromium, or other nickel containing matrix, having therein a separate tungsten-boron phase. The alloy is used as a facing or coating for a number of base materials, and in particular as a piston ring facing for the bearing portion. The alloy is made by utilizing a plasma jet spray technique wherein hydrogen is utilized in combination with nitrogen or argon as a plasma gas.

The hydrogen should be flowed at a rate of 20 to 30 standard ft^3/hr. Another important variable which must be controlled is the distance of the gun from the workpiece. This may be varied from 3.5" to 6.5". Under such conditions the base material is coated with a hard facing comprising the alloy containing a hardened tungsten-boron phase.

Magnesium and/or Calcium to Lower Inert Gas Porosity

Thermally induced porosity, which can lead to cracking in metal powders after consolidation, is believed to be caused by the entrapment of the inert gas used for atomization within the metal powders. *J.M. Larson, I.S.R. Clark and R.C. Gibson; U.S. Patent 4,047,933; September 13, 1977; assigned to The International Nickel Co., Inc.* have found that the addition of an activating agent to the molten alloy prior to atomization with an inert gas, such as argon, serves to substantially reduce the porosity of metal powders. Suitable activating agents are characterized by an ability to rapidly diffuse to the surface of a molten metal particle and an affinity for oxygen. Activating agents that are useful for high nickel alloys and highly alloyed steels include magnesium, calcium, lithium, silicon, and rare earths.

The elements which serve as the activating agent should be added to the molten alloys prior to inert gas atomization in amounts ranging from about 0.001 to about 0.1%. When added in amounts less than 0.001%, there is no significant reduction in the amount of general porosity in the metal powders. With more than about 0.08% of these elements present, deterioration in the hot workability and the strength of these materials is found. It is preferred that from about 0.007 to about 0.08% of the elements used as activating agents be added to the molten alloys prior to inert gas atomization to achieve a residual of at least about 0.007% in the molten alloy.

Magnesium is one of the preferred activating agents. It can be added to the molten metal bath prior to gas atomization in the form of a master alloy. A nickel-magnesium master alloy containing from about 5 to about 20% magnesium, e.g., 15% magnesium, has been used for this purpose. Other magnesium alloys such as silicon-magnesium as well as elemental magnesium are also suitable for this purpose and can be substituted.

The element calcium is also a preferred activating agent and can be added to the molten alloy as a master alloy prior to gas atomization. A nickel-calcium master alloy containing from about 3 to about 15% calcium, e.g., 5% calcium, has been found to be useful for the practice of this process. Other calcium-containing master alloys, e.g., calcium-iron, calcium-silicon, etc., are also of use and can be substituted for nickel-calcium.

Since the elements suitable as activating agents are extremely reactive, it is preferred that the alloys to which they are added be gas atomized immediately after addition. That is, these elements have a strong tendency to react with oxygen, e.g., the oxides of the furnace lining, and to be removed from the melt as a surface slag. Also, since they are highly volatile and are added to the melt in many instances under a partial vacuum, as a result of volatilization they may not be available to interact during the gas atomization process to provide the desired resistance to gas entrapment.

Mixtures of the two preferred activating agents, magnesium and calcium, in various proportions can be used in this process. A mixture of these ingredients can provide beneficial hot working characteristics and mechanical properties since neither element will then be present in an excessive amount in the gas atomized powder.

The pressure of the argon gas should be kept relatively low to provide a desirable particle size distribution. Too low a pressure can lead to insufficient energy to atomize the stream into fine particles and result in a large coarse fraction, whereas when pressures are extremely high, there is a greater tendency for collision of energetic particles which is believed to result in greater porosity. Gas pressures between about 275 and 4,410 kPa (40 and 600 psig) are useful for atomization of molten metal. However, it is preferred to use gas pressures between about 1,030 and 1,240 kPa (150 and 180 psig).

Atomization requires superheating metals to temperatures in excess of their melting points. If insufficient superheat is used, a not entirely desirable large, coarse fraction results with the attendant risk of solidification within the tundish. Superheating to too high a temperature leads to increased entrapment of inert gas. This can be due in part to increased solution of the inert gas, but more likely is due to the increased time that particles are in the liquid state. With high superheats, more frequent collisions between particles result in entrapment of larger quantities of argon. Superheating to temperatures of about 85° to about 135°C above the melting point of the metal being atomized has been found to be effective in providing sufficient superheat for atomization with minimization of the amount of argon entrapped.

Inert gas jets should be aligned to converge at a single point. Such accurate alignment provides fewer particle/particle collisions and hence, less porosity. However, since a single gas impact point can lead to undesirable splash-back of powder onto the gas jets, it is more desirable to use more than one impact point; and consequently, it is preferred to align the gas jet at more than one location. The number of gas jets was found to provide no significant effect on porosity. That is, when four gas jets were used, the amount of porosity was essentially the same as when eight gas jets were used.

Nickel-Lanthanum by Reduction and Diffusion

D.L. Martin; U.S. Patents 3,883,346; May 13, 1975; and 3,918,933; November 11, 1975; both assigned to General Electric Company describes a process for storing hydrogen in particles of nickel-lanthanum alloy ranging in size from about 10 to 50 μ by contacting the particles with hydrogen gas which is selectively absorbed by the particles. The alloy consists essentially of a composition within 15% by weight of the stoichiometric composition of the Ni_5La intermetallic compound.

The alloy particles are formed by providing a particulate mixture of lanthanum oxide, calcium hydride and nickel, the calcium hydride being used in an amount in excess of the stoichiometric amount necessary to reduce lanthanum oxide. The particulate mixture is heated in a nonreactive atmosphere to decompose the calcium hydride and thereby effect reduction of the lanthanum constituent; then the resulting mixture is heated in a nonreactive atmosphere to diffuse the resulting lanthanum metal into the nickel particles to form the alloy. The resulting alloy-containing mixture is left in an oxidizing or moisture-containing atmosphere to oxidize or hydrolyze the excess calcium precipitated therein causing the mixture to disintegrate, and the nickel-lanthanum alloy particles are recovered.

All parts and percentages used herein are by weight unless otherwise noted and where screen size is referred to, it is the U.S. Standard Screen Size.

Example: The formulation was as follows:

	Grams
Nickel powder (−400 mesh, i.e., −38 μ)	58.71
La_2O_3 (1.1 μ)	32.58
Calcium hydride (−40 mesh)	17.68

The nickel powder and lanthanum oxide were mixed together in a jar mill for 15 hours to produce a thoroughly blended mixture. Calcium hydride was then added to the mixture which was then tumbled for an additional 10 minutes under a nitrogen atmosphere in a dry box. The mixture was poured into a nickel container which was then placed in an electric hydrogen furnace and heated to a temperature of about 850°C where it was maintained for 2 hours to reduce the lanthanum oxide. It was then heated rapidly in a period of 35 minutes to 1050°C at which temperature it was held for 2 hours to carry out the diffusion of the lanthanum into the nickel powder.

The product was then placed in the cold chamber of the hydrogen furnace where it cooled to room temperature in hydrogen. The resulting reacted cake was placed in a hydrator, i.e., a high humidity chamber, over the weekend. The cake broke into a flowable particulate material due to the hydration of excess metallic calcium. The calcium hydroxide in the resulting powder was then removed by repeated washes in water and by a final wash with dilute acetic acid. The powder was then washed with alcohol, and dried under vacuum. The powder particles had a size of about 7.2 μ.

A portion of the dried product was subjected to x-ray diffraction analysis using the Debye-Scherrer technique. The x-ray powder pattern obtained indicated that it was practically single phase Ni_5La. Two extremely weak extra lines could not be accounted for as either the Ni_7La or $Ni_{17}La_2$ compound. Standard chemical analysis of this product powder showed it to contain calcium in an amount of 0.38% by weight of the intermetallic compound.

Urea to Control Particle Size

T.C. Quinby; U.S. Patent 4,072,501; February 7, 1978; assigned to the U.S. Department of Energy has developed a method for preparing powders of controlled particle size, powders having an average particle size less than about 15 μ in diameter. Powder mixtures prepared according to this method are very homogeneous, having a composition variance of less than ±1 weight percent at the 95% confidence level.

Finely divided powders are prepared by first reacting an aqueous solution containing dissolved metal values with excess urea. The urea reacts with the water present in the solution, including water of hydration of soluble species, according to the reaction:

$$NH_2CONH_2 + H_2O \leftrightarrows 2NH_3\uparrow + CO_2\uparrow$$

The reaction of urea with water in the solution takes place at a temperature sufficient to melt the urea. Depending upon the metal salt in the solution, the urea/salt mixture melts at about 120° to 132°C, whereupon urea reacts with the water forming volatile products, leaving the urea present in excess of the water content as a molten urea solution containing the metal values.

The aqueous solution can be a solution of any soluble metal salt or mixture of metal salts. Preferred are metal salts such as halides, nitrites, or nitrates whose anions form volatile products upon subsequent heating with molten urea. The preferred method of carrying out the reaction is to add solid urea to a concentrated aqueous solution and heat the resulting mixture to a temperature sufficient to melt the urea. It is preferred that the initial aqueous solution be essentially saturated to minimize the amount of urea reacted.

After the reaction of water in the solution with urea is complete, as evidenced by a volume reduction representative of the original aqueous volume present (volume of water plus stoichiometric urea), the resulting molten urea solution is heated to cause metal values dissolved therein to precipitate. Metal values present in the molten urea solution precipitate homogeneously at about 180°C, forming a mixture of urea and the precipitated metal values. As the precipitation goes to completion, the temperature of the mixture rises, indicating an exothermic reaction.

The chemical composition of the precipitate is believed to be either a hydrated oxide or an amine compound. It is believed that a part of the urea in the molten urea/metal salt mixture is decomposed as it is heated above its melting point, leaving a liquid product. The term "molten urea" is used here to indicate the liquid phase present after the urea/water reaction and after the precipitation reaction, regardless of its actual chemical composition.

After the completion of the precipitation reaction the resulting mixture containing precipitated metal values is heated to evaporate volatile material from the mixture, leaving a dry powder containing the metal values. In actual practice, at the completion of the precipitation step, the product is virtually a solid mass of precipitate bound together with residual molten urea. The volatile products, e.g., NH_4NO_3, are evaporated by heating to about 400°C at atmospheric pressure.

In order to form metal from reducible metal precipitates, the dry powder is heated in the presence of a reducing gas such as hydrogen. In order to provide metal oxide particles, the dry powder can be calcined at above about 650°C. The product of the calcination step is a fine oxide powder or mixed oxide powder. When halide salt solutions are used, it is belived that the calcination step should be performed in the presence of oxygen. When oxysalts are used such as nitrates, sulfates, phosphates, etc., oxygen is probably not needed for the calcination step.

It is not necessary that the heating steps of this process be carried out in a stagewise manner. All that is required is that the water in aqueous solution react with urea prior to the precipitation step and that the precipitation be complete prior to evaporating to dryness. Accordingly, it is a simple matter to devise a continuous or varying heating rate to accomplish these objectives.

Any metal cation or mixture of metal cations can be converted to oxides according to this process. The particle size of the oxide product is a function of the amount of excess urea present (excess with respect to reaction with water). Since excess molten urea is the solvent for the metal values, the greater the amount of excess urea, the smaller will be the precipitated particles.

It is a matter of routine experimentation to determine the oxide/urea ratio needed to provide a particular particle size. For example, oxide particles in the 0.2 to 0.5 μ range are prepared from molten urea solution containing the equivalent of 20 g metal oxide per liter of molten urea. Oxide particles in the 10 to 12 μ range are prepared from molten urea solution containing the equivalent of 50 g metal oxide per liter of molten urea.

Elimination of Carbide Segregation to Prior Particle Boundaries

Carbide segregation to prior particle boundaries generally is undesirable because it reduces workability and mechanical properties. For example, in extrusions such segregations are aligned in the extrusion direction resulting in serious impairment of transverse properties. Thus, whereas longitudinal properties of a powder metallurgical compact, for example, of a nickel-base superalloy, may be excellent, transverse strength properties, particularly ductility, have been found to be very low.

Examination of fractured transverse specimens has shown that failure had occurred through the carbide stringers. Thus the loss in transverse ductility seems to be the direct result of the low strength fracture path afforded by the carbides segregated at prior particle boundaries. It has been found that, in many cases, carbides had formed on the surface of the powder prior to or during consolidation. Also, it has been observed that such segregated carbides are not broken up or dissolved during the working or the subsequent high temperature exposure during article manufacture.

R.E. Allen, J.L. Bartos and P. Aldred; U.S. Patent 3,890,816; June 24, 1975; assigned to General Electric Company describe an alloy powder composition, the interrelationship of the elements of which inhibit formation of undesirable carbides which lead to carbide segregation to prior particle boundaries.

It is a nickel-base alloy powder of appropriate size for use in powder metallurgical techniques and comprises 0.03 to less than 0.3 weight percent C, no more than about 3.6 atomic percent of the M_6C carbide formers Mo and W, and at least about 0.3 atomic percent of the strong MC carbide formers Cb, Ta, Hf and Zr.

The alloy powder is further characterized by the substantial absence of $M_{23}C_6$ carbide. In one form, the powder consists essentially of, by weight, 3-8% Al, 2-8% Ti, 4-30% Cr, 0.03 to less than 0.3% C, at least one of the elements Mo and W with Mo, when selected, being up to 6% and W, when selected, being up to 12% (provided no more than 3.6 atomic percent of the two are present), up to

10% Ta, up to 10% Cb, up to 8% Hf, up to 5% V, up to 1% Zr, up to 1% B, up to 30% Co, with the balance essentially Ni and incidental impurities.

One form of the article includes the alloy powder composition and is characterized by the absence of carbide segregation to prior particle boundaries, the substantial absence of $M_{23}C_6$ carbide and the carbide which is present being predominantly MC-type carbide in the 0.2-1 μ size range.

Example: A large number of alloys were prepared in powder form and consolidated into shapes for further evaluation. The following Table 1 gives the composition of some of these alloys.

The alloy compositions in Table 1 were vacuum induction melted and cast prior to being converted to powder by inert gas atomization. The powder was then sized for use in making test specimens by screening to −60 +325 mesh size. The powder was placed in deformable containers, in this case metal containers of stainless steel, which were evacuated to about 10-15 μ pressure prior to their being sealed. The billets were then extruded to square bars in the temperature range of about 1850°-2250°F. Below that temperature range excessive forces are required to extrude or to obtain full density. Above about 2250°F, the nickel-base superalloy melting point is approached too closely.

After extrusion, a variety of evaluations were conducted on the alloys including heat treatment, mechanical property testing, metallographic studies and calculations of the volume percent of carbides and atomic percent of critical elements included in the alloy composition. The following Table 2 presents certain of these data.

The data of Table 2 has been divided into three groupings of alloys depending on their tendency for carbide segregation to prior particle boundaries. Group I lists alloys the structure of which includes neither $M_{23}C_6$ nor M_6C carbides but only MC carbide. Of these alloys in Group I, none showed carbide segregation to prior particle boundaries. Alloy E is a commercially available nickel-base superalloy used for manufacture of wrought products but not, prior to this process, for use in powder metallurgical compaction.

It should be noted that the elements Mo and W, which tend to result in the formation of M_6C carbide, are maintained within the range of the process of a maximum of 3.6 atomic percent. In addition, the strong MC formers of Cb, Ta, Hf and Zr are included within the range of at least about 0.3 atomic percent.

By way of contrast, the alloys of Group II all were found to have carbide segregation in prior particle boundaries. All of such Group II alloys, except for Alloy 21, included $M_{23}C_6$ carbide within their structure. Except for Alloy 21, the absence of the abovelisted strong MC carbide formers should be noted. Alloys A, B, C and D all are commercially available nickel-base superalloys. Alloy 21 is included in Group II of Table 2 as an alloy having a tendency toward carbide segregation to prior particle boundaries because of its 0.3 weight percent carbon content.

Table 1

Alloy	COMPOSITION (weight percent) BALANCE NICKEL AND INCIDENTAL IMPURITIES											
	Co	Cr	Mo	W	Al	Ti	Ta	Cb	Zr	Hf	B	C	V
1	14.3	9.4	0	6.0	4.2	3.2	0	3.9	.05	0	.02	.19	0
2	8.4	11.7	0	6.1	4.3	3.3	0	2.8	.05	0	.02	.05	0
3	7.8	8.9	1.1	[illegible].6	4.7	3.5	0	2.3	.05	2.0	.02	.15	0
4	15.1	10.7	2.9	5.9	3.8	3.9	0	1.7	.05	2.0	.02	.15	0
5	14.8	10.1	0	9.0	5.2	3.0	0	0	.05	0	.02	.05	0
6	16.0	12.0	3.0	0	5.5	2.5	0	1.5	.05	0	.02	.05	0
7	10.0	9.3	2.0	7.0	4.3	4.0	3.8	0	.05	0	.015	.17	0
8	8.0	12.0	6.0	3.0	4.5	3.5	0	1.5	.05	2.0	.02	.15	0
9	16.0	11.5	5.0	0	5.5	3.5	0	0	.05	0	.02	.15	0
10	13.8	9.6	8.9	0	4.7	3.7	0	0	.05	2.1	.02	.15	0
11	14.5	10.0	6.2	5.8	3.4	4.4	3.6	0.8	.05	0	.02	.15	0
12	14.8	10.2	4.8	5.8	4.3	4.4	0	0.6	.05	0	.02	.15	0
13	12.8	7.9	2.3	5.7	4.6	5.1	0	2.2	.05	0	.015	.15	0
14	10.2	4.7	1.8	5.4	5.3	6.0	0	2.7	.05	0	.015	.15	0
15	12.1	12.3	2.4	7.6	3.7	4.1	2.6	0.9	.05	0	.015	.17	0
16	9.2	7.8	1.8	7.0	3.6	4.6	4.4	0	.05	0	.015	.17	0
17	8.6	7.1	1.7	6.7	4.8	4.7	4.5	0	.05	0	.015	.17	0
18	6.7	4.0	1.3	6.2	5.6	5.6	5.5	0	.05	0	.015	.17	0
19	12.5	7.3	4.6	0	4.3	6.2	0	0	.06	0	.014	.18	1.4
20	11.0	5.7	2.2	0	5.3	5.8	3.2	0.6	.06	0	.014	.18	1.3
21	12.4	7.5	2.3	5.6	4.7	5.3	0	2.3	.05	0	.02	.30	0
23	13.0	8.4	2.6	5.3	4.0	4.1	0	2.0	.05	4.6	.02	.15	0
25	15.5	16.7	3.3	0	6.5	2.7	0	0.7	.06	0	.02	.18	0.3
28	7.8	6.8	4.9	0.8	7.1	2.6	0	0	.06	1.0	.02	.18	0
A	11.0	19.0	9.8	0	1.5	3.2	0	0	0	0	.01	.12	0
B	18.5	15.0	5.0	0	4.2	3.2	0	0	.06	0	.03	.10	0
C	15.2	10.0	3.0	0	5.6	4.6	0	0	.07	0	.01	.17	1.0
D	9.5	14.1	4.0	3.9	3.0	5.0	0	0	.03	0	.02	.18	0
E	8.0	14.0	3.5	3.5	3.5	2.5	0	3.5	.05	0	.01	.07	0

Table 2

Group	Alloy	. . . Carbides (vol %) . . . M_6C	$M_{23}C_6$	MC	x* (%)	y* (%)
I	1	0	0	1.8	2.4	1.9
	2	0	0	0.5	1.7	1.9
	3	0	0	1.4	2.0	1.1
	6	0	0	0.5	0.9	1.7
	7	0	0	1.7	1.2	3.4
	13	0	0	1.4	1.4	3.2
	14	0	0	1.5	1.7	2.8
	16	0	0	0.8	1.5	3.4
	17	0	0	0.8	1.5	3.1
	18	0	0	0.8	1.8	2.8
	20	0	0	0.8	1.4	1.3
	23	0	0	1.5	2.8	3.3
	25	0	0	1.6	0.4	1.8
	28	0	0	1.6	0.4	3.0
	E	0	0	1.4	2.2	3.3
II	5	0	0.6	0.2	0	2.8
	9	0	1.7	0.7	0	2.9
	19	0	2.0	0.4	0	2.6
	21	0	0	2.8	1.4	3.1
	A	2.8	0.4	0.2	0	5.9
	B	0	1.1	0.4	0	2.9
	C	0	1.9	0.8	0	1.7
	D	0.1	2.1	0.8	0	3.6
III	4	0.2	0	1.4	1.8	3.6
	10	2.1	0	1.4	0.7	5.3
	12	1.4	0	1.0	0.4	4.7
	15	0.5	0	1.5	1.4	3.9

*x = Cb, Hf, Ta or Zr; y = Mo or W.

It has been found during the evaluation of this process that alloys including such amounts of C after hot isostatic pressing displayed carbide segregation to prior particle boundaries. Therefore, this process is defined as including an alloy having less than 0.3 weight percent C. The lower limit for C has been defined as about 0.03 weight percent because alloys including less than that amount have insufficient C to form carbides which would create such segregation problems.

Any presence of the $M_{23}C_6$-type carbide is detrimental. It has been found that the complete absence of M_6C-type carbide is not quite as critical. Depending on the composition of the alloy, small amounts of M_6C-type carbide can be tolerated. Typical examples of alloys which can or cannot tolerate M_6C-type carbide in respect to the tendency to carbide segregation in prior boundaries are shown in Group III of Table 2.

Evaluation has shown that Alloy 4 does not show such carbide segregation in prior particle boundaries whereas Alloys 10, 12 and 15 do result in such detrimental carbide segregation. Therefore, this process defines the alloy as including a total content of no more than 3.6 atomic percent of the sum of Mo and W to limit the formation of M_6C-type carbide and thus further limit the tendency of the alloy to carbide segregation in prior particle boundaries.

In Table 2, the volume percent of carbides was calculated with the following assumptions: if the sum of Mo and W is greater than or equal to 3.5 atomic percent, the excess above that amount forms M_6C; if the sum of Cb, Ta, Hf and Zr is greater than or equal to 0.3 atomic percent, the carbon forms MC; and if the sum of Cb, Ta, Hf and Zr is less than 0.3 atomic percent, half of C forms $M_{23}C_6$ and half forms MC.

Representing the specifically preferred form of this process as shown in Table 1 and Table 2 are Alloy forms 4, 7 and 13 within the composition range, by weight, of 3-5% Al, 3.5-5.5% Ti, 7-11% Cr, 0.15-0.2% C, 2-3% Mo, 5-7% W, up to 4% Ta, up to 2.5% Cb, up to 2% Hf, up to 0.1% Zr, 0.01-0.05% B, 10-16% Co, with the balance essentially Ni and incidental impurities.

Freezing of Metallized Salt Solution

D. Tither; U.S. Patent 3,997,331; December 14, 1976; assigned to Crysus (Lancashire) Ltd., England has found that very small particles of metallized salts can be produced if a solution of the salt is frozen in an inert atmosphere under reduced pressure and thereafter the solvent is evaporated also under reduced pressure.

This process produces metallized salts of very small particle or crystal size. It comprises the steps of placing a solution of a metallized salt in a chamber, closing the chamber, evacuating the chamber to remove air therefrom and to reduce the pressure within the chamber, admitting to the chamber a gas which is inert relative to the metallized salt in an amount which is not sufficient to raise the pressure in the chamber to atmospheric pressure, reducing the temperature of the solution so that the solution freezes, and thereafter evaporating the solvent from the frozen solution.

When a solution of a metallized salt, for example ammonium paratungstate, is treated in accordance with the process, not only does the product have a very small particle size which allows the production of tungsten powder in the size range 0.2 to 0.8 μ by a hydrogen reduction process, but in the case of ammonium paratungstate, has a form which is entirely novel.

Hitherto ammonium paratungstate was known to have three hydrated crystal forms, each of which has a characteristic x-ray diffraction pattern. However, when ammonium paratungstate produced according to this process was examined by x-ray analysis no diffracted lines whatsoever where observed. This indicates that the ammonium paratungstate produced by this process is different in form from the previously known forms but the difference was confirmed by scanning electron microscopy.

The process can also be used in conjunction with nickel salts, such as nickel chloride and nickel nitrate solutions and is capable of producing these metallized salts of nickel in a form which permits "submicron" nickel powder to be produced by a hydrogen reduction process.

The apparatus for carrying out the process comprises a chamber, means for evacuating the chamber, means for admitting inert gas to the chamber and means for controlling the temperature with the chamber.

Example 1: Fine particles of ammonium paratungstate were produced. Tungstic acid and ammonium hydroxide were placed in the mixing tank in quantities to form 8% by weight ammonium tungstate and the mixture stirred so as to keep it of uniform consistency.

Filtered ammonium tungstate was delivered to the trays on the top shelf of the apparatus and when those trays were full, the trays on the lower shelves were filled by overflow from the trays on the top shelf through overflow weirs provided on the trays. The depth of solution in each tray was ⅝".

When all the trays were filled, the vacuum pump and Roots blowers were set to operate and the pressure reduced to 500 torrs. A chlorinated fluorinated hydrocarbon sold under the name "Freon" liquid was then admitted to the chamber until the pressure in the chamber rose to 600 torrs.

Cold fluid at a temperature of −50°C was then circulated through the shelves until the solution in the trays was frozen. The total pressure was now lowered to 0.1 torr and the temperature of the shelves raised gradually to 120°C by circulating fluid at that temperature through the shelves.

The solvent from the frozen solution was evaporated off until the product in the trays was not more than 10% by weight of the original charge. The chamber was then opened and the trays removed to recover the product therein which was found to be ammonium paratungstate, more than 50% by weight of which had an agglomerated particle size of less than 53 μ. These multiparticulate agglomerates consisted of many particles of submicron size, and particles less than 10 μ could be recovered by sieving. The multiparticulate agglomerates of size between 76 μ and 53 μ (the usually accepted most suitable size for x-ray diffraction analysis) were examined using an x-ray diffractometer, but no diffracted lines were observed.

The particles of size less than 10 μ were examined by x-ray analysis but no diffracted lines were observed. It was also examined under the electron microscope which showed the particles to have a wholly novel structure.

Example 2: Fine particles of nickel have been obtained by the hydrogen reduction at 400°C of nickel chloride. Nickel scrap was dissolved in dilute hydrochloric acid to produce a solution of metallized salt, i.e., nickel chloride. This solution was treated in the same manner as described in Example 1 except that after the solution was frozen, the total pressure in the chamber was reduced to 5×10^{-3} mm Hg.

The salt produced was found to be hygroscopic and eventually lost its unique structure on prolonged exposure to atmosphere. However, hydrogen reduction of the salt produced extremely fine nickel powder of submicron size.

Example 3: A solution of nickel nitrate was treated in the same manner as described in Example 2. The salt produced was thermally decomposed at 600°C to nickel oxide and then reduced at 400°C to yield a nickel powder of submicron particle size.

Ni-Fe-Cr Oxidation-Resistant, Dispersion-Strengthened Alloy

J.W. Schultz and H.F. Merrick; U.S. Patent 3,912,552; October 14, 1975; assigned to The International Nickel Co., Inc. have developed a process to produce a dispersion-strengthened nickel-base alloy characterized by outstanding oxidation resistance and high temperature strength which can be produced in sheet form and the like, and fabricated into various forms, provided the alloy contains, in addition to a controlled amount of dispersoid, correlated percentages of chromium, aluminum, iron, titanium, etc.

The process comprises an age-hardenable, austenitic nickel-base alloy containing, by weight, about 10 to about 30% chromium, about 10 to about 40 or 50% iron, with the chromium and iron percentages being advantageously interrelated such that they represent a point on or above the line **A-B-C** of Figure 3.2a, aluminum in an amount of at least about 4%, at least one refractory dispersoid in small but effective amount, e.g., 0.2 or 0.3% by volume, sufficient to impart improved stress-rupture strength at temperatures on the order of 1900° or 2000°F, the dispersoid having an average particle size of about 50 angstroms to about 5000 angstroms, and up to about 1% titanium.

Figure 3.2b is a diagram depicting an interrelationship of the chromium, nickel, and iron components of the alloy of this process, wherein the chromium concentration is adjusted to allow for the aluminum and titanium content of the alloys. In the diagram described, the adjusted chromium level, calculated as Cr + 6 + (Al − 3) + Ti, takes into account the effects of aluminum and titanium on the phase relationships in the Fe-Ni-Cr system. The adjusted chromium is referred to as chromium equivalents or Cr_{eq}.

It is most advantageous that the alloy be produced by mechanical alloying, after which it is processed such that it is comprised of relatively coarse grains, elongated in one or two working directions. Preferably the grains have aspect ratios of about 3:1 to about 100:1 with average widths of about 15 to about 2,000 μ, and average lengths of about 150 to about 12,000 μ, the grain thickness being less than the smaller of the width and length dimensions.

Figure 3.2: Iron-Nickel-Chromium Alloy

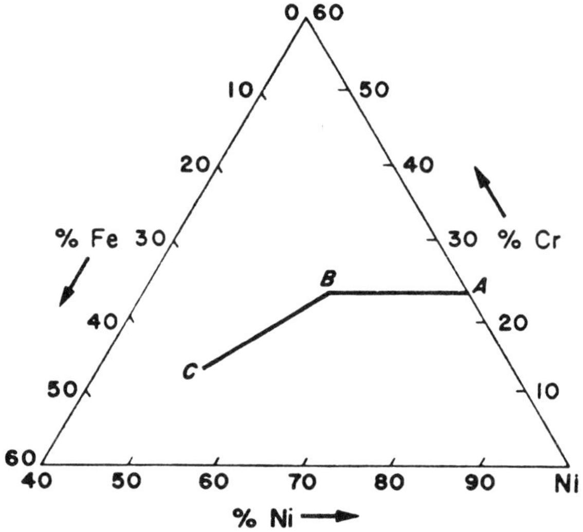

(a) Sulfidation Resistance

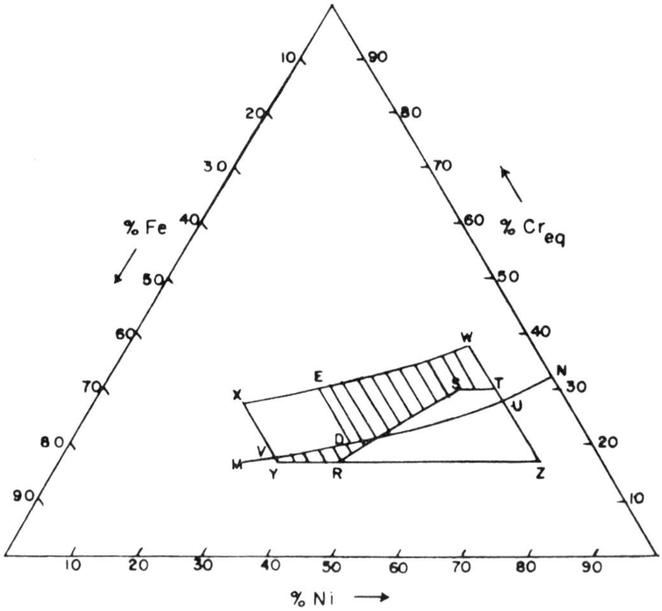

(b) Interrelationship of Alloy Components

Source: U.S. Patent 3,912,552

Grains elongated in two dimensions (i.e., generally elliptical) can be produced by working the alloy in two directions, e.g., cross-rolling at an elevated temperature, the two major axes of the grains being disposed longitudinally and transversely in the rolling plane, while grains elongated in a single direction, i.e., fibrous grains, can be produced by working in one direction.

The alloy combines precipitation and dispersion hardening and provides what is believed to be a most superior combination of high strength and sulfidation and oxidation resistance at temperatures up to at least 2000°F while affording comparatively high ductility which contributes to the ready fabrication thereof.

As mentioned, production of the alloy is advantageously carried out by mechanically alloying, i.e., high energy milling, a powder charge of the desired composition such that the initial powder constituents are interdispersed, comminuted and welded together to provide composite particles. The powder charge can comprise elemental and/or alloyed powders, generally of fine size not exceeding −10 mesh.

The mechanical alloying process is conducted so as to produce composite powder particles characterized by a cohesive internal structure in which the constituents, including the initial dispersoid material, are intimately united to provide a homogeneous interdispersion of comminuted fragments of the starting constituents, the composition of the individual composite powder particles corresponding to the final alloy product.

Mechanical alloying is beneficially carried out such that the composite powders exhibit substantially saturation hardness. Such composite powders have substantial compositional homogeneity, the dispersion particle being distributed substantially uniformly throughout the various composite powder particles at average spacings not exceeding 1 μ.

Mechanical alloying can be conducted under dry conditions in a high energy mill such as the Szegvari attritor, for example. The preferred milling conditions with a 4-gallon Szegvari attritor are about 16 to 24 hours with an impeller speed of about 250 to 350, e.g., 290 rpm, using a nitrogen-0.7% oxygen atmosphere and ⅜" diameter steel balls present in an amount sufficient to provide a ball-to-powder weight ratio of about 15:1 to 20:1.

The mechanically alloyed powder can then be hot consolidated, e.g., by hot extrusion or hot compaction. Generally, powder consolidation can be conducted by extrusion of powder held in a metal, e.g., steel container, at a temperature in the range of about 1800° to 2200°F, e.g., 1950°F, with an extrusion ratio of about 5:1 to 20:1, e.g., 10:1. The consolidated material may be extensively hot worked, e.g., hot rolled over a temperature range of about 1800° or 1900°F to about 2200° or 2300°F to provide hot reductions of up to about 75 or 90% or higher.

The hot workability of the consolidated material readily leads to the production of sheet, strip and other hot reduced mill forms. The hot worked product is then grain coarsened by heating to a sufficiently high secondary recrystallization temperature, e.g., on the order of about 2200° or 2400°F and below the incipient melting point of the alloy to coarsen the grains. It is important that the consolidated product be sufficiently worked to impart to it energy that is adequate to provide secondary recrystallization on subsequent heat treatment. If the product is worked too much, coarse, substantially equiaxed grains will result, whereas too little work results in a relatively fine equiaxed grain structure.

The grain-coarsened product can later be subjected to other hot working and heat treatment operations within the above limits without foregoing the secondarily recrystallized grain structure that confers good high temperature properties upon the alloy.

Mechanical Alloying of High-Strength, Corrosion Resistant Alloy

J.S. Benjamin and J.W. Schultz; U.S. Patent 3,926,568; December 16, 1975; assigned to The International Nickel Co., Inc. have found that nickel-base alloys containing controlled percentages of chromium, aluminum, titanium, molybdenum, tungsten, tantalum, zirconium, boron, carbon and a dispersoid such as yttria afford good stress-rupture properties at intermediate temperatures, e.g., 1400°F, and at more elevated temperatures, e.g., 1900°F, together with good oxidation and/or control of nitrogen important for enhancing sulfidation resistance.

The chromium, titanium and tungsten impart a most potent effect in achieving optimum resistance to sulfidation. By the same token, titanium and tungsten, for example, tend to detract from 1900°F stress-rupture strength, although not at 1400°F. Chromium and aluminum are indispensible in conferring resistance to oxidation yet aluminum at 1900°F is deemed somewhat detrimental whereas chromium impairs stress-rupture properties at both 1400° and 1900°F. Such considerations are indicative of the careful balance of chemistry required, and this balance is further required to develop a proper grain structure.

With regard to yttria, it confers high stress-rupture strength, particularly at the higher temperature levels (1900°F). A small amount of this constituent, e.g., 0.4%, is effective in imparting this benefit, though it is preferred that about 0.5 to 1.5% be present. Percentages of yttria above 2% are unnecessary. Tantalum enhances strength characteristics especially in the 1200° to 1400°F temperature region. From 1.5 to 3% of this element is deemed beneficial.

In referring to nickel as constituting the "balance" or "balance essentially" in respect of the above composition, the presence of other elements in amounts which do not adversely affect the basic characteristics of the alloys is not excluded. In this connection, the alloys may contain up to 3% columbium (although they are preferably columbium free); up to 10% cobalt; up to 3% hafnium; up to 0.75% or 1% oxygen; up to 3% iron (as a contaminant); and up to 0.3% nitrogen. But it is most preferred, that nitrogen not exceed about 0.1% particularly in seeking the optimum of sulfidation resistance.

Cobalt has a negative effect at 1900°F and should, if present, be held to less than 8%. In addition, while yttria is very much preferred as the added dispersoid constituent, other refractory dispersoids, e.g., thoria and lanthana, can partially replace or be used in lieu thereof where a lesser combination of overall properties can be tolerated. Other refractory dispersoids can be present aside from the aforementioned yttria, lanthana and thoria, including the oxides, carbides, borides, and nitrides (provided the total nitrogen content does not exceed about 0.1%) of one or more materials such as thorium, zirconium, hafnium, titanium and oxides of aluminum, yttrium, lanthanum, cerium, etc. It is to be also understood that a given range for a particular element of the alloys can be used with a given range for any other element.

The alloys should be produced by mechanical alloying which involves the dry, intensive high energy milling of a powder charge whereby the initial powder constituents are simultaneously fragmented and cold bonded to provide a homogeneous, intimate interdispersion of alloying elements in the form of composite particles corresponding to the alloy composition desired. In this connection, the impacting elements used are preferably hard and impact resistant such as through hardened 52100 steel.

Using a 4 gallon attritor mill, for example, a ball-to-powder ratio of about 15:1 to 25:1 by volume is preferred, the impeller speed being suitably conducted over the range of 250 to 300 rpm for a period of about 12 to 24 hours. For processing in a 10 gallon attritor a speed of 150 to 200 rpm can be used, a period of 15 to 40 hours being satisfactory. Larger attritors permit reduced impeller speeds.

Subsequent to mechanical alloying the composite alloy product particles can be compacted as by hot extrusion. A temperature of about 1800° to 2125°F is suitable and an extrusion ratio of about 10:1 to 25:1 is quite satisfactory. Thereupon, the extruded piece can be hot worked if desired.

It is most important that the alloys then be treated to develop a germinative grain growth such that a microstructure is produced characterized by coarse elongated grains having a high aspect ratio, e.g., 2:1 to upwards of 100:1. This comprises subjecting the alloys to a heat treatment within the temperature range of approximately 2125° to about 2300°F. If lower temperatures are used such as on the order of 2100°F, the alloys will retain the fine grain of the extruded structure and this is undesirable. On the other hand, should the temperature appreciably exceed 2300°F, then incipient liquation will develop and this is also undesirable. This treatment was followed by aging within the temperature range of about 1250° to 2000°F, benefically from 1450° to 1600°F, for a period of from about 16 to 30 hours.

Heat- and Wear-Resistant Alloy

H. Doi and K. Nishigaki; U.S. Patent 3,916,497; November 4, 1975; assigned to Mitsubishi Metal Corporation, Japan have developed a heat resistant and wear resistant alloy of the carbide-dispersion and precipitation-hardening type, to be used for cutting tools, etc. It has a basic composition consisting, in weight percent based on the total weight of the alloy, of from 10 to 90% dispersed particles composed of one or more kinds of carbides or composite carbides of transition metals from Groups IV-A, V-A and VI-A and the balance being from 50 to 70% Ni, from 2 to 10% Ti, from 0.5 to 10% Al, and one or more kinds of alloy elements selected from the group consisting of from 1 to 10% Fe, from 1 to 20% Co, and from 1 to 20% Cr.

The alloy also contains one or more kinds of alloy elements selected from the group consisting, in weight percent based on the weight of a Ni-base matrix, of no more than 5% Nb, no more than 10% Ta, no more than 20% Mo, no more than 20% W and no more than 5% V.

Alloys according to this process provide excellent cutting characteristics, particularly in the low cutting speed range, in which high speed steel finds its principal application. High speed steel, in general, contains MC, M_6C and $M_{23}C_6$ type carbides dispersed therein. The hardness of M_6C and $M_{23}C_6$ type carbides does not go as high as 1,800 to 2,100 in terms of Vickers hardness. In contrast thereto, the hardness of the principal carbides of this process, such as for instance TiC, goes as high as 3,000 to 3,200 (Vickers hardness). Also, they have good wettability to the matrix to Ni base superalloys as well as high oxidation resistance. The result is that the excellent properties of TiC are well reflected in the cutting performance of tools which contain it.

When the alloys of this process are used for cutting tools to be used in a low speed cutting range, the total amount of one or more kinds of carbides or composite carbides contained in the alloy should preferably be in the range of from 20 to 70%.

Example 1: 10% WC of a size of 1 μ was added to 20% TiC powder of 3 μ in size which had been obtained by crushing commercially available TiC of −100 mesh in a wet type ball mill. Then the following were added to this TiC-WC mixture: 35% Ni, 10% Co, 5% Fe, 10% NiAl (Ni:Al = 7:3), 2% Ti, 5% Mo and 3% Cr. These added powdery elements form a binding phase for the carbide particles. The powder mixture thus prepared was wet-mixed, compacted, sintered under vacuum of 10^{-1} mm Hg at 1350°C for 1.0 hour. The sintered product was subjected to solution treatment at 1150°C for 4 hours and thereafter to aging treatment at 750°C for 4 hours. The hardness of the alloy thus obtained was 63 RC (Rockwell C scale).

Example 2: The following were added to TiC powder of 3 μ in size; 5% Co, 40% Ni, 7.9% NiAl (Ni:Al = 7:3), 2% Ti, 5% Fe, 5% Cr, 5.09% Mo and 0.01% B. These powdery elements form a binding phase for the carbide particles. The powder mixture thus prepared was wet-mixed, compacted, sintered under vacuum at 1350°C for 1.0 hour, subjected to solution treatment under vacuum at 1150°C for 4 hours, then oil-quenched and tempered at 760°C for 3 hours. The hardness of the thus obtained alloy after measurement was on the Rockwell hardness C scale, 57 after sintering, 55 after quenching, and 62 after tempering, respectively. The transverse rupture strength was 180 kg/mm^2.

The cutting test results of such tool alloys listed in the table below show the superiority of the alloys of this process in wear resistance and smoothness of finished surface. The cutting conditions were as follows: work material, AISI 4340; feed, 0.1 mm/rev; depth of cut, 1.0 mm; cutting time, 2 minutes.

	. . . . . Speed (m/min).				
	20	50	80	125	
Sample	. . . . Flank Wear (mm).				**Finish**
Process	0.05	0.08	0.1	0.2	good finished surface
Control	0.2	0.5	0.7	1.0	poor finished surface

Magnetic Powders

IRON

Stabilization of Pyrophorous Powder in Liquid Medium

J. Wegener; U.S. Patent 4,030,913; June 21, 1977; assigned to U.S. Philips Corp. has found that the stabilization of pyrophorous iron powder in a liquid medium can be carried out in a very short time when the powders are contacted under careful conditions with organic compounds which contain nitrogen bound to oxygen, for example, nitromethane, nitrobenzene or nitrosobenzene.

$NO-C_6H_5$ and $C_6H_5-N=N-C_6H_5$ could be demonstrated as a reaction product in the passivation with $NO_2-C_6H_5$. In the conversion with NO_2CH_3 inter alia gaseous reaction products, such as N_2 and C_2H_6, are formed. Since the gaseous reaction products and also excessive NO_2CH_3 do not interfere at all in the further processing, nitromethane is preferred.

The ratio of a molar concentration of iron to N−O containing compound is preferably at most 20 in a compound comprising a nitro group. It is even more favorable to choose $[Fe]/[NO_2-R]$ to be smaller than 20 (R is the organic radical in question). The lower limit of the ratio is determined only by the price of the $N-O_x$ containing passivating agent ($x = 1$ or 2). In a passivation with pure $N-O_x$ containing compound such a vehement reaction may occur that a fire starts. For NO−R compounds as a passivating agent a suitable maximum ratio is 10. For molecules having several N−O containing groups corresponding suitable ratio values apply.

Example 1: 0.72 mol (40 g) of Fe_{active} were ground in 250 ml of toluene in a PVC flask with 0.063 mol (7.8 g) of nitrobenzene and approximately 50 g of steel balls (ϕ = 3 mm) for 30 minutes. The powder and the balls were then separated from the liquid, washed several times with toluene and dried in an N_2 flow. The possibly occurring heat tone (temperature increase) was then established by means of a thermo element present in the powder and addition of O_2 to the N_2. Even with the N_2:O_2 ratio of air, no temperature increase occurs. The powder is stable in air.

Example 2: 0.36 mol (20 g) of Fe_{active} were stirred in 200 ml of benzene in a polynecked flask with 0.018 mol (3.2 g) of dinitrobenzene under N_2 by means of a KPG stirrer. Duration 1 hour. Filtering was then carried out succeeded by several washings with benzene. The powder was dried in N_2 flow and tested for its stability against oxygen in the manner described in Example 1. The powder is stable in air.

Example 3: 0.111 mol (6.2 g) of Fe_{active} were stirred in a polynecked flask in 50 ml of benzene with 0.010 mol (1.1 g) of nitrosobenzene under N_2. Duration 45 minutes. The method was then continued as described in Example 2 and the powder was tested for its stability in air. The powder showed a good stability.

Example 4: 0.082 mol (4.6 g) of Fe_{active} were reacted while stirring in a polynecked flask with 40 ml of benzene and 0.0082 (0.5 g) of nitromethane. Duration of the test 45 minutes. The method was then continued as described in Example 2. The powder is stable in air.

In the following table, static magnetic properties of powders are recorded which had been stabilized according to Example 1 to 4. For comparison, the table also states values of N_2/O_2-treated powders. The table comprises values for σ_S, the magnetic moment per kg in a field of 10^6 A/m (expressed in Wbm/kg); σ_R, the remanent magnetic moment per kg after magnetization in a field of 10^6 A/m (expressed in Wbm/kg); σ_S/σ_R, the ratio between the two moments; H_c, the magnetization coercive force (expressed in A/m); H_R, the remanent coercive force (expressed in A/m); and H_c/H_R, the ratio between the two coercive forces.

	σ_S	σ_R	σ_R/σ_S	H_c	H_R	H_c/H_R
Example 1	1.86	0.88	0.47	8.91	11.10	0.80
N_2/O_2-treated	2.06	0.98	0.48	9.44	11.47	0.82
Example 2	1.87	0.90	0.48	9.71	11.92	0.81
N_2/O_2-treated	1.80	0.85	0.47	9.83	12.10	0.81
Example 3	1.98	0.94	0.47	9.52	11.80	0.81
N_2/O_2-treated	1.80	0.85	0.47	9.83	12.10	0.81
Example 4	1.93	0.92	0.48	9.71	11.98	0.81
N_2/O_2-treated	1.80	0.85	0.47	9.83	12.10	0.81

Note: σ_S, σ_R in 10^{-4} Wbm/kg
H_c, H_R in 10^{-4} A/m

Crystalline Silica Coating

G. Matsui, K. Toda, A. Takedoi and N. Horiishi; U.S. Patent 4,133,677; January 9, 1979; assigned to Toda Kogyo Corp., Japan have developed a process for producing acicular magnetic metallic particle powder which has improved magnetic properties, namely, both high σ_S and H_c values. It is commercially valuable as high-power and high-density magnetic recording media. The process comprises the following:

1. Preparing an aqueous suspension of acicular particles of at least one metal oxide selected from the group consisting of ferric oxide, iron (III) oxide hydroxide and each of these containing Co and/or Ni therein while adjusting the pH of the suspension at a desired value higher than 8.
2. Adding an amorphous water-soluble silicate into the suspension while agitating under a nonoxidizing atmosphere to thereby coat homogeneously and densely the acicular particles of metal oxide with the amorphous silicate.
3. Water-washing, collecting and drying the resulting particles to obtain the acicular particles of metal oxide homogeneously and densely coated with the crystalline silica.
4. Subsequently heating them in a stream of a reducing gas at temperatures between 400° to 800°C to obtain acicular magnetic metal particle powder.

An alternative method for producing acicular magnetic metallic particle powder comprises the steps of:

1. Preparing an aqueous suspension of acicular particles of at least one metal oxide selected from the group consisting of ferric oxide, iron (III) oxide hydroxide and each of these containing Co and/or Ni therein while adjusting the pH of the suspension at a desired value higher than 8.
2. Adding an amorphous water-soluble silicate into the suspension while agitating under nonoxidizing atmosphere to coat homogeneously and densely the acicular particles of metal oxide with the amorphous silicate.
3. Neutralizing the resultant amorphous silicate on the particles by adding an acid into the suspension to obtain the acicular particles of metal oxide homogeneously and densely coated with the crystalline silica.
4. Water-washing, collecting and drying the resultant particles, and subsequently heating them in a stream of a

reducing gas at temperatures between 400° to 800°C to obtain acicular magnetic metal particle powder.

Example 1: *Preparation of Starting Material Particles*—42 liters of aqueous solution containing 52 mols of $FeSo_4$ was added into 18 liters of 15.2 N NaOH aqueous solution previously placed in a reactor, while introducing N_2 gas thereinto at a flow rate of 200 l/min. This procedure was continued for 15 minutes at pH 12.2 and at a temperature of 48°C to form ferrous hydroxide colloid. Thereafter, air was introduced at a flow rate of 300 l/min into the resulting colloidal solution at 40°C to carry out oxidizing reaction for 20 hours, thereby producing acicular iron (III) oxide hydroxide particles. Subsequently, the obtained particles were washed with water, filtered, dried, and ground in usual manner. These resultant particles were acicular α-FeOOH by a chemical analysis and were acicular particles having 0.8 to 1.0 μ of long axis and 15:1 axial ratio of long axis to short axis by electromicroscopic observation.

Example 2: *Preparation of Particles Coated with SiO_4*—360 g of acicular α-FeOOH particles obtained in Example 1 was dispersed in water to prepare 10 liters of suspension. This suspension had a pH value of 7.8 and a viscosity of 4.7 poises. NaOH solution was poured into the suspension for adjusting the pH value thereof to 8.6. Thereafter 38.9 g of sodium silicate (No. 3 on J.I.S. K-1408) containing 28.55 weight percent of SiO_2 was added into the suspension, followed by stirring and dispersing the particles while preventing oxidizing gas such as air from intermixing as far as possible. The resulting suspension had a pH value of 10.2 and a viscosity of 2.7 poises.

The obtained particles were then washed with water, filtered in usual manner and then dried at 110°C to thereby obtain acicular iron (III) oxide hydroxide particles coated with 4.44 mol percent of SiO_2 based on the amount of iron therein. Accordingly, 96% of the initial sodium silicate calculated as SiO_2 amounts was coated on the particles.

Example 3: *Reduction Treatment Process*—300 g of the resultant particle powder obtained in Example 2 was reduced in a reductor by introducing H_2 gas at a flow rate of 3 liters/min and at 600°C for 3 hours. The reduced acicular SiO_2 coated metallic iron particles were preliminarily dipped in toluene solution so as not to bring about a radical oxidation thereof when the particles were taken out in the air. Thereafter, the particles were coated with stable oxidized film by evaporating toluene from the surfaces of the particles. The physical properties of acicular magnetic iron particles thus obtained are: long axis, 0.60 μ; ratio of long axis to short axis, 10; saturated magnetic flux density, 165 emu/g; coercive force, 620 Oe.

Comparative Examples: The resultant particles of Example 1 were reduced to acicular magnetic iron particles without coating the surface thereof with silica by the same procedure of Example 3 at a temperature of 350°C for 2 hours. The physical properties of the particles so obtained are: long axis, 0.65 μ; ratio of long axis to short axis, 8; saturated magnetic flux density, 95 emu/g; coercive force, 425 Oe.

Oxidation Resistance by pH Control

N. Takahashi, S. Ozaki, T. Naruse, H. Isono and T. Nishihara; U.S. Patent 4,167,582; September 11, 1979; assigned to Victor Company of Japan, Ltd, Japan have developed a finely divided metallic powder comprising iron, and optionally, one or more other metals, which exhibits a saturation magnetization great enough to serve as the material of the magnetic recording media for high density recording purposes and is excellent in stability, particularly in the resistance to oxidation by air.

The magnetic powder is a metallic powder comprising iron and, optionally, at least one other metal, which powder is obtained by the reduction of a metal oxide powder with a pH value not smaller than 6.0 whose metal component corresponds to the composition of the metallic powder. The saturation magnetization of the metallic powder is greater than about 150 emu/g at room temperature.

A metal oxide powder obtained from a basic solution is useful for the metallic powder. This oxide powder may optionally be treated in an acidic liquid phase environment before the reduction as long as the pH value of the acid-treated powder remains in the above specified range. A metal oxide powder precipitated from an acidic solution and thereafter treated in a basic liquid phase environment so as to have the above required pH value also is useful.

The metallic powder may contain less than about 15 weight percent of oxygen and preferably has an acicular particle shape with a mean length of about 0.3 to 0.5 μm and a mean length-to-width ratio of about 7:1 to about 10:1. Preferred examples of metal useful for alloying with iron are cobalt, nickel, manganese and chromium.

In the production of a magnetic recording medium, the metallic powder is dispersed in an organic liquid together with an organic polymer which is soluble in the organic liquid and serves as a binder. Optionally use will be made of a dispersing agent and other commonly used additives in preparing this dispersion, a magnetic paint. A thin magnetic layer is laid on a base of a nonmagnetic material by the application of this magnetic paint to the base, followed by drying and calendering procedures.

A magnetic recording medium thus produced does not differ in construction from conventional magnetic recording media: in the dried magnetic coating, the metallic powder is dispersed in the solidified polymer binder. This recording medium, however, is excellent both in its magnetic characteristics represented by the maximum magnetic flux density and coercive force and in its chemical and physical stability represented by resistance to oxidation (rusting) in a high temperature, high humidity atmosphere and the rate of variation in the maximum magnetic flux density B_M with the lapse of time.

In this process, this rate of variation is made less than 5% of an initial value for

B_M after the maintenance of the recording medium for 340 hours in a 90% relative humidity atmosphere at a temperature of 80°C. Accordingly this recording medium is fully practicable and can serve high density recording purposes.

The process is based on the finding that the chemical stability, i.e., resistance to air-oxidation, of a finely divided iron powder greatly depends on the pH of an oxide powder the reduction of which gives the iron powder. Of course, the pH of the oxide powder is principally determined by the acidity or basicity of a liquid phase environment employed in the preparation of the oxide powder.

Thus, a metallic powder featuring a satisfactory resistance to oxidation can be obtained only when an oxide powder to be reduced has a pH value not smaller than 6.0, and it is requisite to the pH measurement that the oxide powder is thoroughly washed and heated in advance of the measurement.

Example 1: A powder of gamma hematite γ-Fe_2O_3 was obtained from goethite α-$Fe_2O_3 \cdot H_2O$ prepared in a basic liquid phase environment (a sodium hydroxide solution was added to a ferrous sulfate solution) through reduction and a succeeding mild oxidation. The pH of γ-hematite powder was 8.2. The reduction of this γ-hematite powder in a hydrogen gas stream at about 350°C gave a finely divided metallic powder, which had an acicular particle shape and contained about 13% by weight of oxygen. The saturation magnetization σ_S of this metallic powder was 158 emu/g at room temperature.

A magnetic paint was prepared by dispersing 100 parts by weight of the metallic powder together with 25 parts by weight of a conventional polymer binder, 2 parts by weight of a dispersing agent and 3 parts by weight of other commonly used additives for lubrication and endurance purposes in 300 parts by weight of an organic solvent. This paint was applied to a polyester film, followed by drying, to give a 1.4 μm thick magnetic coating. For the produced magnetic recording medium, the maximum magnetic flux density B_M was 3,350 G and the coercive force Hc was 600 Oe.

This magnetic recording medium was kept placed in an environmental test chamber maintained at a relative humidity of 90% and at a temperature of 80°C to examine its resistance to oxidation by air. The measurement of the magnetic properties was repeated after the lapse of 170 hours and 340 hours from the start of the environment test, with the result that substantially no change occurred in any of the above presented characteristics. The appearance of the magnetic coating too remained unchanged when the test was continued for 340 hours.

Example 2: A Co-containing γ-Fe_2O_3 powder was prepared using a basic liquid phase environment. The amount of Co in this oxide powder was 5.4% by weight of the total of Fe and Co, and the pH for this powder was 8.6. The reduction of this material in a hydrogen gas stream gave a metallic powder, which was essentially an Fe-Co alloy with a saturation magnetization value σ_S of 152 emu/g at room temperature.

Using this alloy powder, a magnetic recording medium was produced by the procedures described in Example 1. The maximum magnetic flux density B_M for this recording medium was 2,970 G and the coercive force Hc was 720 Oe. When this recording medium was maintained for 170 hours in a 90% relative humidity atmosphere at a temperature of 80°C, neither the magnetic characteristics nor the appearance of the magnetic coating exhibited any substantial variation.

Ureas and Isocyanates for Passivation

E.A. Chandross, M. Robbins and H. Schonhorn; U.S. Patent 3,964,939; June 22, 1976; assigned to Bell Telephone Laboratories, Inc. have found a class of compounds which, without polymerization, passivate fine particles of oxidizable metals. These compounds are ureas, thioureas, isocyanates and isothiocyanates containing at least one organic substituent with at least two carbons. For passivation these compounds are applied to the essentially oxide-free metal powders by immersing the powders in a solution of the protective species in a nonreactive organic solvent. It is considered that corrosion protection is achieved in this method by some modification of the surface properties of the particle since it has been found that the degree of protection is insensitive to the molecular weight of the substituents.

The amount of organic material incorporated in the final device can be minimized by washing the powders in pure solvent after treatment in the protective solution with little or no effect on the degree of protection. Iron powders, suitable for such uses as transformer cores and magnetic recording tape, and Co_5Sm powders, suitable for the production of permanent magnets, have been protected by this method and have shown little degradation after long term aging at room temperature and accelerated aging at high temperatures in air or moist oxygen.

Iron powders whose average least dimension was 0.3 μm were produced by hydrogen reduction of γ-ferric oxide. The ferric oxide particles were placed in a ceramic crucible and heated to 400°C while maintaining a flow of hydrogen gas through the reaction vessel. The powders were cooled to room temperature and, while still in a hydrogen atmosphere, were immersed in a 5 weight percent solution of the protective compound in benzene. The protected powders were filtered from the solution, rinsed in fresh benzene, and then dried at 60°C at a reduced pressure of approximately 100 torr.

The saturation magnetization of the powders was measured soon after treatment and again after aging. The results of these measurements and the aging method used are indicated in the table for several exemplary protective materials. For comparison the saturation magnetization of pure iron is indicated. While the saturation magnetization of the protected powders is less than that of pure iron it is significantly greater than the saturation magnetization reported for powders protected by encapsulation in polymers (*Journal of Electrochemical Society,* 117, p. 138, 1970).

Protective Material	Aging History	Saturation Magnetization (σ_S in emu/g)
N,N'-diheptylthiourea	as prepared	151
	one day in air at 100°C	139
	10 days in air at 100°C	136
N,N'-diethylthiourea	as prepared	169
	10 days in air at 100°C	142
	>1 year in air at 35°C	150
Octadecylthiourea	as prepared	176
Octadecylisothiocyanate	as prepared	159
	10 days in air at 100°C	139
N,N'-diheptylurea	as prepared	173
	>1 year in air at 25°C	165
Octadecylurea	as prepared	177
	>1 year in air at 25°C	152
Pure bulk iron	—	218

Co_5Sm powders were prepared in an essentially oxide-free state by grinding of arc melted pieces while immersed in a 5% solution of N,N'-diheptylthiourea in benzene, rinsed and dried. No significant weight increase was observed after accelerated aging by flowing water saturated oxygen gas over the powders at 60°C for more than 100 hours.

A magnetic recording band was made by mixing together 145 g of iron particles, protectively treated with N,N'-diheptylthiourea together with 131 g of commercial, polymer based binder mixture. The mixture was cast in a recording band mold and cured at 150°C for 15 minutes. The recording response of the band was satisfactory.

Gaseous Reducing Agent for Titanium-Containing Ion

P.T. Greene, H. Rau and J.R. Wegener; U.S. Patent 3,910,785; October 7, 1975; assigned to U.S. Philips Corporation describe a method of preparing a metal powder consisting mainly of iron by reduction of finely divided acicular iron oxide or iron oxide hydrate with a gaseous reduction agent.

The iron oxide particles or the iron oxide hydrate particles contain at least 0.1% and at most 10% (atomic) of titanium relative to the iron. The good dispersibility of the metal particles thus prepared is evident from the directivity ratio of a magnetic tape manufactured with the particles. Furthermore, the needle shape of the particles is also maintained during the reduction when high reduction temperatures are used.

Presumably the iron oxide particles or the iron oxide hydrate particles contain at least 0.2 atomic percent and at most 4 atomic percent of titanium relative to the iron, since good results have been realized within this range.

The iron oxide particles or the iron oxide hydrate particles can be doped with titanium. During the preparation of the particles, a precipitate is formed from a solution of an iron salt and a titanium salt. In this manner iron and titanium are united in a simple manner.

In another case the iron oxide particles or the iron oxide hydrate particles are covered with a layer of a titanium compound. For that purpose the finely divided iron oxide or iron oxide hydrate is suspended in a dilute titanium salt solution.

Iron Oxide Hydrate Doped with Nickel or Cobalt

In other work by *H. Rau, J.R. Wegener and P.T. Greene; U.S. Patent 3,837,839; September 24, 1974; assigned to U.S. Philips Corporation* for preparing a metal powder which consists mainly of iron by reduction of finely divided acicular iron oxide hydrate with a gaseous reduction agent, powdered iron oxide, preferably in hydrated form, is mixed with an organic nickel salt or organic cobalt salt, which salts are decomposable at temperatures between 300° and 425°C. The mixture is heated for 2 or 3 hours in a reducing atmosphere at a temperature between 300° and 425°C. This reduction time is sufficient and is shorter than the time necessary for the reduction in the absence of nickel and/or cobalt, which under equal circumstances is at least 4 hours.

A short reduction time is important in preparing finely divided powders because otherwise an undesirable sintering takes place. It has been found that a considerable shortening of the reduction time can be achieved if a special iron oxide hydrate is used.

According to the process, the iron oxide hydrate particles are doped with at least 0.3% of at least one metal which is catalytic for hydrogen reactions and the reduction takes place at a temperature below 350°C. Metals which are catalytic for hydrogen reactions are Ni, Co, Ru, Pt and Pd. Owing to the fact that the iron oxide hydrate particles are acicular, there is an upper limit of the doping. The resulting powders show properties (coercive force, rectangularity of the hysteresis loop) which make them suitable for magnetic recordings. The reduction process occurs at temperatures below 350°C and often even considerably lower than 350°C at a rate suitable for practical purposes. The iron oxide hydrate particles are in particular doped at least with Ni and/or Co.

Flowing Hydrogen to Effect Reduction

H. Rau and K.-G. Knauff; U.S. Patent 3,967,986; July 6, 1976; assigned to U.S. Philips Corporation have developed a method of preparing iron powder for recording in which iron oxide hydrate powder is reduced by flowing hydrogen in intimate contact with the iron oxide hydrate powder, preferably in a fluidized bed.

The gaseous reduction agent is intimately contacted with the material to be reduced, and the flowing quantity of gaseous reduction agent in relation to the material to be reduced corresponds at least to the formula V = 0.4 x T − 120, where V is the volume in liters of the quantity of gaseous reduction agent at 0°C and 1 atm which flows per hour and per gram of Fe contents of the material to be reduced, and T is the reduction temperature in °C, T being between 300° and 600°C.

The intimate contact between gaseous reduction agent and material to be reduced results in a shortening of the reduction time with the temperature, the quantity of flowing reduction agent and Fe contents being maintained constant. The upper limit of the volume of reducing agent is determined only by economical considerations.

In one embodiment of the method which is advantageous in the range between 300° and 400°C, reduction may be carried out in a single-stage fluidized bed.

On the basis of the abovementioned formula, such large flowing quantities occur at higher temperatures and in certain circumstances, the range of the fluidized bed is left (the possibility exists that the powder is blown out of the reaction vessel). In order to remain within the range of the method the filling mass is reduced at higher temperatures.

The very short reduction times occurring then make it useful to change from discontinuous to continuous operation; otherwise filling and emptying will consume more time than the reduction itself. The continuous method may of course also be used at temperatures below 400°C.

A favorable embodiment of continuous operation is in a tilting furnace, such as is illustrated in British Patent Specification No. 1,104,852.

Porous Ferromagnetic Powders

R.J. Deffeyes; U.S. Patent 4,146,504; March 27, 1979; assigned to Graham Magnetics Inc. has produced small particles characterized by high porosity and structures or particles formed from the porous particles. Porosity is quantitatively measured by bulk density of the powder. The powders also tend to have a very high internal surface area. The internal surface area is suitably characterized by the nitrogen absorption method. In general, particles prepared according to the process are less than about 5 μ in average diameter. A particular advantage of the process is that particles of extremely small particle size can be prepared.

If the particles are to be loaded into a polymeric matrix, particle size is relatively important. In some applications, e.g., where magnetic properties are important, it will be the magnetic particle size which will be important. Superparamagnetic elemental particles below 100 angstroms can dominate the magnetic characteristics of agglomerate particles far larger.

The materials are inorganic powders characterized by surface areas in excess of 5 m^2/g, especially an effective pore size of from 25 to 200 angstroms in diameter and formed by the thermal decomposition of organometallic salts, e.g., metal oxalates, metal formates, and the like, so that the resultant products have a bulk density of less than 0.4 g/cc when the specific gravity of the material from which the powders are formed is 8 g/cc. In general, and in allowing for a range of materials of different specific gravity, the ratio between bulk density and specific gravity should be less than about 0.08 to 1, advantageously less than about 0.05 to 1. The most porous materials have such ratio below 0.02 and even as low as 0.002 and lower.

The most advantageous of the particulate materials have very high surface areas, i.e., over about 25 m^2/g. This high surface area may be utilized directly as in catalysis. Lower surface area material, say those between 5 and 25 m^2/g may be advantageously utilized indirectly as in facilitating comminuting the powder to a powder of reduced particle size, i.e., a powder approximating the elemental particle size.

Example 1: A material was prepared as follows. A first solution was formed of 4 lb of a cobalt nitrate solution (aqueous) containing 15% cobalt by weight; 7 lb, 13.5 oz of citric acid; 2 lb, 7 oz of a 67% sucrose solution (aqueous); 3 lb, 10 oz of deionized water; and 2 lb, 2 oz of denatured ethyl alcohol.

A second solution was formed of 2 lb of 8 oz of oxalic acid dihydrate in 7 lb, 4 oz of 67% sucrose solution; 7 lb 4 oz deionized water and 7 lb, 4 oz of denatured alcohol.

These two solutions were pumped at rates of 230 ml/min and 345 ml/min, respectively, into a ⅛" stainless steel "T" coupling and through 7′ of ¼" (i.d.) tubing into a receptacle. A very thick pink dispersion of cobalt oxalate crystals resulted. This dispersion will be called a "seed" mixture.

Thereupon 53 lb of an iron-bearing solution was prepared by dissolving 12.5 parts $FeSO_4$, 194 parts of water, 0.5 parts of citric acid, and 3 grams of sodium sulfite.

A metal-bearing solution was then prepared containing 16 lb of cobalt nitrate solution (15% cobalt by weight); 8 lb 6 oz of nickel nitrate solution, containing 10.2% weight of nickel; 53 lb of iron-bearing solution as described above and 93 lb of water.

An oxalic acid solution was prepared by dissolving 17 lb 8 oz of oxalic acid dihydrate in 270 lb of water. When solution of the oxalic acid solution was complete, the seed solution was added to it and mixed in.

The metal solution is sprayed, at a rate of 3.7 gal/hr, into a baffled and very-well agitated mixer holding the seed dispersion. After the precipitation has been completed the resulting oxalate crystals were stirred for an additional 30 minutes and then filtered, washed with 50 gallons of hot water, then 15 gallons of ethanol. The crystals were then coated, from isopropanol, with 7% of polyamide resin (e.g., a material sold by AZ Products Company of Eaton Park, Florida, under the trademark Azamide 325) based on the weight of metal in the oxalate, dried. This material was divided into a large number of samples, reduced under several sets of firing conditions and evaluated.

The data presented below shows how various properties of the metal changed with respect to reducing conditions. In this data:

H_c = Coercivity of the powder
SA = Surface area (BET)
Sq = Squareness
M = Magnetic moment in oersteds
OA = Linseed oil absorption (cc/g)
Wr = In a hysteresis loop showing an unintegrated value of the magnetic moment with time (a dM/dt curve), the peak-spread width at 50% of the peak height over the distance from origin of the curve to a vertical line from the center of the peak.
BD = Bulk density

Reduction was carried out until the CO_2 level in the off gas fell below 0.02%.

Ex. No.	Percent H_2/°F	H_c	SA	Sq	M	Wr	OA	BD
1	0/700	760	34.7	0.82	192	0.55	–	0.045
2	0/800	797	–	0.83	177	–	–	–
3	5/600	765	70	0.59	169	1.38	–	0.044
4	5/650	772	–	0.69	156	0.99	–	0.043
5	5/700	850	43	0.78	191	0.70	1.4	0.06
6	5/750	831	29	0.77	163	0.61	–	0.047
7	5/775	744	23	0.80	170	0.59	–	0.063
8	5/800	747	39	0.80	166	0.65	–	–
9	10/800	686	–	0.77	180	–	–	–
10	20/650	887	–	0.71	173	0.82	–	0.046
11	20/700	890	–	0.63	–	–	–	–
12	20/750	862	–	0.72	152	0.70	–	–
13	20/800	470	9	0.81	156	0.77	–	–
14	30/700	813	–	0.68	162	0.83	–	0.055
15	30/750	538	–	0.84	170	0.59	–	–
16	30/800	442	9	0.83	149	0.90	–	–
17	40/650	850	24	0.69	167	–	–	–
18	40/700	833	12	0.76	171	0.52	–	–
19	40/750	507	7.0	0.81	151	0.65	1.1	–
20	40/775	462	3.2	0.82	164	0.76	–	–
21	40/800	353	2.5	0.82	159	–	–	–
22	5/850	756	37	0.77	200	0.66	–	0.05
23	40/725	793	24	0.72	162	0.69	0.8	0.067
24	40/850	–	4.4	–	–	–	1.46	0.11
25	50/750	–	5.3	–	–	–	–	0.078

Example 2: A quantity of 600 oersted coercivity powder being comprised of 60% cobalt, 20% iron and 20% nickel was found to have a magnetic moment of 140 gauss in its powder form.

This material was then smeared out vigorously with a linseed oil. A drop of linseed oil was placed on a glass surface; 1 g of metal powder was incrementally added to the linseed oil and kneaded with a spatula. Enough oil and powder was added alternately to maintain a stiff metal-like strip on the plate as the powder is broken down. As a stiff paste in linseed oil, it manifested a magnetic moment of 1,100 gauss.

Next, a sample of the original powder material was dispersed in magnetic tape with about 15% of its weight of cobalt-doped iron oxide being subjected to compounding with matrix resins in a sigma blade mixer. The mix containing about 25% by weight of pigment was charged into a glass tube, oriented in a 60-cycle BH meter and found to have a magnetic moment of 4,500 oersteds, and a squareness of 0.86. (The original powder had a squareness of about 0.6.)

The increasing magnetic moments are indicative of the decreased bulk density of the material as it is broken up by the relatively mild shearing actions described. The increased squareness indicates that smaller particles, more acicular in nature are formed during the shearing procedures.

The total increase in bulk density achieved by these techniques is typically from 10 to 150 times; it is usually in excess of 30 times the original bulk density.

That little damage is done to a magnetic recording powder by these shearing actions is demonstrated by the facts that the processing procedure does not increase coercivity less than about 10% and there is no substantial increase in strain anisotropy introduced into the powder.

Vacuum Evaporation Treatment

R.F. Bush and J.W. Mitchell; U.S. Patent 4,113,521; September 12, 1978; assigned to International Business Machines Corporation have found that by employing evaporation techniques, small particles, of the order of 100 A, of Fe_3O_4 or FeO having relatively high magnetic moments can be fabricated.

The evaporator consists of a chamber or bell jar on an inert support. Inserted into a chamber is a ceramic or other inert and highly refractory support on which is placed an Al_2O_3 boat. The latter could be comprised of any other material that is inert at very high temperatures. Iron having a purity of 99.999%, is inserted as a pellet or group of pellets into the boat. 100 amperes is supplied to the Al_2O_3 boat, heating the iron to approximately 2250°C. Prior to the actual heating of the iron, the chamber is evacuated down to 10^{-6} mm of pressure and then backfilled to 1.0 mm of argon by conventional means.

Located within the evacuated chamber are pyrex cylinders behind which are placed magnets. The evaporated iron particles were captured by the pyrex cylinders. These particles oxidize immediately upon exposure to air, when removed from the evacuated chamber. The backfilled argon pressure can be altered from 1.0 mm to 0.1 mm during the evaporation process. In general, the greater the argon pressure, the larger are the iron particles that are evaporated.

After the Fe_3O_4 particles were retrieved from their capturing cylinders they were magnetically evaluated. The particles averaged in size from 100 A to 200 A and had values of magnetic moments that were in a range of 82 to 111 emu/g. The relatively high magnetism is believed due to the fact that the evaporation of iron permits very small discrete particles of pure iron to be formed and the final particle is covered with an outer shell of Fe_3O_4, but such outer shell is only 30% of the total volume of Fe° (Fe_3O_4) particles. The particles can be readily separated.

It is believed that the evaporation process, using the parameters set forth above, permits 100 to 200 A sized particles to form where the larger volume (about 70%) of the particles is a core of pure Fe° surrounded by the oxide Fe_3O_4.

Hydrogen Reduction of Treated Acicular Oxides

E. Koester, G. Wunsch, P. Deigner, W. Stumpfi and H.H. Schneehage; U.S. Patent 4,017,303; April 12, 1977; assigned to BASF AG, Germany have developed a process for the manufacture of acicular ferromagnetic metal pigments containing iron, having a reduced field strength distribution and a steeper residual magnetization curve, by applying an alkaline earth metal cation selected from the group consisting of Ca, Ba, and Sr cation, and an organic compound which contains at least one group which can form chelates of the Formula 1 to appropriate acicular iron oxides and then reducing the treated oxides.

(1)

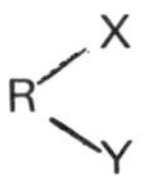

In the formula X and Y are:

$-OH$, $=O$, $-N{<}^{R'}_{R'}$, $=NH$, $=NOH$ or $=N-R''$,

wherein R′ is a terminal short chain carboxylic acid linked to nitrogen by means of an amino bond, R′′ is alkyl or aryl and R is a two-membered or three-membered carbon chain or a part of an aromatic ring, whereby the organic compound forms a chelate with the alkaline earth metal cation in the solvent used to form the solution.

The solvent is removed by evaporation and the so-treated acicular iron oxide is reduced with hydrogen at temperatures of up to 400°C to form metal pigments which retain their acicular shape.

Example 1: 100 g of α-FeOOH needles of length 0.5 μ and 30:1 length:thickness ratio are dispersed in 2,000 cc of methanol in which 3.7 g of barium acetylacetonate have been dissolved. After evaporating off the methanol, the powder is dried at 100°C. After reduction with 200 liters of hydrogen per hour at 275°C for 8 hours an acicular iron powder is obtained.

Example 2: The procedure followed is as in Example 1 except that 7.5 g of barium acetylacetonate are dissolved in the methanol and the reduction is carried out at 300°C.

Example 3: 100 g of α-FeOOH of Example 1 are dispersed in a solution of 3.2 g of strontium acetylacetonate in 2,000 cc of methanol, and after evaporating off the liquid, the material is reduced to the metal with hydrogen for 8 hours at 175°C.

Example 4: 100 g of α-FeOOH of Example 1 are dispersed in a solution of 2.7 g of calcium acetylacetonate in 2,000 cc of methanol, and after evaporating off the liquid, the material is reduced to the metal with hydrogen for 8 hours at 175°C.

Example 5: 100 g of α-FeOOH of Example 1 are dispersed in 1.5 liters of water in which 6 g of cobalt chloride hexahydrate have been dissolved. Cobalt hydroxide is precipitated on the surface of the α-FeOOH by adding dilute ammonia solution until the pH is 10, and the pigment is then dried. 100 g of the resulting α-FeOOH containing 1.5% of Co are dispersed in a solution of 3.8 g of barium acetylacetonate in 1.5 liters of methanol and, after evaporating off the liquid, the material is reduced to the metal with hydrogen for 8 hours at 325°C.

Example 6: 100 g of α-FeOOH of Example 1 are dispersed in 1.5 liters of ethanol to which 3.6 g of barium hydroxide octahydrate dissolved in 50 cc of water are added, and a solution of 2 g of triethylenetetramine in 50 cc of ethanol is added to the dispersion. After evaporating off the liquid, the material is reduced to the metal with hydrogen for 8 hours at 300°C.

Comparative Experiment 1: 100 g of α-FeOOH of Example 1 are dispersed in 2,000 cc of methanol. After evaporating off the methanol, the dried powder is reduced for 8 hours at 275°C with 200 liters of hydrogen per hour.

Comparative Experiment 2: 100 g of α-FeOOH of Example 1 are dispersed in 2,000 cc of ethanol in which 0.38 g of silver nitrate are dissolved. After evaporating off the alcohol, the powder is dried. The reduction is carried out at 250°C with 200 liters of hydrogen per hour and is complete after 8 hours.

Experiment Number	Reduction Temperature (°C)	Needle Length (μ)	Length: Thickness Ratio	Magnetic Values . . . at 800 kA/m . . . H_c . (kA/m) .	H_R	h_5/h_{25}
Ex. 1	275	0.2-0.4	15-25:1	105	132	1.45/0.49
Ex. 2	300	0.2-0.4	15-25:1	107	132	1.35/0.47
Ex. 3	275	0.2-0.3	15-20:1	104	128	1.48/0.50
Ex. 4	275	0.2-0.3	15-20:1	105	131	1.42/0.48
Ex. 5	325	0.2-0.3	15:1	110	135	1.40/0.47
Ex. 6	300	0.2-0.4	15-25:1	104	125	1.49/0.52
1*	275	0.1-0.2	5-8:1	90	122	1.85/0.62
2*	250	0.07-0.15	5-8:1	98	121	1.65/0.55

*Comparative

In related work *E. Koester, G. Wunsch, E. Schoenafinger and H. Schneehage; U.S. Patent 4,050,962; September 27, 1977; assigned to BASF AG, Germany* produce acicular ferromagnetic metal pigments containing iron by applying equivalent amounts of an alkaline earth metal cation and of a monobasic, dibasic or tribasic aliphatic carboxylic acid of up to 6 carbon atoms onto appropriate acicular iron oxides and then reducing the treated oxides.

Carboxylic acids, for the purposes of the process, are saturated or unsaturated aliphatic carboxylic acids of up to 6 carbon atoms and with up to 3 acid groups, in which carboxylic acids one or more hydrogen atoms of the aliphatic chain may be substituted by hydroxyl or amino radicals. Hydroxydicarboxylic acids and hydroxytricarboxylic acids, such as tartaric acid or citric acid, are particularly suitable.

Iron oxides which are usable are the acicular iron oxides from among α-FeOOH, Fe_3O_4, Fe_2O_3, and γ-Fe_2O_3, and the same iron oxides modified with cobalt and/or nickel. These oxides may also be used for the manufacture of ferromagnetic metal pigments if they contain other elements, provided that the incorporation of other foreign elements does not disturb the acicular form. Particularly preferred ferromagnetic metal pigments contain iron and up to 25 atom percent of cobalt.

The alkaline earth metal cation applied simultaneously with the carboxylic acid to the iron oxide is chosen from amongst calcium, barium and strontium.

Reduction of Phosphorus Oxyacid and Carboxylic Acid Treated Iron

W. Steck, H.H. Schneehage, C. Jaeckh, W. Huebner, M. Ohlinger and W. Loeser; U.S. Patent 4,155,748; May 22, 1979; assigned to BASF AG, Germany have developed a method of producing acicular ferromagnetic metal particles which are distinguished by a narrow particle size spectrum coupled with a pronounced acicular shape and which therefore exhibit a narrow field strength distribution, a very steep residual induction curve and only slight temperature dependence of the magnetic properties by reducing a finely divided acicular iron compound selected from the group consisting of iron oxide and iron oxide hydrate with a gaseous reducing agent at a temperature of from 230° to 500°C.

Deposited on the iron oxide or iron oxide hydrate, prior to reduction is (a) a hydrolysis-resistant substance selected from the group consisting of oxyacids of phosphorus, their esters and inorganic salts in such an amount that 0.2 to 2% by weight of phosphorus is present, and (b) a compound selected from the group consisting of aliphatic monobasic, dibasic and tribasic carboxylic acids of from 1 to 6 carbon atoms in such an amount that 0.1 to 2.1% by weight of carbon is present.

The examples which follow illustrate the process. The coercive force H_C [kilo-amps/m], the specific remanence $M_R/\rho[nTm^3/g]$ and the specific saturation magnetization $M_S/\rho[nTm^3/g]$ of the powder samples were measured in a vibrating sample magnetometer at a field strength of 160 kA/m. The coercive force H_c is calculated on the basis of a tap density, ρ, of 1.6 g/cm^3.

Example 1: 50 g of goethite having a particle length of 0.82 μm and a length-to-width ratio of 35:1 are suspended in 750 ml of water, with intensive stirring. 1 g of oxalic acid ($C_2H_2O_4 \cdot 2H_2O$) followed by 0.35 ml of 85% strength phosphoric acid is added to this suspension. After continuing the stirring for 10 minutes, the solid is filtered off and the filter cake is dried in air at 120°C. Reduction of the goethite, treated in this way, for 8 hours, at 310°C in 30 liters of hydrogen per hour gives an acicular iron powder.

The magnetic properties and analytical values of the resulting powder are given in Table 1.

Table 1

	Content of PO_4	C	H_C	H_C at $\rho = 1.6$	M_S/ρ	M_R/ρ	M_R/M_S*
	. . .(%) . . .						
Example 1	1.4	0.04	77.6	66.8	154	84	0.55
Example 2	1.2	0.08	83.9	71.0	146	80	0.55
Comparative Experiment 1A	0	0	73.3	62.9	130	75	0.587
Comparative Experiment 1B	1.3	0	75.5	66.5	131	66	0.50
Comparative Experiment 1C	0	0.06	64.4	58.9	142	81	0.57

*M_R/M_S = relative remanence.

Example 2: The procedure described in Example 1 is followed except that phosphoric acid and oxalic acid are added simultaneously to the suspension.

The magnetic properties of the resulting iron powder, and the analytical values, are shown in Table 1.

Comparative Experiment 1: 50 g of goethite are suspended in 750 ml of water as described in Example 1, and the procedure of Example 1 is then continued (A) without additives, (B) after adding 1 g of oxalic acid, or (C) after adding 0.35 ml of 85% strength phosphoric acid.

The magnetic properties of the iron powders obtained in these Comparative Experiments, and the analytical values, are also shown in Table 1.

In 3 parallel batches A, B and C, 50 g portions of α-FeOOH having an average needle length of 0.51 μm and a length-to-width ratio of 28.3:1 are suspended in 750 ml of water.

Batch A is filtered off as in Example 1 and the filter cake is dried at 120°C. After reduction for 8 hours with 30 l/h of hydrogen at 350°C, an acicular iron powder is obtained.

0.35 ml of 85% strength H_3PO_4 are added to Batch B and the reduction is carried out at 350°C.

0.35 ml of 85% strength H_3PO_4 and 1 g of $C_2H_2O_4 \cdot 2H_2O$ are added simultaneously to Batch C. The reduction is carried out at 350°C.

The magnetic properties of the metal pigments are summarized in Table 2.

Table 2

	Content of PO_4 . . . (%) . . .	Content of C . . . (%) . . .	H_c	H_c at $\rho = 1.6$	M_S/ρ	M_R/ρ	M_R/M_S*
Dispersion A	–	–	73.0	62.5	127	70	0.55
Dispersion B	1.8	–	77.0	66.9	121	61	0.50
Dispersion C	1.3	0.08	82.2	71.6	133	72	0.54

Acicular Particles from Reduction of Goethite Particles

M. Oshimura, K. Yokoyama, Y. Hayashi and A. Miyake; U.S. Patent 4,133,676; January 9, 1979; assigned to Hitachi Maxell, Ltd., Japan describe a process for producing acicular ferromagnetic metal particles having a particle size of 0.1 to 1 μm and a dimension of crystallite-size of not more than 215 A in the effective thickness of the crystallite in the direction perpendicular to the reflecting plane (110), which have a high coercive force, preferably not less than 1,400 oersteds. They are useful for a high-density magnetic recording tape, video mother tape, or permanent magnet material.

The preparation comprises reducing with heating goethite particles which are prepared by treating an aqueous solution of a ferrous salt with a large amount of a basic agent, followed by oxidation of the resulting reaction mixture.

That is, the goethite particles are prepared by adding with agitation an aqueous solution of a ferrous salt to an aqueous solution of a basic agent, so that the pH value of the mixture becomes not less than about 12, preferably not less than about 13.5, whereby ferrous hydroxide is precipitated, and then blowing an oxygen-containing gas (e.g., air) into the reaction mixture at room temperature or an elevated temperature (e.g., at 20° to 80°C), whereby ferrous hydroxide is oxidized to produce goethite, and isolating the precipitated goethite particles, washing with water and then drying.

The starting ferrous salt includes ferrous sulfate, ferrous chloride, ferrous bromide, ferrous acetate and is usually used in an amount of 0.2 to 0.5 mol/l based on the whole volume of the reaction mixture after adding the basic agent. The basic agent includes alkali metal hydroxides or carbonates (e.g., sodium hydroxide, potassium hydroxide, lithium hydroxide, sodium carbonate, potassium carbonate, lithium carbonate), alkaline earth metal hydroxides (e.g., calcium hydroxide, magnesium hydroxide, strontium hydroxide) and ammonium hydroxide, but alkali metal hydroxides are the most preferable from the viewpoint of the solubility and the pH value of the solution.

The basic agent may be used in an amount of not less than 6 mols, preferably not less than 8 mols, more preferably not less than 10 mols, per 1 mol of the ferrous salt. The upper limit of the amount may be restricted by the solubility of the basic agent, and hence, the amount of the basic agent is usually in the range of 6 to 80 mols, preferably 8 to 30 mols, more preferably 10 to 20 mols, per 1 mol of the ferrous salt.

In the preparation of the goethite particles, a small amount of salts of other metals may be added to the aqueous solution of a ferrous salt. Suitable examples of other metal salts are sulfate, chloride, bromide or acetate of nickel, chromium, cobalt or copper, and these salts may usually be used in an amount of a few to several percents by weight based on the weight of the ferrous salts. The goethite particles containing other metal components can give the ferromagnetic iron particles having excellent antioxidation properties by treating it as described below.

The goethite particles thus obtained are dehydrated with heating at about 200° to 800°C, whereby α-ferric oxide is produced, and the resulting ferric oxide is reduced with a reducing gas (e.g., hydrogen gas) at a temperature of about 340° to 420°C, by which the desired ferromagnetic iron particles having a high coercive force can be obtained.

Example: To a 5 liter glass-made vessel were added NaOH (800 g, 20 mols) and water (2 liters) and thereto was added with vigorous agitation a solution of ferrous sulfate ($FeSO_4 \cdot 7H_2O$, 278 g, 1 mol) in water (2 liters) to precipitate white green ferrous hydroxide.

While keeping the solution containing the precipitates at 40°C, air was blown into the solution at a rate of 20 l/min for 10 hours in order to oxidize the ferrous hydroxide. The resulting yellow precipitates were separated by filtration, washed well with water and then dried at 100°C to give acicular goethite particles having a particle size (average length in long axis) 0.4 μm and an axis ratio (long axis:short axis) 15:1.

The goethite particles obtained above were dehydrated by heating at 500°C to give α-ferric oxide (α-Fe_2O_3). The α-ferric oxide (5 g) was uniformly developed onto a quartz board. The board was set within an electric furnace and thereto was passed through hydrogen gas in a rate of 1 liter/minute at 360°C for 6 hours and thereby the ferric oxide was reduced to give ferromagnetic iron particles (Product No. 1).

The particles had almost the same particle size and axis ratio as those of goethite particles and had a D_{110} of 140 A which was measured by an x-ray diffraction. The particles had also a coercive force (H_c) of 1,700 oersteds which was measured at a maximum magnetic field of 10,000 oersteds by using a vibrating sample magnetometer, a maximum magnetization (σ_S) of 146 emu/g and a square ratio (residual magnetization/maximum magnetization: σ_R/σ_S) of 0.50.

In the same manner as described above excepting that the amount (molar ratio to that of ferrous sulfate) of NaOH was varied, various ferromagnetic iron particles (Product No. 2 to 9) were prepared, and the particle size, D_{110} and coercive force of these particles were measured. The results are shown in the following table.

Product No.	Amount of NaOH (molar ratio)	Particle size (μm)	D_{110} (Å)	Hc (oersted)
2	1	0.2	410	660
3	4	0.3	310	1000
4	6	0.6	250	1250
5	7	0.5	225	1320
6	8	0.6	200	1420
7	10	0.3	180	1500
8	11	0.3	170	1540
9	12	0.2	145	1620

Gaseous Reduction of Silicon Compound Coated Raw Material

M. Amemiya and S. Asada; U.S. Patent 4,043,846; August 23, 1977; assigned to Hitachi, Ltd. and Hitachi Maxell, Ltd., Japan developed a method of producing ferromagnetic iron powder which has both a high coercive force and a high remanence ratio. It comprises dipping iron compound powder in a solution in which silicone oil is dissolved, removing solvent from the iron compound powder, reducing the iron compound powder in a reducing atmosphere at an elevated temperature and forming oxide layers on whole surfaces of the individual particles of the iron compound powder.

As a starting raw material, oxalate, formate, oxide or oxyhydroxide of ferromagnetic metal or their mixture, may be used; the iron compound is preferable because of the high value of the magnetization in the obtained metal; and the oxide and the oxyhydroxide are preferable too because of the facility of obtaining needlelike shaped powder. Another reason why the oxide is preferable is that the number of atoms made free at the reducing step is few whereby the particles are hard to break down. Among the iron oxides usable as the starting material, magnetite and ferric oxide such as maghemite are desirable because of the facility of obtaining needlelike shaped powder. It is to be desired that the particles of the raw material are needlelike. In addition, iron oxide containing Co, Ni, Al, Cr, Mg, B or the like may be used as the starting material to improve corrosion-resistance property.

The starting material powder is reduced by heating it in a reducing atmosphere after the starting material powder is coated with a solution of inorganic silicon compound or a solution of organic silicon compound, which has a boiling point higher than 250°C, and then the solvent of the solution is removed by drying. The drying step may be omitted when the solvent can be removed by natural drying, in another step. As the inorganic silicon compound or organic silicon compound, there is usually used a silicate such as sodium silicate or a silicon compound selected from the group consisting of sodium silicate, potassium silicate, silicic acid, potassium metasilicate, magnesium silicate, cobalt silicate, calcium metasilicate, metasilicic acid, sodium metasilicate or silicone oil, respectively.

The latter is used more usually. A more desirable result can be obtained when a solution containing silver ions is used with the silicon compound solution. Both of them may be used as a mixed solution or as two separate solutions. The starting material powder is dipped once in the former case, but twice in the latter, namely, it should be dipped in a silicon compound solution and in a silver ion solution separately. As the silver compound used for the solute, there can be used many silver salts such as silver nitrate, silver sulfate and silver chloride.

For the solvent of silicone oil, many organic solvents, such as ketone, formamide, aromatic compound can be used. Water, aqueous acidic solution and aqueous alkaline solution are used for solvents for inorganic silicon compounds. In the solution containing silver ions, water, liquid ammonia, strong inorganic acid or alcohol are employed as solvents. However, the latter can be used only for silver nitrate.

In this method, the amount of inorganic silicon compound or organic silicon compound, such as silicone oil, in the solution is more than 0.05 g/l. When the amount of silicon compound in the solution is less than 0.05 g/l, it is difficult to expect a better result than the prior art method. The preferable amount of the silicon compound in the solution is more than 0.1 g/l and the most preferable amount is more than 0.2 g/l. Furthermore, the silver compound may be contained in the solution in an amount of more than 0.02 g/l and preferably more than 0.05 g/l. The upper limit value thereof may be determined by the concentration in saturated solution.

Any reducing atmosphere containing reducing gas such as H_2 and CO may be used at the heating step for reduction; additionally H_2 gas and town gas can be obtained easily and therefore are preferable.

The suitable ratio of H_2 to H_2O, CO to CO_2 or the like for reducing Fe_3O_4, which is the rate-determining step in reducing every kind of iron oxide, is well known and may be applied here.

The suitable reducing temperature ranges from 250° to 600°C in the case that the solute of the solution in which the starting material powder is dipped is only silicon compound, especially silicone oil or sodium silicate. It ranges from 200° to 700°C in the case of using both silver salt and silicon compound as the solutes. It ranges preferably from 300° to 550°C in the former case, and from 250° to 600°C in the latter case.

According to the composition of the reducing atmosphere and the heating temperature, various heating times are necessary for the reduction process. The composition of the discharged gas of the heating atmosphere is somewhat different from the composition of the charged gas during the reduction process, but it becomes almost the same as that of the charged gas in the range of the compositional intrinsic fluctuation after the reduction process is completed. Accordingly, the completion of the reduction process can be recognized by detecting the discharged gas composition, and the necessary heating time becomes apparent naturally.

The produced powder can become corrosion-resistant and be convenient to handle, by forming surface layers such as oxide layers on the surface of the produced metal particles according to the well-known method concerning the nonactivation treatment of pure iron, namely by making the particle surfaces nonactive.

Oxides Surface Treated with Cobalt

S. Umeki, K. Aue and Y. Tokuoka; U.S. Patent 4,067,755; January 10, 1978; assigned to TDK Electronics Co., Ltd., Japan have developed a method of producing a powder magnetic material for use as a magnetic recording medium. More particularly, the disclosed powder magnetic material is adapted to provide a magnetic recording medium of low electrical resistance. In particular, an alloy of cobalt and at least one element selected from the group consisting of vanadium, molybdenum, copper, manganese, nickel, zinc and iron are deposited onto the surface of particles of magnetic iron oxides such as Fe_3O_4 or γ-Fe_2O_3. Powdered magnetic material of high coercivity, high magnetic flux density and low electrical resistance is obtained.

Example 1: About 25 g of spiculate or needle-shaped (acicular) magnetite whose major axis is about 0.5 and whose axis ratio is about 6 to 8 is dispersed (slurried) into a solution so as to be reacted at 100°C for an hour, the composition of the solution being as follows:

$CoCl_2 \cdot 6H_2O$	6 g
$Na_2C_4H_4O_6 \cdot 2H_2O$	50 g
H_3BO_4	20 g
$NaHPO_2 \cdot H_2O$	24 g
H_2O	500 ml
$CuCl_2 \cdot 2H_2O$	2 g

The pH value of the solution is adjusted to 9.0±(0.1) in 5N NaOH solution. After the product is washed with water and dried, the substance is heat-treated in a nitrogen atmosphere at 300°C for an hour. The magnetic properties and electrical resistance of the product obtained were measured.

Table 1 below shows the measured values of various materials, wherein the column labeled No. 1 denotes pure magnetite Fe_3O_4 before treatment according to this process, column No. 2 denotes magnetite containing cobalt only, and column No. 3 denotes magnetite containing deposited alloy of cobalt and copper according to this process.

Table 1

		.......Material........		
		No. 1	**No. 2**	**No. 3**
Coercive force	H_c (Oe)	390	544	520
Residual magnetic flux density	B_R/ρ (G-cm^3/g)	485	525	514
Saturated magnetic flux density	B_M/ρ (G-cm^3/g)	970	1,072	1,040
Rectangular ratio	B_R/B_M	0.50	0.49	0.49
Electrical resistance	R (Ω)	1×10^8	1×10^{10}	1×10^6
Deposited metals	Co (wt %)	0	5.0	3.0
	Cu (wt %)	0	0	0.4

As shown in Table 1, the electrical resistance of material of No. 3 is lower than that of materials No. 1 and 2 because of the addition of copper. The coercive force and magnetic flux density are also much improved by the addition.

Though the amount of deposited metals is shown individually as Co or Cu in Table 1, it denotes chemically analyzed results, and Co and Cu were actually deposited in the form of an alloy. Such expression is used in the same manner in the other examples.

The measurement of electrical resistance is carried out by the following method. Powder magnetic material is packed into a glass tube having a diameter of 4 mm and a length of 60 mm. The powder is packed to a density of about 0.7 g/cm^3. Mercury electrodes are attached to both ends of the glass tube, and the resistance measurement is carried out by an electrometer. Magnetic characteristics are measured by a self-recording flux meter under the condition that H max was 5,000 Oe. These measuring methods are used in other examples in the same manner.

Example 2: In this example, manganese is utilized instead of copper in Example 1, and the magnetic properties and electrical resistance of the resultant powder magnetic material (No. 4) produced by the same method as the method of Example 1, are shown in Table 2 below.

Table 2

		Material No. 4
Coercive force	H_c (Oe)	608
Residual magnetic flux density	B_R/ρ (G-cm^3/g)	543
Saturated magnetic flux density	B_M/ρ (G-cm^3/g)	1,083
Rectangular ratio	B_R/B_M	0.50
Electrical resistance	R (Ω)	1×10^7
Deposited metals	Co (wt %)	1.5
	Mn (wt %)	0.3

As shown in Table 2, the electrical resistance is decreased by the addition of manganese, in comparison with the characteristics of the materials No. 1 and No. 2 as shown in Table 1; and high coercive force and high magnetic flux density of material No. 4 are obtained. Thus, the effect of manganese is substantially the same as that of copper.

Example 3: About 50 g of spiculate magnetite, whose major axis is about 0.5 μ and whose axis ratio is about 6 to 8, is dispersed into a solution to be reacted at 70°C for an hour, the composition of the solution being as follows:

$CoCl_2 \cdot 6H_2O$	5 g
Rochelle salt	13 g
Hydrazine hydrate	25 g
H_2O	500 ml
$ZnCl_2$	2 g

In this example, zinc is used instead of copper (Example 1) or manganese (Example 2). The pH value of the solution is adjusted to be about 12.0 in 5N NaOH.

Table 3 shows the measured values of magnetic characteristics and electrical resistance of the obtained powder magnetic material.

As shown in Table 3, the electrical resistance of the materials No. 5 and 6 is smaller than that of materials No. 1 and 2. The decrease in electrical resistance results from the addition of zinc. The electrical resistance of the material No. 6, however, shows a rather high value, 1×10^9, and hence not more than 3 weight percent zinc should be added.

Table 3

	Material	No. 5	No. 6
Coercive force	H_c (Oe)	560	450
Residual magnetic flux density	B_R/ρ (G-cm^3/g)	453	430
Saturated magnetic flux density	B_M/ρ (G-cm^3/g)	914	900
Rectangular ratio	B_R/B_M	0.50	0.48
Electrical resistance	R (Ω)	1 x 10^7	1 x 10^9
Deposited metals	Co (wt %)	4.0	1.0
	Zn (wt %)	3.0	4.0

Prevention of Deterioration of Magnetic Properties

In order to prevent deterioration of characteristics of ferromagnetic metal or alloy particles, *Y. Tokuoka and K. Fukuda; U.S. Patent 4,113,528; September 12, 1978; assigned to TDK Electronics Co., Ltd., Japan* add an oxidizing agent to an aqueous slurry containing ferromagnetic metal or alloy particles which are obtained by a reduction in a liquid phase.

The ferromagnetic metal ions can be Fe^{2+}, Fe^{3+}, Co^{2+}, Ni^{2+} and a mixture thereof and the ions are in the form of the water-soluble salts such as sulfate, nitrate, chloride etc. The concentration of the ferromagnetic metal salt is usually in a range of 0.1 to 2 mols/l and especially about 1 mol/l.

Suitable reducing agents include sodium or potassium boron hydride and the concentration of the reducing agent is usually in a range of 0.1 to 2 mols/l especially about 1 mol/l.

Suitable oxidizing agents include hydrogen peroxide, potassium permanganate, sodium chromate, sodium nitrite etc.

The concentration of the oxidizing agent can be enough to oxidize the surface of the ferromagnetic metal or alloy particles and is dependent upon the rate of the addition of the oxidizing agent, the volume of the washing tank and the velocity of the formation of the ferromagnetic metal or alloy particles, and is usually in a range of 1 x 10^{-3} to 10^{-1} mol/l, preferably 1 x 10^{-3} to 9 x 10^{-3} mol/l.

When a mixture of ferrous sulfate and cobalt chloride at a ratio of 1:9 (cobalt chloride/ferrous sulfate) is used as the metal salt and sodium boron hydride is used as the reducing agent and the reactor is disposed in the field of 1,000 Oe, where each solution of equimol is fed from the tanks at rates of about 0.25 cc/sec, the yields per 1 liter of the solution are plotted as a straight line.

The ferromagnetic particles are recovered after the reaction and are washed with ethanol and acetone and dried at 200°C in hydrogen gas atmosphere.

The typical characteristics are shown in Table 1.

Table 1

Sample No.	Temp. for drying (° C)	Saturation magnetization (emu/g)	Coercive force (Hc)
1	200° C	150 emu/g)	560 Oe
2	room temp.	134 emu/g	748 Oe

Examples: Sample I was prepared by feeding the mixture of the reaction solution and ferromagnetic particles at a rate of 5 cc/min and feeding water at a rate of 2 l/min, and feeding hydrogen peroxide solution at a rate of 10 cc/min with stirring by the disperse-mixer at 10,000 rpm, whereby the sample was washed at pH of 7.

Sample II was prepared in accordance with the process for preparing the Sample I except feeding 0.3 M solution of potassium permanganate at a rate of 5 cc/min.

Sample III was prepared in accordance with the process for preparing the Sample I except rotating the disperse-mixer at 8,000 rpm in pH of 6 without adding an oxidizing agent.

In the reference example, the sample was prepared in accordance with the process for preparing the Sample I except feeding no air and no oxidizing agent. The magnetic field for measurement was 5,000 Oe. The results are shown in Table 2.

Table 2

Sample No.	Coercive force (Oe)	Saturation magnetization (emu/g)	Squareness ratio
I	950	142	0.55
II	1025	136	0.55
III	1025	105	0.56
Reference	986	92	0.51

Particle Recovery Using a Magnetic Drum

In a related process by *Y. Tokuoka, K. Fukuda and A. Hosaka; U.S. Patent 4,097,313; June 27, 1978; assigned to TDK Electronics Co., Ltd., Japan* the ferromagnetic metal or alloy particles are also prepared by reducing ferromagnetic metal ions in the solution with a reducing agent to obtain a slurry of ferromagnetic metal or alloy particles. Then the particles are separated by attracting the particles on a magnetic drum which is rotated under contacting with the slurry, and washing the particles and separating the particles from the magnetic drum.

Example: An aqueous solution of 1 mol of sodium boron hydride and an aqueous solution of 0.9 mol of ferrous sulfate and 0.1 mol of cobalt sulfate were respectively fed into a reactor at a rate of about 30 cc/min. The ferromagnetic particles which were precipitated at a rate of 1 to 1.5 g/min were continuously recovered and stored.

The magnetic characteristics of the ferromagnetic particles were measured at the time of production and after storing them. Sample I was prepared without a lubricant coating on the surface of the magnet drum. Sample II was prepared by a silicone grease coating. The results are shown in the table below.

Sample No.	Sample I			Sample II		
Days for storing	0	7	14	0	7	14
Saturation magnetization (emu/g)	115	115	114	141	141	140
Coercive force (OE)	1191	1193	1193	1075	1079	1077
Squareness ratio	0.56	0.56	0.56	0.56	0.56	0.56

The saturated magnetic moment of the conventional ferromagnetic iron oxide particles is lower than 90 emu/g. Samples I and II maintained high saturated magnetic moment for a long time.

Sample II had a higher saturated magnetic moment compared with that of Sample I. This fact shows that the silicone grease coating is effective for improving it. It is considered to be caused by the following reason. Since the surface of the magnet drum is coated with the silicone grease as a water-repellent lubricant, the separation of water from the ferromagnetic particles is effectively attained to prevent the oxidation of the ferromagnetic particles with water.

In accordance with the process, the ferromagnetic particles in the solution can be recovered speedily and the magnetic characteristics of the ferromagnetic particles can be maintained for a long time.

COBALT

Acicular Powders Having High Squareness Ratios

J.E. Ehrreich and A.R. Reti; U.S. Patent 3,855,016; December 17, 1974; assigned to Graham Magnetics, Inc., describe a process for providing small acicular ferromagnetic cobalt and cobalt-containing particles of uniform shape which do not tend to agglomerate during their fabrication. These particles are usually composed predominantly of cobalt metal and always comprise at least 43% be weight of cobalt. These needle-like particles have a typical diameter of 0.02 to 0.5 μ and a typical length of 0.1 to 2.0 μ. They may consist substantially entirely of cobalt or may contain cobalt alloyed with other metals.

Measurements on a bulk sample of unoriented cobalt particles of this process give coercivities greater than 400 Oe with typical values ranging from 600 to 1,000 Oe and more. The samples also have high saturation magnetization, in excess of 60 emu/g. Still further, the bulk material is characterized by a high degree of squareness, e.g., ratios on the order of 0.25 to 0.5.

Thus, the acicular cobalt and cobalt-containing particles of this process have magnetic and physical properties which make them especially suitable as the magnetic pigment in magnetic recording members. Furthermore, these desirable characteristics do not materially degrade with time.

Example 1: A quantity of 238 g of $CoCl_2$ is dissolved in a container to form a 1 liter aqueous solution. Then 100 g of sodium hydroxide is dissolved in a second container to form a second 1 liter aqueous solution. The cobalt chloride and sodium hydroxide solutions are mixed together, using a magnetic agitator bar, in a 4 liter beaker for 3 minutes to form a cobalt hydroxide precipitate. The 135 g of oxalic acid are dissolved in a separated container to form a 1.5 liter aqueous solution. The oxalic acid solution is mixed thoroughly into the cobalt hydroxide precipitate for 5 minutes to form a cobalt oxalate precipitate.

The resultant mixture is filtered in a Buchner funnel and washed several times with water. Then it is rinsed several times with acetone and air dried. The resulting cobalt oxalate particles are acicular (that is "needle-shaped"). Following this, 0.05 g of silicone polymer oil (G.E. RTV910 Diluent) is mixed with 0.95 g of the acicular cobalt oxalate particles using sufficient tetrahydrofuran to insure uniform coating of the silicone oil on the particles. Next, the coated particles are air dried. Then the coated particles are charged into a 1" diameter glass tube in a tube furnace, heated to 360°C in a hydrogen atmosphere, and held at this temperature for 1 hour, thereby reducing the metal salt to cobalt.

The material is then cooled to room temperature while still in a hydrogen atmosphere, and subjected to a 2 minute argon purge. Next the product is washed several times with acetone. A magnet is used to effect separation from the wash, and the separated material is air dried. The resulting product is a strongly magnetic powder having a coercivity of 705 Oe, a sigma value of 82 emu/g, and a squareness figure 0.40. The particles are composed mainly of cobalt, and have little or no silicone oil on their surfaces. Also the cobalt particles are acicular with an average particle size of 0.3 μ in diameter by 1 μ in length.

Example 2: Cobalt particles are prepared as described in Example 1, except that the cobalt oxalate crystals are not coated with silicone oil during their reduction step. Now the resultant cobalt particles have a coercivity of 313 Oe, a sigma value of 101 emu/g, and a squareness characteristic of only 0.21.

Exchange Anisotropy Properties

E. Koester, G. Wunsch, P. Deigner, D. Graw, K. Uhl and W. Steck; U.S. Patent 3,959,032; May 25, 1976; assigned to BASF AG, Germany have developed a process for producing magnetic materials which essentially consist of a phase A with a ferromagnetic spin system and a phase B with an antiferromagnetic spin system, phases A and B interacting by exchange coupling, and the manufacture of such materials. The materials consist essentially of two-phase magnetic single-domain particles, with the alloy $[Co_xNi_{1-x}]$ as the ferromagnetic phase A and its oxidized form $[(CoO)_x(NiO)_{1-x}]$ as the antiferromagnetic phase B, with values of x of 0.04 to 0.96. They are manufactured by surface oxidation of a pigment layer, which at least covers the surface but may extend deeper, having the composition of phase A, to form an oxide layer, at least 40 A thick, of the phase B, followed by heating in an inert gas atmosphere, or by reduction of a corresponding layer of phase B to a layer of phase A.

Magnetic single-domain particles of this composition have a Neel temperature which varies with x. As a result, materials are obtained which have an asymmetrical hysteresis loop, attributable to exchange anisotropy, at temperatures which can be varied within wide limits in accordance with the intended use of the material.

In one embodiment, the materials consist of a core of phase B, namely the oxide material, and an outer shell of the ferromagnetic metal phase A.

A further advantageous structure of the materials according to the process comprises a system of several layers of the two phases A and B. In such materials, the two phases alternate and here again the core can consist of phase A or phase B. This structure makes it possible to increase the effective area over which the coupling of the ferromagnetic and antiferromagnetic system occurs.

In a preferred embodiment, only a part of the magnetic material exhibits the exchange interaction, in particular as an asymmetrical hysteresis loop.

The materials can contain additional elements in each phase. Possible modifiers are any metals present in such amounts that they do not interfere with the exchange interaction between the metallic phase A and the oxide phase B. In particular, iron, manganese, zinc, chromium, aluminum and other metals can be added. The amounts in which such additives do not interfere with the manufacture of the materials according to the process is easily determined experimentally and depends on the incorporation of the particular material into the particular crystal lattice.

The examples which follow illustrate an advantageous selection of materials according to the process and their manufacture. Among parameters measured are the specific induction of the material, $B_S/\sigma[nTm^3/g]$, and the specific remanence, $B_R/\sigma[nTm^3/g]$. j_{er} is shown as a measure of the proportion of magnetic pigment exhibiting exchange anisotropy in the total composition of the material. j_{er} is determined by saturating the sample in a field of 160 kA/m, heating it in the remanent state to above the particular Neel temperature T_N and again cooling it. The specimen is then placed in an alternating field of 200 kA/m which slowly decreases to zero. j_{er} is now obtained by measuring the residual remanence and expressing it as a percentage of the remanence of the untreated material after saturation of 160 kA/m.

Example 1: 200 g of NaOH were dissolved in 2,750 ml of water in a 6 liter four-neck flask, the solution was warmed to 80°C and 356 g of $CoCl_2 \cdot 6H_2O$ and 118 g of $NiCl_2 \cdot 6H_2O$ dissolved in 1,250 ml of water were added dropwise in the course of 40 minutes while stirring (at about 300 rpm) and passing nitrogen (120 liters of N_2/hr) through the flask. After completion of the addition, the mixture was stirred for a further 2.5 hours at 80°C. The hydroxide precipitate $Co_xNi_{(1-x)}(OH)_2$, wherein x = 0.75, was filtered off, washed with distilled water and dried in a vacuum drying oven at 60°C.

80 g of the dried product were reduced for 14 hours in a rotary furnace at 300°C by means of 100 liters of H_2/hr. The oxygen content of the resulting product was 1.7% by weight. The product was oxidized by treating it with a 1:12 mixture of air and nitrogen at room temperature for 7 hours and was then heated under nitrogen at 300°C for 4 hours. The oxygen content was then 5.4%.

A saturation induction B_S/σ of 83.5 nTm^3/g, a remanence B_R/σ of 38.0 nTm^3/g and a coercive force H_C of 53 kA/m were found, in a magnetic field of 160 kA/m. The Neel temperature T_N was about 75°C. j_{er} was about 1.3%.

Example 2: The method followed was as in Example 1, but only 238 g of $CoCl_2 \cdot 6H_2O$, and 238 g of $NiCl_2 \cdot 6H_2O$, were dissolved in 1,250 ml of water. The solution was added dropwise in the course of 1 hour and the mixture was then stirred for a further 3 hours at 80°C.

The hydroxide precipitate $Co_xNi_{(1-x)}(OH)_2$ in which x = 0.5, contained in a part of the mother liquor, was introduced into a 3 liter tumbler autoclave and kept at 200°C under autogenous pressure for 24 hours. The precipitate was then filtered off, washed with distilled water and dried in vacuo at 50°C.

70 g of the dried product were dehydrated for 2 hours in a rotary furnace at 250°C by means of 100 liters of N_2/hr and were then partially reduced for 2 hours at 300°C by means of 100 liters of H_2/hr. The product was then oxidized for 40 minutes in a rotary furnace at 200°C by means of a 1:10 mixture of air and nitrogen. After heating for 4 hours under nitrogen, the sample was cooled to room temperature and postoxidized at this temperature for 2 hours by means of a 1:10 mixture of air and nitrogen. Air was then passed over the materials for 10 minutes. The oxygen content was 8.4%. The following values were measured at 160 kA/m:

$$B_S/\sigma = 63 \text{ nTm}^3/\text{g}$$
$$B_R/\sigma = 27.4 \text{ nTm}^3/\text{g}$$
$$H_c = 31.5 \text{ kA/m}$$
$$j_{er} = 14\%$$
$$T_N = 120°\text{C}$$

Improved Dispersability by Sulfuric Acid Treatment

Magnetic cobalt-phosphorus particles formed by chemical oxidation-reduction are found to agglomerate and not to disperse homogeneously when mixed with organic resin binders. *A. Simonetti; U.S. Patent 3,905,841; September 16, 1975; assigned to International Business Machines Corp.* has found that when the same type of cobalt-phosphorus particles are treated with a sulfuric acid containing solution and then incorporated into an organic resin binder, the resulting mixture exhibits homogeneous dispersion characteristics.

The sulfuric acid treatment can vary as to time, temperature, acid concentration, and the presence or absence of other constituents or water in the treating solution. Percentage weight loss of the dry cobalt-phosphorus particles is considered indicative of the degree of treatment. Amounts of weight loss of 10% or less begin to provide particles with improved dispersion characteristics.

Particles treated according to this process which show a weight loss of from about 35% to about 50% exhibit preferred improved dispersion characteristics in an organic resin binder. Greater amounts of weight loss achieve a similar result, but provide a greater loss of raw materials without any significant improvement in dispersion characteristics. Small uniform cobalt-phosphorus alloy particles having improved dispersion characteristics in organic resin binders are formed by this technique.

In preferred embodiments a stoichiometric amount of sulfuric acid containing solution is determined and utilized to achieve a desired percentage weight loss. Where the stoichiometry is thus controlled, the reaction can be allowed to proceed to completion without any concern as to the reaction time and temperature, although about 13 minutes at about 45° to about 52°C is considered an adequate time and preferred temperature to complete such a reaction.

Controlled Chemical Reduction in DC Magnetic Field

T.J. Beaulieu, M.A. Marchese, F.T. Plante and R.H. Tlaskal; U.S. Patent 3,954,520; May 4, 1976; assigned to International Business Machines Corporation describe metallic magnetic cobalt-phosphorus particles produced by chemical reduction having improved characteristics for use in magnetic recording media and having a W/H_C of about 1.2 or less.

W/H_C represents a figure which can be determined with mathematical precision and which is normalized to take into account the average coercivity of each sample. It is derived by plotting a derivative curve of coercivity and magnetic field.

The full width or range of coercivity at half (50%) the maximum height of the derivative curve is measured. This full width at half the maximum height has been designated as "W." W, in Oe, is by itself a meaningful representation of the uniformity of coercivity in the sample. Unfortunately, however, W by itself does not take into account the variations in W which are to be expected as the average coercivity H_C of the sample varies. It has thus been determined to normalize W for the variation in coercivity for each sample by dividing W by H_C. Therefore, as used herein, the figure of merit is:

$$W/H_C = \frac{\text{full width of the derivative curve at half its maximum height}}{\text{average coercivity}}$$

They are formed by preparing a bath including a soluble cobalt salt and then chemically reducing the cobalt cations to cobalt metal in a dc magnetic field of at least about 200 gauss using hypophosphite anions as a reducing agent. Other bath parameters and additives may be selected and controlled in accordance with any of the techniques known in the art. The operating ranges for the dc magnetic field are broad and require no control other than that they produce the desired improvement in W/H_C.

The reaction time during which the magnetic particles are produced is not necessarily critical in so far as obtaining the desired range of W/H_C is concerned. However, reaction sequence, timing, and time can affect other physical and magnetic characteristics. Reaction time can be controlled, e.g., by quenching the reaction with large volumes of water. The presence or absence of other constituents in the reaction mixture can also be varied. The figure of merit W/H_C of the resulting cobalt-phosphorus particles is considered to be indicative of the success or failure of treatment.

Example: A cylindrical beaker containing an aqueous 3,520 ml solution including 140 g cobalt sulfate ($CoSO_4 \cdot 7H_2O$), 280 g sodium citrate dihydrate, and 160 g sodium hypophosphite ($NaH_2PO_2 \cdot H_2O$) was prepared, placed in a solenoidal magnetic field of 1,000±100 gauss intensity, and heated to boiling. The solution was subjected to mechanical stirring during its preparation and heating. To the hot solution was added 80 ml of 0.1% palladium chloride ($PdCl_2$)-hydrochloric acid (HCl) solution followed in about 10 seconds by 400 ml of 28% ammonium hydroxide (NH_4OH). Stirring was stopped and within several seconds a vigorous reaction occurred in which the reaction mixture darkened and gas evolved from the mixture. Finely divided dark gray particles were formed and precipitated within the beaker.

After about 6 minutes, the reaction was quenched with about 4,000 ml of cold water. The magnetic field of the solenoid was then turned off, the dark precipitate allowed to settle, and the reaction mixture decanted. The particles were then washed 3 times with water, 3 times with acetone, and dried as completely as possible under nonoxidizing conditions. The particle yield was about 86%, based on the amount of available cobalt.

An unweighed portion of the resulting particles was dispersed in an organic binder, coated on a narrow substrate, the substrate placed within a 1,000 Oe orienting magnetic field of a solenoid, and the binder dried. The resulting film was then used for determination of the W/H_C of the particles by the vibrating sample magnetometer (VSM). A series of minor hysteresis loops and the major hysteresis loop of the coating were run. Other magnetic properties were determined by packing a measured amount of dry particles in a glass cylinder for measurement by the VSM.

The particles were found to exhibit an average coercive force of 510 Oe and a squareness ratio of 0.82. The full width at half maximum W was found to be 260 Oe, and the W/H_C 0.51. Electron micrographs of the powder indicated that it consisted of particles having an average length of about 0.3 μ with a length to width ratio of about 3 to 1. Its chemical composition was about 92.2% cobalt, 2.4% phosphorus, 0.6% palladium, with the balance believed to be oxygen, the oxygen being present mainly at the surface of the particles.

Using the technique of this example, samples having average coercivities in the range of about 500 to about 550 Oe and W/H_C of less than 1.2 can be consistently prepared. Materials in this coercivity range having good magnetic uniformity are formed into recording media which are easily written and read by contemporary recording equipment. However, such media produces three times the signal output of the best iron oxide media now available over a frequency range of 0 to 10,000 flux changes per inch.

Any significant variation from the procedure of this example results in a substantial change in the average coercivity and in the W/H_C of the cobalt-phosphorus produced.

The process of *D.R. Craig, G.M. Lederle and F.T. Plante; U.S. Patent 3,899,369; August 15, 1975; assigned to International Business Machines Corporation* is similar to the preceding one except that the dc magnetic field is 1,000 gauss instead of 200.

ALLOYS

Enhancement of Decomposition Products by Seeding

In making metal-bearing powders comprising the formation of an organometallic salt and the subsequent decomposition of the salt to a metal-bearing powder, e.g., the thermal reduction of metal oxalates to metal *R.J. Deffeyes; U.S. Patent 3,965,046; June 22, 1976; assigned to Graham Magnetics Incorporated* uses the following procedure: reacting a carboxylic acid with a metal reactant to form an organometallic salt seed; reacting a carboxylic acid with a second metal reactant, in a liquid medium with the first carboxylic salt, and growing a second organometallic salt material on the seed, the seed causing a modification in the normal growth pattern of the second salt; recovering a solid organometallic salt product comprising both the seed and the second organometallic salt; and decomposing the salt product to a metal-bearing powder.

The seed comprises, as its primary metallic component, cobalt and is needle-like in shape and forms means to modify the morphology of the second salt such that, on reduction to a metal which is derived from a second metal reactant, the magnetic properties of the metal are substantially enhanced over the properties achievable when the seed crystal is not utilized.

Example 1: A metal-containing solution was prepared with 200 ml water containing 29.0 g $Co(NO_3)_2 \cdot 6H_2O$. A solution of 25.2 g of oxalic acid dihydrate in 400 ml of water was prepared in a 1,000 ml baffled Erlenmeyer flask equipped with a magnetic stirring bar. Then the previously prepared metal-containing solution was added to the flask to form a seed-bearing dispersion.

Thereupon a precipitating solution of 25.2 g of oxalic acid dihydrate in 400 ml of deionized water was prepared in a baffled 1,000 ml Erlenmeyer flask. A quantity of 100 ml of the seed dispersion was mixed into the precipitating solution. Next, a metal-bearing solution was prepared as follows:

$FeSO_4 \cdot 7H_2O$	6.0 g
$Co(NO_3)_2 \cdot 6H_2O$	13.2 g
$Ni(NO_3)_2 \cdot 6H_2O$	6.0 g
Water	200 ml

The metal-bearing solution was added slowly to the precipitating solution bearing the seed. This addition was carried out over a period of five minutes with the agitator running at full speed. The crystal products were filtered, washed with hot water, dried by washing in isopropanol and coated with a polyamide resin Azamide 325 (AZ Products, Inc.). The resin was applied in about a 5% solution of isopropanol. The resin was coated in a quantity of resin equivalent to 2.3% by weight based on dry oxalate. The oxalate was dried.

The resultant oxalate crystal was placed into a boat and reduced in a tube furnace at 800°F with 2.4 l/min dry nitrogen passing over the sample. The tube was cooled to room temperature and exposed to a 3% O_2, 97% N_2 atmosphere for 3 hours. Then it was exposed to dry air for another hour.

The resultant product was metal powder primarily comprised of an alloy of cobalt with the following properties.

Sigma, emu/g	153
Coercivity	512
Squareness	0.74
Bulk density, g/ml	0.3
Surface, m^2/g	25.4

Example 2: Using the same general procedures disclosed in Example 1, a copper-oxalate seed was used to initiate growth of an equant particle composed of a second organometallic salt of the 60:20:20 cobalt:nickel:iron type. Copper, all appearing in the seed portion, equalled 1½% of the total metal in the completed oxalate.

The reduced oxalate product was an equant metal powder exhibiting a squareness of 0.5. Such a powder finds particular application in magnetic disc storage devices.

Co-Ni-Fe Organometallic Oxalates

R.J. Deffeyes and W.R. Tyler; U.S. Patents 3,909,240; September 30, 1975; 4,004,917; January 25, 1977; and 4,061,824; December 6, 1977; all assigned to Graham Magnetics Incorporated prepared highly acicular organometallic salts from organic acids and a metal cation by initiating the precipitation or growth of the salt in the presence of a quantity of complexing agent. The resulting metal complex (or metal complexes) apparently controls the morphology and shape assumed by the embryonic and growing organometallic salts. The method is especially advantageous when used to promote the growth of highly acicular salts.

A visual feature of the acicular organometallic salts is the pearlescence or iridescence of the precipitate as it forms with stirring. A distinct pinkish color seems to predominate in cobalt-based salt systems whereas in the prior art such systems tended to exhibit a brownish tinge. It is believed these visual features are primarily due to the uniformity of the organometallic salt's particle size and the acicular nature of the particles.

The initial precipitation of seed crystals can take place in the same medium in which the primary growth or organometallic salt is to occur. In such a situation, it is usually desirable to utilize a liquid medium wherein the solubility of the organometallic salt is less than its solubility in water. This is conveniently achieved by adding an organic solvent to the water. Such water-soluble solvents as acetone, lower alkyl alcohols like ethanol, and the like, are entirely satisfactory.

In another embodiment, seed crystals may be formed in a first medium containing the complexing agent and, advantageously, a water-solvent medium and then transferred to the primary crystal-growing solution. One advantage of this latter procedure is that, when the complexing agent is properly selected, the population density of seeds or seed-promoting complexes in the first medium can be so high that a very small volume of the medium containing the seeds or seed-promoting complex can be used to initiate the growth of crystals in the primary crystal-growing solution. This is true even when the medium for the primary crystal-growing is entirely aqueous. Thus, it is seen that this embodiment can be carried out using a relatively small amount of organic solvent and complexing agent.

The following working examples are given as illustrative embodiments of the process and products produced thereby. Unless otherwise specified, the examples are carried out according to the following general procedure.

The precipitation is carried out in a 1,000 ml Erlenmeyer flask with four baffle-type indentations molded at 90 angular degree intervals around the base of the flask. Agitation is achieved by a magnetic stirring bar about 1.6″ long and 0.5″ in diameter.

The addition of solutions to the flask is accomplished through a funnel. The times of addition are longer as the viscosity of the solution being added increases.

Cobalt nitrate is used as the cobalt source but other such water-soluble cobalt salts as the acetate, chloride or sulfate are generally useful. The commercial varieties of these salts normally contain 3% of nickel based on the weight of cobalt. Experiments have indicated there is little or no advantage to the use of pure cobalt salts.

The water is used as deionized water processed through a standard commercial laboratory deionizer. A colorless commercial syrup is used as a source of sugar complexing agent (Karo, Best Foods, Englewood Cliffs, NJ). It is formed principally of the type that digests to dextrose units. When sucrose is specified, ordinary sugar from any grocery store is used. Other reactants are utilized in their ordinary commercial form. The methanol is reagent grade, but such purity is not required. The ethanol is denatured ethanol. The glycerine is USP, but again this purity is probably not required.

The organometallic salts, the preparation of which is disclosed below are coated with a resin and decomposed as follows.

The organometallic salt filter cake is air dried after washing and rewetted with isopropanol. A polyamide resin, Resymide 1125, is dissolved in isopropanol. A quantity of the resin is used which is about 7% of the weight of the metal in the organometallic compound to be coated. The isopropanol suitably contains a concentration of about 10% resin. A thin slurry of the organometallic salts in the resulting liquid is formed and the isopropanol is evaporated as the slurry is stirred. The resulting dry powder is a mass of resin-coated particles ready for reduction.

In general, the reduction is carried out by placing the powder in sample boats containing eight different compartments. (Usually a different sample is in each compartment to facilitate comparative testing.) The sample boats are sealed in tubes having gas inlet and gas outlet ports, then placed in an oven, with conduits to both the outlet and inlet port extending from the oven to a gas-supply source and a gas exhaust system, respectively.

The tubes are purged with N_2; then the oven temperature is set for 650°F and a gas mixture of 20% hydrogen and 80% nitrogen is passed through the tubes at the rate of 40 cc/min. Within about 45 minutes, CO_2 evolution starts. After the rate of CO_2 evolution peaks and drops to less than 0.01% by volume of the off-gas, the tube is purged with nitrogen, and removed from the furnace. After cooling to room temperature, this usually takes an hour or so, a mixture of 3% oxygen and 97% nitrogen is passed through the tube. This gas mixture is usually allowed to run for several hours but the exact time, or even the exact composition, is not particularly critical.

Then the reactor is purged with air for two hours and opened. The samples are removed and are ready for analysis. Usually a sample from a control batch of oxalate is fired with each eight-sample load as a means to confirm the reproducibility of the reduction process.

Example 1: A quantity of 25.2 g of oxalic acid dihydrate is dissolved in a mixture of 250 g of Karo and 150 g of water to form a first solution. To this first solution is added a second solution of 29.1 g of cobalt nitrate hexahydrate in 100 g of Karo and 50 g of water. This addition is made over a period of about 30 minutes.

After the addition is completed, stirring is continued for an additional 30 minutes. The mixture was filtered and washed two times with distilled water at 60°C. The filter cake, beige initially, turned pinkish in color on being washed with the water. Next the precipitate is washed in acetone. A sample is collected for a microscopic inspection; the remainder of the precipitate is resin coated and reduced to metal.

Example 2: A procedure similar to that of Example 1 was followed excepting the first solution was comprised of 25.2 g oxalic acid dihvdrate, 200 g of sucrose and 200 g of demineralized water. The solution added to the first solution, through a funnel and over a period of about 13 minutes, was formed of 29.1 g of cobalt nitrate hexahydrate in 75 g sucrose and 75 g of water.

Samples were taken as indicated in Example 1. After the precipitation of the cobalt oxalate salt, it was filtered, washed and reduced to form a metal more fully described in Table 1 below.

Table 1

	Powders of Examples					
Properties	**1**	**2**	**3**	**4**	**5**	**6**
Specific magnetic moment, emu/g	125	117	118	119	114	119
Coercive force, Oe						
RT	770	783	799	1,020	1,020	730
Liquid N_2 temp.	1,344	1,480	1,400	1,690	1,848	–
Ratio liquid N_2/RT	1.75	1.89	1.75	1.66	1.81	–
Squareness	0.78	0.78	0.81	0.79	0.80	0.76
Particle size, μ						
Length, average	10.6	9.3	7.5	3.5	6.3	7.9
Standard deviation	4.6	4.1	3.9	1.0	1.6	3.3
Width, average	0.82	0.59	0.41	0.20	0.41	1.34
Standard deviation	0.44	0.36	0.19	0.04	0.07	0.67
L/D ratio, average	14.6	18.4	20.2	17.9	15.8	7.95
Standard deviation	5.8	10.5	11.4	5.0	4.9	4.6
Minimum value	6.2	6.7	7.1	10.0	10.6	3.75
Maximum value	29.4	54.0	55.4	33.3	29.6	14.9

The following examples were carried out, according to the general procedures of Example 2, to form metallic powders. Properties of these metals and the organometallic salt precursor are given in Table 1.

Example 3:

Component	Grams
Solution 1	
Oxalic acid dihydrate	25.2
Ethylene glycol	200
Water	200
Solution 2	
Cobalt nitrate hexahydrate	29.1
Ethylene glycol	75
Water	75

Addition time was 5 minutes and 45 seconds.

Example 4:

Component	Grams
Solution 1	
Oxalic acid dihydrate	25.2
Ethanol	125
Water	125
Karo	150
Solution 2	
Cobalt nitrate hexahydrate	29.1
Ethanol	50
Water	50
Karo	50

The addition took 10 minutes and 4 seconds. The mixing action was relatively poor because of the viscosity of the mix.

Example 5:

Component	Grams
Solution 1	
Oxalic acid dihydrate	25.2
Acetone	125
Water	125
Karo	150
Solution 2	
Cobalt nitrate hexahydrate	29.1
Acetone	50
Water	50
Karo	50

Again, the addition took about 6 minutes.

Example 6:

Component	Grams
Solution 1	
Oxalic acid dihydrate	25.2
Water	375
Karo	25
Solution 2	
Cobalt nitrate hexahydrate	29.1
Water	140
Karo	10

Solution 2 was added to Solution 1 in a little less than 5 minutes. The acicularity of the oxalate increased and, consequently, the bulk density dropped.

Example 7: A seed crystal mixture was prepared according to the following procedure.

Solution	Grams
Solution 1	
Cobalt acetate	25
Citric acid	25
Water	50
Karo	50
Ethanol	50
Solution 2	
Oxalic acid dihydrate	25.2
Karo	150
Water	150
Ethanol	150

Solution 2 is added slowly at 18°C into Solution 1 in a 1 liter baffled Erlenmeyer flask equipped with a magnetic stirring bar. Initially stirring is so vigorous that bubbles of air are whipped into the oxalic solution.

When about a third of the Solution 2 is added, the material will be so viscous that it will no longer vortex. The agitator is stopped momentarily to clear the solution of air bubbles, then stirring is resumed. The cobalt is introduced about ⅛" above the magnetic stirring bar. As the remainder of the cobalt is added the mix becomes a very viscous suspension. A 200 ml sample of this suspension is filtered on a Buchner vacuum filter using Whatman #5 filter paper.

The crystals were washed with 400 ml of hot water and then with 200 ml of acetone. The crystals lift off the filter paper as a paper-like sheet. These crystals were difficult to photograph and measure, but size data are included in Table

2, based on scanning electron microscope (SEM) pictures taken at 10,000X. A 5 g sample of these crystals was coated with 0.113 g of Resymide in isopropyl alcohol. The alcohol was evaporated and the powder placed in a tube furnace sample boat. It was reduced in 20% hydrogen and 80% nitrogen until the CO_2 in the vent gas dropped below 0.04%. It was cooled to 20°C under nitrogen and then exposed to 3% oxygen in nitrogen for 16 hours. The sample was next exposed to dry air for two hours and then analyzed magnetically. Data are given in Table 2.

Table 2

	Control	Ex. 7
Seeded added, ml	none	(all seed)
Coercive force, Oe	805	1,087
Squareness	0.75	0.74
Specific magnetic moment	115	106
Length, average, μ	8.7	2.5
Diameter, average, μ	1.5	0.19
L/D ratio, average	6.4	13.3

Fe-Ni-Co Magnetic Powder for Tape

R.J. Deffeyes; U.S. Patent 4,054,530; October 18, 1977; assigned to Graham Magnetics Incorporated provides a cobalt-based magnetic powder of excellent magnetic properties and excellent resistance to chemical deterioration, even when the powder is in the submicron size range.

The magnetic properties are insensitive to small variations in metallic ratios as may occur from batch-to-batch (and indeed from particle-to-particle in a batch) under production conditions.

The metallic ratios involved are defined as follows on a weight percentage basis: iron, 10 to 33%; nickel, 8 to 30%; and cobalt, 50 to 82%.

In the more advantageous embodiments the nickel is raised to at least about 12% level and no more than about 24% iron is utilized. Moreover, it is found that, in many applications, manganese can be substituted for all or some of the nickel; the resultant powders have relatively low magnetic moment but they retain most of the other desirable properties.

The nickel-bearing powders of the process are characterized by the following specific properties:

Coercive force—500 to 900 Oe
Specific magnetic moment—120 to 190 emu/g
Squareness—0.5 to 0.85
Corrosion resistance—Ability to maintain at least 50% of the specific magnetic movement (sometimes called the sigma value) when exposed to 50% relative humidity at 166°F for 100 hours.

Example: The following procedure is particularly useful in providing acicular particles for use, say, in magnetic tape. A seed crystal of cobalt oxalate was prepared by adding to a solution of oxalic acid (Solution 1) a cobalt-bearing solution (Solution 2).

Solution 1—

25.2 g of oxalic acid dihydrate
150 g of a clear polysaccharide syrup digestible to dextrose (Karo)
150 g of denatured ethyl alcohol
150 g of deionized water.

Solution 2—

25 g of cobaltous acetate tetrahydrate
25 g of citric acid, USP Grade
50 g of Karo syrup
50 g of denatured ethyl alcohol
50 g of water.

Solution 2 is added slowly to the Solution 1 while the resultant mixture is stirred magnetically in a baffled Erlenmeyer-type flask of a 1 liter capacity. As the addition proceeds, a precipitate forms and a resultant suspension becomes so thick that the stirring bar may stop and require restarting. In fact, it is sometimes advantageous to stop the bar and remove air bubbles therefrom to decrease the resistance of the bar to rotation. As the mix becomes even thicker, it is desirable to lower the tip of a funnel through which the addition is being made to a point about ¼" above the stirring bar. The resultant suspension is used as a seed crystal material in the following procedure:

A quantity of 25.2 g oxalic acid dihydrate is dissolved in 400 ml of deionized water to form a third solution. To this third solution is added 100 ml of the seed crystal suspension described above and which results from the addition of Solution 2 to Solution 1. The result of this addition is the formation of a seeded oxalic acid solution.

A fourth solution is then prepared as follows:

18 g of cobalt nitrate hexahydrate
5 g of ferrous sulfate heptahydrate
5 g of nickel nitrate
150 g of water.

This fourth solution is added to the seeded oxalic acid solution with vigorous stirring. The addition takes about 6 minutes. After the addition of the solution is complete, the resulting mix is agitated for an additional 30 minutes before being filtered.

The filter cake is washed with 600 ml of hot water, then 200 ml of acetone.

After being air dried at about 80°F the resultant metal oxalate was wetted with some isopropanol treated with a 10% solution of a resinous polymeric material, a polyamide resin (Resymide 1125). The amount of resin was selected to equal 7% by weight of the metal content of the oxalate. This type of treatment has been found to be useful in the thermal reduction of organometallic salts to ferromagnetic powders. Some ethanol was added to the oxalate-resin mixture to make a thin slurry and facilitate distribution of resin on the surfaces of the oxalate. The oxalate was dried by evaporation.

The reduction was carried out by placing the powder in sample boats containing eight different compartments. The sample boats are sealed in tubes having gas inlet and gas outlet ports, then placed in an oven, with conduits to the outlet and inlet ports extending from the oven to a gas-supply source and a gas exhaust system, respectfully. The tubes are purged with N_2; then the oven temperature is set for 650°F and a gas mixture of 20% hydrogen and 80% nitrogen is passed through the tubes at the rate of 4 cc/min. Within about 45 minutes, CO_2 evolution starts.

After the rate of CO_2 evolution peaks and drops to less than 0.01% by volume of the off-gas, the tube is purged with nitrogen, and removed from the furnace. After cooling to room temperature, this usually takes an hour or so, a mixture of 3% oxygen and 97% nitrogen is passed through the tube. This gas mixture is usually allowed to run for several hours but the exact time, or even the exact composition, is not particularly critical.

Then the reactor is purged with air for 2 hours and opened. The samples are removed and are ready for analysis.

Specific magnetic moment—138 emu/g
Coercive force—659 Oe
Squareness—0.84.

When samples of the powder were exposed for 100 hours at 166°F and 50% relative humidity, 83% of the original specific magnetic moment was present after the exposure.

The product was an acicular powder having an average particle size of about 0.5 μ by about 0.15 μ. It had a metal content of about 18 parts nickel, 19 parts iron, and 64 parts cobalt.

Samples of this material were stored in a humidity oven at 66°C and 50% relative humidity for 160½ hours. After this exposure, the specific magnetic moment was found to be 112 emu/g. The material comprises about 18% nickel, 18% iron and 64% cobalt and has an extraordinary combination of corrosion-resistance and magnetic properties.

Organic Medium to Minimize Agglomeration

J.E. Ehrreich and A.R. Reti; U.S. Patent 3,843,349; October 22, 1974; assigned to Graham Magnetics Incorporated have found that hyperfine, uniformly sized metal particles are reduced from decomposable salts by immersing the salt particles in an organic medium prior to reduction and keeping them in the medium until reduction is completed. The medium keeps the particles from sintering together so that the resulting metal powder is characterized by minimal interparticle agglomeration. The organic medium is then largely removed so that it does not degrade the characteristics of the metal particle.

Since the separation medium is an organic material, the reactant particles are immersed in or thoroughly mixed with this material which thus coats the individual particles. During reduction, the coatings may be largely removed or deteriorated by thermal decomposition, evaporation, sublimation or the like processes, but their residues serve to provide the desired inter-particle separation. In many cases, the coating residues will not materially interfere with the desired use of the metallic particles. However, where their removal is preferred, they can largely be washed away with suitable solvents. In appropriate cases, magnetic techniques can also be used at this point to aid in separation.

In certain cases some organic coating material may remain on the particles. This material may bind metallic particles together in random orientations, which tends to detract from the desirable characteristics of the metallic powder, particularly in magnetic applications. However, the extent of this effect is much less than the corresponding degradation resulting from inter-particle sintering when this process is not practiced. Indeed, the degradation products of the coating material can be of material benefit even during the advanced stages of the reduction process by radiating heat throughout the porous mass of powder being subjected to reduction.

Polymers and other organic chemicals that have been used as the coating material include such film-forming materials as silicone oil, silicone polymer, silanes, epoxy resin, polyacrylic esters, and polyurethane. Usually, the organic material is applied in a suitable solvent to ensure complete coverage of the oxalate crystals.

Example 1: A quantity of 238 g of $CoCl_2$ is dissolved in a container to form a 1 liter aqueous solution. Then 100 g of sodium hydroxide is dissolved in a second container to form a second 1 liter aqueous solution. The cobalt chloride and sodium hydroxide solutions are mixed together, using a magnetic agitator bar, in a 4 liter beaker for three minutes to form a cobalt hydroxide precipitate. Then 135 g of oxalic acid are dissolved in a separate container to form a 1½ liter aqueous solution. The oxalic acid solution is mixed thoroughly into the cobalt hydroxide precipitate for 5 minutes to form a cobalt oxalate precipitate.

The resultant mixture is filtered in a Buchner funnel and washed several times with water. Then it is rinsed several times with acetone and air dried. The resulting cobalt oxalate particles are acicular (i.e., needle-shaped). Following this, 0.05 g of silicone polymer oil (G.E. RTV910 Diluent) is mixed with 0.95 g of the acicular cobalt oxalate particles using sufficient tetrahydrofuran to insure uniform coating of the silicone oil on the particles. Next, the coated particles are air dried. Then the coated particles are charged into a 1″ diameter glass tube in a tube furnace, heated to 360°C in a hydrogen atmosphere, and held at this temperature for 1 hour, thereby reducing the metal salt to cobalt.

The material is then cooled to room temperature while still in a hydrogen atmosphere, and subjected to a two-minute argon purge. Next the product is washed several times with acetone. A magnet is used to effect separation from the wash, and the separated material is air dried. The resulting product is a strongly magnetic powder having a coercivity of 705 Oe, a sigma value of 82 emu/g, and a squareness figure of 0.40. The particles are composed mainly of cobalt, and have little or no silicone oil on their surfaces. Also, the cobalt particles are acicular with an average particle size of 0.3 μ in diameter by 1 μ in length.

Example 2: A quantity of 238 g of $CoCl_2 \cdot 6H_2O$ is dissolved in a container of 500 ml of denatured alcohol and 500 ml of water. Also 80 g of sodium hydroxide is dissolved in a separate container in 500 ml of water and 500 ml of denatured alcohol. Then the solution is cooled to room temperature. Following this, the cobalt chloride and sodium hydroxide solutions are mixed together in a 4 liter beaker in a magnetic mixer for 5 minutes. During this period, a cobalt hydroxide precipitate forms in the beaker. In a separate beaker, 135 g of oxalic acid are dissolved in 750 ml denatured alcohol and 750 ml of water. The oxalic acid solution is mixed into the cobalt hydroxide precipitate for 5 minutes to form a cobalt oxalate precipitate.

The precipitate is filtered in a Buchner funnel and the oxalate cake is washed with 2 liters of acetone. Then the oxalate is redispersed in 1 liter of acetone and then filtered. The acetone-wet cake is mixed with 75 g of 10% solution of Estane 5702, a polyurethane-type rubber, manufactured by B.F. Goodrich & Co. in tetrahydrofuran. This mix is spread on a polyethylene sheet to air dry. Coated oxalate salt is then charged into an aluminum boat which is approximately 16″ long and 2½″ wide and 1.9″ high. The interior of this boat is separated into four elongated compartments by three equally spaced and centered fins.

The filled boat is placed in a sealed 2⅞″ diameter stainless steel tube. The tube in turn is placed in a 3″ diameter 24″ long, tube furnace so that the filled aluminum boat is equally distant from each end of the furnace. Argon is fed through the stainless steel tube at a rate of 800 cc/min. At the same time, heat is applied to the stainless steel tube and in 1 hour the contents of the boat reaches 360°C. After 30 minutes, the atmosphere is changed to hydrogen, which is passed through the tube at a flow rate of 800 cc/min for a period of two hours. The stainless steel tube with the filled aluminum boat therein is taken out of the furnace and cooled externally with ice while still maintaining the boat in a hydrogen atmosphere and the contents allowed to reach room temperature.

Finally, a 10-minute argon purge is passed through the tube at a rate of 800 cc/min. The contents of the boat are then transferred to a polyethylene bag filled with argon. Care is taken to prevent contact with air, because at this point the material is extremely pyrophoric. After four days, two pinholes are pierced in the bag so that the contents of the bag are exposed to a very slowly increasing oxygen concentration. The product is then removed from the bag after four more days.

The resulting product obtained is ferromagnetic metallic powder having a coercivity of 813 Oe, a sigma value of 86 emu/g and a squareness characteristic of 0.46. The powder is nonpyrophoric and is composed of acicular cobalt particles having cobalt oxide coatings and exhibiting minimal interparticle sintering.

Example 3: A quantity of 19.9 g $FeCl_2 \cdot 4H_2O$ is dissolved in 50 ml H_2O. Separately, a quantity of 8.0 g sodium hydroxide is dissolved in 50 ml denatured alcohol and 50 ml H_2O. These two solutions are mixed together in a magnetic stirrer for 2 minutes to form an iron hydroxide precipitate. Then a solution of 13.5 g oxalic acid in 75 ml denatured alcohol and 75 ml H_2O is mixed into the precipitate for 2 minutes. The resultant precipitate is filtered in a Buchner funnel and washed several times in acetone and air-dried. Then, 1 g of the material in an open bottle is coated with 5% Estane 5702 polyurethane as described in Example 2 and charged into a 1" diameter glass tube in a tube furnace along with a second 1 g quantity of the oxide material uncoated.

Both samples are heated for 30 minutes at 360°C in argon. They are then held at this same temperature in hydrogen for 90 minutes. They are then cooled to room temperature under hydrogen followed by a 2 minute argon purge. The resultant product from the first bottle is ferromagnetic and composed of small, unagglomerated iron particles. The powder has a coercivity of 690 Oe, a sigma value of 38 emu/g and a squareness factor of 0.47. The mass of particles from the second bottle (uncoated) has a coercivity of 560 Oe, a sigma value of 50 emu/g and a squareness of 0.36.

Air-Stable Magnetic Materials

J.G. Smeggil and R.J. Charles; U.S. Patents 3,856,580; December 24, 1974 and 3,856,581; December 24, 1974; both assigned to General Electric Company describe a process which comprises providing particles of magnetic transition metal-rare earth alloy, heating an organometallic compound to decompose the compound and produce a metal vapor, and contacting the metal vapor with the particles to deposit a coating of metal thereon. These metal-coated alloy particles can be magnetized and pressed into a compact to form a magnet. They can also be magnetized before or after distribution in a nonmagnetic matrix and the resulting mixture pressed into a compact useful as a magnet; these metal-coated particles are annealed in air to increase their intrinsic coercive force by at least 10%.

Briefly stated, this process comprises providing particles of a magnetic transition metal-rare earth alloy, heating an organometallic compound to decompose the compound and produce a metal vapor, contacting the metal vapor with the particles to deposit a coating of metal thereon and heating the coated alloy particles at a temperature ranging from about 50° to 200°C for a period of time sufficient to increase their intrinsic coercive force by at least 10%.

In this process a magnetic transition metal-rare earth alloy, e.g, TRE, where T is a transition metal and RE is a rare earth metal, is used in particle form. The transition metal is selected from the group consisting of cobalt, iron, nickel, manganese and alloys thereof.

The rare earth metals useful in this process are the fifteen elements of the lanthanide series having atomic numbers 57 to 71 inclusive. The element yttrium (atomic number 39) is commonly included in this group of metals and, in this discussion, is considered a rare earth metal. A plurality of rare earth metals can also be used to form the process intermetallic compounds, which, for example may be ternary, quaternary or which may contain an even greater number of rare earth metals as desired. Mischmetal, an abundant common alloy of rare earth metals, is particularly advantageous.

Representative of the cobalt-rare earth compounds useful are cobalt-cerium, cobalt-praseodymium, cobalt-neodymium, cobalt-promethium, cobalt-samarium, cobalt-europium, cobalt-gadolinium, cobalt-erbium, cobalt-thulium, cobalt-ytterbium, cobalt-lutecium, cobalt-yttrium, cobalt-lanthanum and cobalt-mischmetal. Examples of specific ternary compounds include cobalt cerium-praseodymium-mischmetal.

Transition metal-rare earth intermetallic alloys or compounds exist in a variety of phases and each phase may vary in composition. A material substantially comprised of the T_5RE single phase is particularly preferred since this phase has shown the most desirable combination of magnetic properties.

In this process, an organometallic compound is used which can be solid, liquid or gas at room temperature and which decomposes at temperatures lower than 500°C to produce a metal vapor. The metal vapor is contacted with the process TRE alloy powder to deposit metal thereon to form a continuous coating of metal which protects the powder from the atmosphere. The coating process yields a coherent substantially uniform layer of metal and is accomplished in an atmosphere in which the reactants are inert. Typical inert atmospheres which are suitable for the coating process include argon, nitrogen or a vacuum. No water vapor or oxygen gas is present to degrade the magnetic properties of the alloy materials.

There are a number of useful organometallic compounds which decompose at temperatures below 500°C. Typical of these is triisobutylaluminum as a source of aluminum. Specifically, the metal coating of aluminum can be deposited according to the following reactions:

$$Al[CH_2CH(CH_3)_2]_3 \xrightarrow{100°C} AlH[CH_2CH(CH_3)_2]_2 + CH_2C(CH_3)_2$$

$$AlH[CH_2CH(CH_3)_2]_2 \xrightarrow{250°C} Al + 1.5H_2 + 2CH_2C(CH_3)_2$$

This operation must be accomplished at reduced pressures because the triisobutylaluminum cannot be successfully distilled above 10 mm Hg.

For the vapor deposition of copper phenylcopper C_6H_5Cu is useful. It affords advantages similar to those listed for the Al deposition, and in addition, it takes place at a very low temperature.

In the following examples the intrinsic coercive force of each sample was measured at room temperature. Specifically, a specimen of the powder was prepared for magnetic measurement by introducing it into a body of molten paraffin wax in a small glass tube and cooling the wax in an aligning magnetic field of 20,000 Oe to align the particles along their easy axis until the paraffin solidified. The intrinsic coercive force of the sample was then measured after applying a magnetizing field of 30,000 Oe. All parts and percentages used herein are by weight unless otherwise noted.

Example 1: A sintered body of compacted CoSm alloy powder was ground to a powder using a jaw crusher and a jet mill. The alloy powder was comprised substantially of Co_5Sm phase and a minor amount of Co_7Sm_2 phase. Approximately 5 g of the alloy powder having a size of −325 mesh (U.S. Standard Screen Size), e.g., an average particle size of about 6 μ, were placed in a U-tube along with about 1 g of copper acetylacetonate, present as the hydrate $Cu(CH_3COCHCOCH_3)_2 \cdot 2H_2O$, and in the powder form which was calculated to yield copper in an amount of about 2% by weight of the alloy powder.

A slow stream of argon gas was passed through the U-tube across the mixture which was mixed by moving a magnet beneath it. After a few minutes, a period of time considered suitable for the removal of the majority of oxygen present by the argon stream, the mixture was first gently heated with a Meeker burner and occasionally during the entire heating process a magnet was used to stir up the mixture to assist both in driving off the water of hydration and in obtaining a more uniform metal coating on the alloy particles. The temperature of the mixture was then raised slowly to about 400°C to accelerate the decomposition of the copper acetylacetonate.

After a minute or two a brown coating was observed being deposited on the alloy particles and on the walls of the U-tube. Heating was continued for several more minutes then stopped. The coated alloy particles were allowed to cool to room temperature with argon continuing to flow over them. After the powders were cool a magnet was used to separate the coated alloy particles from a small amount of nonmagnetic material present, presumably unreacted copper acetylacetonate and elemental Cu metal.

The alloy particles were examined under an optical microscope and by scanning electron microscopy. They appeared to be enveloped by a continuous uniform brown coating of copper metal.

The intrinsic coercive force, H_{ci}, of a portion of these copper-coated particles, as well as of a portion of the uncoated particles of the same composition and size, was determined and the results are shown in Run No. 1 of Table 1.

Table 1

Run No.	Time (hr)	H_{ci} Uncoated Particles	H_{ci} Cu-Coated Particles
1	0	7,550	9,960
2	16	6,360	10,660
3	24	6,060	10,730
4	40	6,150	10,930
5	112	6,260	11,450

As shown by Table 1, the copper-coated particles had an intrinsic coercive force initially higher than that of the uncoated particles, and after being heated at 92°C in air for periods of time as long as 112 hours, the intrinsic coercive force of the copper-coated particles improved significantly whereas that of the uncoated particles deteriorated significantly. This indicates that the copper coating was continuous and provided an effective barrier to the atmosphere to prevent atmospheric oxygen from reacting with and degrading the magnetic properties of the Co_5Sm powder. In addition, the significant increase in intrinsic coercive force of the copper-coated particles after heating may be due to removal of sites for reverse domain initiation on the now protected surface of the alloy particles.

Example 2: In this example the cobalt-samarium alloy powder used was substantially the same as that set forth in Example 1. The intrinsic coercive force of portions of the powder was determined before and after annealing in an air oven at 150°C for 30 minutes and the results are shown in Run No. 6 of Table 2.

Table 2

Intrinsic Coercive Force, H_{ci} (oersteds)	Run No. 6 Control	Run No. 7 1% Cr	Run No. 8 5% Cr	Run No. 9 2% Al
Original	7,080	7,420	6,420	8,250
½ hr at 150°C	5,210	8,350	7,650	–
16 hr at 175°C	–	–	–	9,900

Portions of the alloy powder were coated with metal substantially as set forth in Example 1. Specifically, chromium hexacarbonyl, $Cr(CO)_6$, in an amount of 4% by weight of the alloy powder, was admixed therewith and the mixture placed in a U-tube where it was maintained under a stream of argon and substantially continuously mixed with a magnet. It was calculated that in this mixture chromium hexacarbonyl would yield chromium in an amount of 1% by weight of the alloy powder. At a temperature of 400°C the chromium hexacarbonyl decomposed, and after about a minute, a silvery coating was observed on the alloy powder and on the walls of the tube. Heating and mixing was continued for about 10 additional minutes to insure complete coating of the particles with chromium and then stopped.

The chromium-coated alloy particles were cooled to room temperature under argon. A magnet was then used to separate the coated particles from a small amount of nonmagnetic material present. The intrinsic coercive force of a portion of the coated powder was determined and is shown in Run No. 7. The remaining coated powder was heated in an air oven at 150°C for 30 minutes, then cooled to room temperature and its intrinsic coercive force determined as shown in Run No. 7 of Table 2.

The procedure used in Run No. 8 was the same as that of Run No. 7 except that chromium hexacarbonyl was used in an amount of 20% by weight of the alloy powder which was calculated to yield chromium in an amount of 5% by weight of the alloy powder.

In Run No. 9 triisobutylaluminum was used in an amount which coated the alloy particles with aluminum in an amount of about 2% by weight of the alloy powder. The coating procedure differed from Run No. 7 in that a double U-tube was used wherein the triisobutylaluminum was placed in one U-curve and the alloy powder was placed in the second U-curve. The triisobutylaluminum was decomposed at a temperature of 250°C and the resulting aluminum vapor was carried by the argon stream into contact with the alloy powder where it condensed on the powder which was being continuously mixed with a magnet to insure the deposition of a continuous uniform coating of aluminum.

The metal-coated particles of Runs 7-9 were examined under an optical microscope and a scanning electron microscope. They appeared to be completely and substantially uniformly coated by metal.

The intrinsic coercive forces of the uncoated and coated powders, before and after annealing in air, are given in Table 2.

As illustrated by Table 2, the intrinsic coercive force of Run No. 6, the uncoated control powder, deteriorated significantly after 30 minutes at 150°C whereas the intrinsic coercive force of the metal coated particles of Run Nos. 7-9, which illustrate the process, increased significantly after the annealing treatment in air. This indicates that the process metal-coated particles provided an effective barrier to the atmosphere to prevent atmospheric oxygen and moisture from reacting with and deteriorating the magnetic properties of the process alloy powder.

In further work by *J.G. Smeggil and R.J. Charles; U.S. Patent 3,856,582; December 24, 1974; assigned to General Electric Company*, the fabrication of matrix bonded transition metal-rare earth alloy magnets is described. Briefly, this comprises admixing particles of a transition metal-rare earth alloy with an organometallic compound which decomposes at a temperature below 500°C and pressing the mixture to form a green body. The green body is heated to decompose the organometallic compound to produce a nonmetallic product and a metal vapor. The metal vapor deposits an interconnecting continuous coating of metal on the exposed surfaces of the pressed alloy particles thereby preventing penetration by the atmosphere, and the nonmetallic product is outgassed from the body leaving the resulting coated compact porous.

Organometallic Lubricants for Sintering

In work related to the three preceding patents *J.G. Smeggil; U.S. Patent 3,853,640; December 10, 1974; assigned to General Electric Company* describes lubricants which will not have any significant deteriorating effect on the sintered product and which can be selected to enhance the magnetic properties of the sintered product. In accordance with this process an organometallic compound is used as the lubricant for die-pressing magnetic transition metal-rare earth alloy powders.

The organometallic compound is a lubricant for the powder being pressed and it is inert with respect to the powder being pressed. Specifically, the organometallic compound is a solid or liquid at room temperature and it is air-stable at room temperature. It thermally decomposes at a temperature below 500°C to yield a metal vapor and a nonmetal specie or species of decomposition, at least one of which is organic, and which are stable to interaction with the pressed transition metal-rare earth alloy powders at the temperatures needed for this decomposition to occur.

Specifically, the decomposition of the organometallic compound below 500°C produces a metal vapor which condenses metal on the exposed surfaces of the pressed powder and also produces an organic specie, or nonmetal species at least one of which is organic, which diffuse out of the body. No water vapor or oxygen is present to degrade the magnetic properties of the pressed powder.

The particular metal deposited by the decomposition reaction can be chosen to fulfill a number of purposes. For example, the metal remanent can be inert and not partake or interact in any significant manner, either chemically or magnetically, with the transition metal-rare earth alloy material, e.g., Zr. Or else the metal can be chosen to coat the transition metal-rare earth alloy powders and offer protection against later oxidation. Or the metal can be chosen to change the chemical composition of the transition metal-rare earth alloy material, e.g., Co deposited from cobalt acetylacetonate and cobalt-rare earth alloy pressed material in some desired manner.

Generally, the organometallic compound is used in an amount ranging from about 0.5 to 10% by weight of the particles of transition metal-rare earth alloy to be pressed. The specific amount of organometallic compound depends largely on the lubricating characteristics of the organometallic on the particular properties desired in the final sintered product as imparted by the particular metal deposited by the decomposition reaction. Generally, such a metal is deposited in the sintered product in an amount ranging from a detectable amount to about 2% by weight of the transition metal-rare earth alloy.

Thus, this process comprises admixing an organometallic compound and particles of a transition metal-rare earth alloy to form a substantially intimate mixture, the organometallic compound decomposing at a temperature below 500°C, pressing the mixture to form a green body, heating the green body to decompose the organometallic compound to yield a metal vapor and a nonmetal specie of decomposition, the metal vapor depositing metal in the body and the nonmetal specie of decomposition diffusing out of the body, and sintering the body at a temperature of at least 900°C to produce a sintered product having a density of at least 87% of theoretical.

Surface Nitriding

In the course of the reduction of metal salts powder such as a ternary oxalate of Fe, Co and Ni by heating in a reducing atmosphere at a temperature of about 300°C for a period of 2 to 6 hours, *N. Takahashi, K. Chiba, M. Nakamura, S. Ozaki, H. Isono, T. Naruse, T. Mishihara and I. Fukushima; U.S. Patent 3,904,448; September 9, 1975; assigned to Victor Company of Japan, Ltd., Japan* introduce a nitrogen-containing compound gas such as NH_3 into the reaction system atmosphere at a temperature of 250° to 400°C for a period of 15 to 60 minutes in order to form a layer of metal nitride on the surface of the powder. The nitride layer protects the resulting alloy powder against air-oxidation and deterioration in the magnetic properties.

Example 1: A ternary oxalate of iron, cobalt and nickel prepared by a conventional coprecipitation method with the composition of 50, 48 and 2% by molar ratio, respectively, was used as a starting material. 50 mg of the oxalate powder was put into a boat and placed in a reaction tube, and heated at 300°C for 5 hours in a 0.5 l/min stream of hydrogen. Then ammonia gas was introduced into the tube mixed with the hydrogen stream in an amount of about 10% of hydrogen, continuing the 300°C heating for another 1 hour.

The obtained ternary alloy powder had the saturation magnetization value of 211 emu/g and coercive force of 650 Oe.

Example 1A (Prior Art): The procedures of Example 1 were repeated except that the heating was carried out in the stream of hydrogen gas alone throughout the total period of 6 hours.

Example 2: Example 1 was repeated except that the heating temperature was 290°C.

Example 2A (Prior Art): Example 1A was repeated except that the heating temperature was 290°C. The saturation magnetization and coercive force of the alloy powders prepared in the above examples are tabulated below.

	Temperature (°C)	Heating Time ..... (hr) ..... in H_2	in NH_3	Saturation Magnetization (emu/g)	Coercive Force (Oe)
Ex. 1	300	5	1	211	650
Ex. 1A	300	6	0	199	640
Ex. 2	290	5	1	202	630
Ex. 2A	290	6	0	160	640

The effect of the nitrification according to the process on the saturation magnetization will be apparent from the table. The coercive force was substantially constant and satisfactory in these examples because of moderate reaction temperatures. If a higher temperature is employed to raise the saturation magnetization value above the result of Example 1A according to the prior art, a decrease in the coercive force is inevitable.

In the thus formed nitride layer, iron nitride takes the most important part, and it can exist as various kinds of structure ranging from Fe_8N to Fe_3N. The magnetic properties of iron nitride do vary depending on the structure; e.g., Fe_8N is ferromagnetic. It is, therefore, possible to obtain various properties of alloy powders, ranging from nonmagnetic to ferromagnetic ones, from the same starting material by the control of the extent of the nitrification.

The nitrification in the method of this process is affected by the reaction temperature, the duration of the reducing heating before introduction of the nitrogen-containing compound gas, the flow rate of that gas and the duration of the nitrifying heating. The heating temperature should be between about 250° and 400°C and preferably above 300°C. Both the reduction and nitrification are difficult to proceed at below 250°C, but excessive nitrification and/or sintering of the powder may result at above 400°C.

To achieve the initial reduction of the metal salts, at least a part of the heating in a reducing atmosphere should be carried out prior to the start of, e.g., the ammonia gas supply. Hence, it is preferable to commence the nitrifying heating after 2 to 6 hours of the reducing heating. When the initial reducing heating time exceeds 6 hours, the almost completely reduced particles cannot be easily nitrified, and sintering of the particles may be accompanied. From this view point, the nitrogen-containing compound gas is preferably mixed with the reducing gas rather than used alone after completion of the initial reduction. The mixing ratio of the nitrogen-containing compound gas to the reducing gas should be at least 10% by volume and preferably from 20 to 80%. The minimum heating time for the nitrification is considered about 15 minutes.

Aluminum Crystalline Oxalate

In order to produce a magnetic alloy powder having a large coercive force and high residual magnetization *N. Takahashi, M. Nakamura, S. Ozaki, H. Isono and I. Fukushima; U.S. Patent 3,925,114; December 9, 1975; assigned to Victor Company of Japan, Ltd., Japan* allow Fe, Co and/or Ni ions to react with oxalic acid, in the presence of Al ion, to precipitate a crystalline oxalate powder containing Al and at least one of the above three ferromagnetic metals.

The amount of Al is preferably from 1 to 20 mol percent of the total metal mols. The obtained crystalline oxalate powder is then reduced into a magnetic alloy powder in the usual way, e.g., by H_2, preferably at a temperature range between about 300° to about 400°C.

The introduction of Al into a starting solution is adaptable for the preparation of either any individual metal powder of Fe, Co and Ni with aluminum or any of their mixed alloys.

In this process, water-soluble salts such as chlorides of Fe, Co and Ni are dissolved in water together with a similar salt of Al. The aqueous solution is adjusted to a suitable acidity, preferably to about pH 1.5, and is then mixed and stirred with a solution of oxalic acid $C_2H_2O_4 \cdot 2H_2O$ in a water miscible solvent, preferably in acetone containing a small amount of toluene. The stirring is continued usually for about 10 minutes at room temperature or at a slightly elevated temperature to precipitate crystalline multicomponent Fe, Co, Ni and Al oxalates. The precipitate is considered not a simple mixture of four kinds of oxalates, but a crystalline four-component oxalate.

A crystalline powder of this oxalate obtained through filtration, washing and drying of the precipitate is heated for several hours in a reducing atmosphere, usually in a stream of hydrogen, at a temperature ranging from about 300° to 400°C to give a magnetic alloy powder. The amount of Al in the aqueous solution should be at least about 1 mol percent of the total metals. Existence of only a trace of Al, e.g., up to 0.1 mol percent has little effect on the magnetic properties of a resulting alloy powder.

Example: A solution was prepared by dissolving 25.2 g of $C_2H_2O_4 \cdot 2H_2O$ in 200 ml of acetone followed by the addition of 5 ml of toluene. The temperature of the solution was kept at 35°C in a constant temperature bath.

Meanwhile, 8.8 g of $FeCl_2 \cdot nH_2O$, 8.6 g of $CoCl_2 \cdot 6H_2O$, 0.4 g of $NiCl_2 \cdot 6H_2O$ and 0.2 g of $AlCl_3 \cdot 6H_2O$ were dissolved in 75 ml of water. The aqueous solution was adjusted to pH 1.5 and was heated to 35°C.

Then, the aqueous solution was poured into the oxalic acid solution all at once with vigorous stirring, and stirring was continued for 10 minutes, keeping the liquid temperature at 35°C, to allow the metal ions in the solution to completely react with the acid. The resulting precipitate was filtered, washed and dried to give a crystalline powder of mixed Fe, Co, Ni and Al oxalates.

The obtained oxalate powder was reduced into an alloy powder under the following conditions: 80 mg of the oxalate powder was put in a platinum boat and was heated in a 0.5 liter per minute stream of hydrogen at 310°C for 6 hours.

The resulting alloy powder showed the following magnetic properties:

Coercive force	880 Oe
Saturation magnetization	191 emu/g
Residual magnetization	64 emu/g

BORON AND HYPOPHOSPHITE REDUCTION

Iron and Lead

M. Aonuma and Y. Tamai; U.S. Patent 4,007,072; February 8, 1977; assigned to Fuji Photo Film Co., Ltd., Japan have prepared a powdery ferromagnetic metal material for a magnetic recording medium containing at least Fe, B and Pb as the main components.

The powdery ferromagnetic metal material is produced by reducing in the presence of a lead compound a metal salt capable of forming a ferromagnetic substance and containing at least Fe in an aqueous solution thereof using a borohydride compound and/or a derivative thereof.

It can also contain ferromagnetic metals such as Co and Ni, or, as an additional element, at least one metal selected from Al, P, Sc, Ti, Cr, Mn, Cu, Zn, Ga, As, Se, Sr, Y, Zr, Nb, Mo, Rh, Pd, Ag, Sn, Te, La, Ce, Pr, Nd, Pm, Sm, W, Re, Os, Ir, Pt, Au, Hg, Bi, and the like.

A most preferred composition comprises about 30 to 95% by weight of Fe, about 0.5 to 20% by weight of B, about 0.1 to 20% by weight of Pb, and about 0 to 70% by weight of Co and/or Ni.

In this process, B is contained in the borohydride or derivative thereof used as a reducing agent, and it is introduced into the powder in the reduction. Borohydride compounds such as borane, borazane, borohydride, sodium borohydride, potassium borohydride, dimethylamino borane, diethylaminoborane, and the like, or the derivatives thereof can be used.

The concentration of the borohydride compound or derivative thereof ranges preferably from about 0.0002 to 10 mols/l, and it has been found that the molar ratio of the reducing agent to the metal ion (reducing agent/metal ion) desirably ranges from about 0.1:1 to 5:1.

The material has excellent antioxidation properties. It can be handled with ease during production because it is not combustible in air. Furthermore, the powder is excellent in orientation in magnetic field when used in preparing a tape because the excellent dispersibility, and particularly, a high squareness ratio (B_R/B_S) can be obtained.

Example 1:

	Quantity (mol/l)
M_1 liquid	
Ferrous chloride	0.36
Cobalt chloride	0.04
Lead acetate	0.002
R_1 liquid	
$NaBH_4$	1.60

Note: Water used as the solvent in both the M_1 and R_1 liquids, hereinafter are the same.

80 parts of the M_1 liquid was charged to a nonmagnetic vessel and a direct current magnetic field of 1,000 Oe was applied to all the vessel. 20 parts of the R_1 liquid was added to the M_1 liquid over a 10 second period while stirring to effect the reaction, and thus a ferromagnetic powder was obtained. The powder was washed, and then replaced with acetone, and dried in a drying oven at 60°C, and a ferromagnetic powder was obtained. This sample is designated No. P-1.

The following composition containing powder No. P-1 as produced above was charged to a ball mill and sufficiently mixed and dispersed to thereby prepare a magnetic coating composition.

	Quantity (parts)
Ferromagnetic powder (No. P-1)	300
Polyester polyol*	40
Polyisocyanate compound**	20
Nitrocellulose***	20
Chromium oxide (Cr_2O_3; particle size about 0.5 μ)	5
Lecithin	3
Methyl ethyl ketone	450
Methyl isobutyl ketone	400

*Condensate of adipic acid and butanediol, having terminal hydroxy groups and a molecular weight of about 1,800.
**Desmodur L; 75 weight percent ethyl acetate solution; adduct of 3 mols of toluene diisocyanate and 1 mol of trimethylol propane.
***Molecular weight of about 800.

The composition was coated on one side of a polyethylene terephthalate film having a thickness of 25 μ in a dry thickness of 5 μ while applying a magnetic field and dried by heating. The thus prepared wide magnetic web was calendered and slit to a ½" width, and thus a video tape was obtained. The surface property of the video tape was quite excellent. This sample is designated No. T-1.

Example 2:

	Quantity (mol/l)
M_2 liquid	
Ferrous chloride	0.40
Lead acetate	0.02
Sodium tartrate	0.10
R_2 liquid	
$NaBH_4$	1.60

The procedure of Example 1 was repeated using the M_2 and R_2 liquids to thereby obtain a ferromagnetic powder. This sample is designated No. P-2. Also, the procedure of Example 1 was repeated using the powder (No. P-2) to thereby prepare a video tape. The surface property of the thus produced video tape was quite excellent. This sample is designated No. T-2.

Example 3:

	Quantity (mol/l)
M_3 liquid	
Ferrous chloride	0.36
Cobalt chloride	0.04
Lead acetate	0.01
Zinc chloride	0.005
R_3 liquid	
$NaBH_4$	1.60
Diethylaminoborane	0.20

The procedure of Example 1 was repeated using the R_3 and M_3 liquids to thereby obtain a ferromagnetic powder. This sample is designated No. P-3. Also, the procedure of Example 1 was repeated using the powder (No. P-3) to thereby prepare a video tape. The appearance of the thus produced video tape was substantially the same as that of Example 1. This sample is designated No. T-3.

Comparison Example:

	Quantity (mol/l)
Mc_1 liquid	
Ferrous chloride	0.36
Cobalt chloride	0.04
Rc_1 liquid	
$NaBH_4$	1.60

A ferromagnetic powder was produced using the Mc_1 and Rc_1 liquids under the same conditions and in the same manner as in Example 1. This sample is designated No. P-4. A video tape was produced using the powder (No. P-4) in the same manner as in Example 1. This sample is designated No. T-4.

The properties of the 4 samples as obtained in Examples 1 to 3 and the Comparison Example were compared and the results obtained are shown in Table 1 below.

Table 1

Example	Sample	H_c (Oe)	B_R/B_S	Surface Property*	Dispersibility*	Output of 5 MHz (dB)**
1	No. T-1	1,050	0.83	Good	Good	+3.5
2	No. T-2	780	0.86	Excellent	Good	+3.9
3	No. T-3	1,200	0.80	Good	Good	+0.5
Comparison	No. T-4	1,100	0.80	Average	Average	0

*The surface property and dispersibility were measured with No. T-4 as a standard.

**The output of 5 MHz was measured with the output of No. T-4 as 0 dB.

Furthermore, the composition, the particle size, and the B_S value of the powders are shown in Table 2.

Table 2

Example	Sample	B_S Value (G)	Particle Size (A)	Composition (%) Fe	B	Pb	Co	Zn
1	No. P-1	11,000	250	72.2	5.5	0.3	7.0	–
2	No. P-2	12,000	400	81.5	5.5	2.8	–	–
3	No. P-3	10,500	300	73.0	5.8	1.3	7.4	0.5
Comparison	No. P-4	11,000	250	72.0	5.6	–	7.0	–

pH Control

In related work, *M. Aonuma and Y. Tamai; U.S. Patent 4,020,236; April 26, 1977; assigned to Fuji Photo Film Co., Ltd., Japan* prepared a ferromagnetic material having a high coercive force and good squareness ratio.

It is prepared, e.g., by mixing an aqueous solution containing at least a salt of a metal capable of forming a ferromagnetic material with a solution of a reducing agent containing a boron hydride compound, and allowing an oxidation-reduction reaction to occur preferably in a magnetic field. After the reaction, the resulting powder is separated from the mother liquor, and washed with water. Here, the surface of the powder is preferably treated with an anodic surface active agent, preferably a fatty acid salt having 11 to 18 carbon atoms.

Examples of fatty acid salts include alkali metal or alkaline earth metal salts of saturated or unsaturated fatty acids comprising an alkali metal such as Na, K, etc., an alkaline earth metal such as Mg, Ca, etc., a saturated fatty acid such as lauric acid, myristic acid, palmitic acid, stearic acid, etc., and an unsaturated fatty acid such as undecyclic acid, oleic acid, elaidic acid, linoleic acid, linolenic acid, etc. Then, the powder is subjected to a dehydration and subsequently dried in a nonoxidizing atmosphere or under vacuum at a temperature of 100° to 200°C.

As another process, water is replaced by acetone, e.g., and the powder is maintained in the acetone. Further, the acetone is evaporated to dry the magnetic powder, and it is used as a magnetic material for magnetic recording. Or acetone can further be replaced by, e.g., n-butyl acetate, and the powder can be used as a magnetic material for magnetic recording without drying.

The concentration of the ferromagnetic metal ion in the aqueous solution of the salt of the metal capable of forming a ferromagnetic material is about 0.002 to 2 mols/liter, preferably 0.01 to 0.5 mol/liter.

The concentration of the other metal ions is about 0 to 15 mol percent, preferably 0 to 5 mol percent.

Desirably, the concentration of the boron hydride compound or the derivative thereof used as the reducing agent is about 0.0002 to 10 mols/liter, and it has been found to be especially desirable to perform the reaction while maintaining the molar ratio of the reducing agent to the metal ion at about 0.1:1 to 5:1, preferaby 0.25:1 to 4:1.

When the surface of the ferromagnetic metal powder is treated with an anodic surface active agent, the concentration of the anodic surface active agent preferably ranges from about 0.001 to 0.3 mol/liter, and the amount thereof employed ranges from about 10 to 500 parts by weight, preferably about 50 to 200 parts by weight, per part by weight of the ferromagnetic metal powder.

The magnetic field employed can be either a direct current field, an alternating current field, or a pulsating current field. When the reaction is carried out in the absence of a magnetic field, the saturation characteristics of the powder become poor, and at the same time, the squareness ratio is degraded considerably. However, such a ferromagnetic powder can be used for special uses.

The pH of the reaction mixture at the end of the oxidation-reduction reaction needs to be at least about 3.0. If the pH is less than about 3.0, the ferromagnetic powder is again dissolved as a result of the reaction, and the yield of the magnetic powder decreases. It is frequently observed that even when washing with water is fully carried out, the saturation magnetization (σ) may sometimes decrease. By adjusting the pH of the reaction mixture at the end of the reaction to at least about 3.0, the time required for washing the reaction product with water after the reaction can be shortened, and it is possible to obtain a ferromagnetic powder having a high coercive force (H_C) and a high residual magnetic flux density (B_R).

Temperature Control

M. Aonuma, Y. Tamai and F. Kodama; U.S. Patent 4,101,311; July 18, 1978; assigned to Fuji Photo Film Co., Ltd., Japan have developed a ferromagnetic metal powder having a narrow distribution of coercive force produced.

The process comprises: Adding a reducing agent to an aqueous solution of a salt of Fe, Co, Fe-Co, Fe-Ni, Co-Ni or Fe-Co-Ni at a rate of 0.1 to about 100 mols/sec over an addition time of one minute or less, the temperature of the salt solution being −10° to 40°C at the start of the addition. During the course of the reduction the temperature of the aqueous solution is maintained at a temperature not more than about 20°C higher than the temperature of the salt solution at the start of the addition.

Usually, the reducing reaction in accordance with this process is carried out in the presence of a reducing agent. Examples of suitable reducing agents are hypophosphorous acid and the alkali metal and alkaline earth metal salts thereof, boron hydride compounds and derivatives thereof. Boron hydride compounds are especially preferred.

The reducing agent is dissolved in water, methanol or a like solvent, preferably water, when used, and it is preferred that hydroxide ion be present in the solution. In this case, the amount of the hydroxide ion affects the weather-resistance of the formed magnetic material. More precisely, when the concentration of the hydroxide ion is in the range of about 0.001 N to 0.6 N, the weather-resistance of the magnetic substance formed is high. Suitable alkaline materials which can be used to obtain this hydroxide concentration are water soluble, produce hydroxide ions in solution and are capable of generating a pH of higher than about 8. Preferred inorganic hydroxides are the hydroxides of Group I and Group II elements such as sodium hydroxide, potassium hydroxide, barium hydroxide and calcium hydroxide, and suitable examples of organic alkaline materials include n-butylamine, isopropylamine, hydrazine, etc.

When hypophosphorous acid or a derivative thereof is used as a reducing agent, if desired, the reaction solution can further contain a complexing agent and a pH buffer, for example, monocarboxylic acids, dicarboxylic acids and hydroxycarboxylic acids. Furthermore, boric acid, carbonic acid, sulfurous acid, hydrochloric acid, sulfuric acid, phosphoric acid, nitric acid, ammonium hydroxide, and alkali metal hydroxides can be employed as pH adjusters or pH buffers. Also, if desired, the reaction solution may contain a reaction initiator such as a noble metal ion such as gold, silver, palladium or platinum, a soluble protein, a carbohydrate, or an organic solvent.

The application of ultrasonic vibration during the reaction is also effective for improving the characteristics of the ferromagnetic powder produced.

The preferred reaction pressure is about 0.5 to 5 atmospheres. The reaction temperature (the temperature at the initiation of the reaction) is usually about $-10°$ to $+100°C$ (e.g., about 60°C or less, preferably about $-10°$ to $+40°C$ when a boron hydride compound is used as a reducing agent), preferably $-10°$ to $+30°C$, although the temperature will vary depending upon the type of the reducing agent employed. The temperature can be reduced during the reaction from the initial temperature, for example, by positively cooling the reaction solution using a cooling medium such as liquid nitrogen.

Preferably, the pH of the reaction solution at the immediate beginning of the reaction is about 12 or less. When a hypophosphite compound is used as a reducing agent, the pH of the solution at the immediate beginning of the reaction ranges from about 5 to 12 and in using a boron hydride compound or a derivative thereof as a reducing agent, the pH of the solution ranges from about 0.5 to 5.

A magnetic field can be applied during the reaction, if desired, and a suitable field strength is about 10 Oe or higher, preferably as high as possible. Usually, a suitable field strength ranges from about 500 to 3,000 Oe. The concentration of the metal ion is usually about 0.002 to 2 mols/liter, preferably 0.01 to 0.5 mol/liter. The concentration of the reducing agent is preferably about 0.0002 to 10 mols/liter. A preferred molar ratio of the reducing agent to the metal ion is about 0.1:1 to about 5:1, preferably 0.25:1 to 4:1.

The addition time (the time span over which the reducing agent or aqueous solution containing reducing agent is added to the metal salt solution) is a very important factor in this process. The addition time should be about one minute or less and preferably is 15 seconds or less. It has been confirmed that inferior effects such as low coercive force, inferior squareness ratio, broad distribution of coercive force and the like are obtained when the reducing agent or aqueous solution containing reducing agent is added over a period of time greater than one minute. Accordingly the adding time should be short.

Chromium Layer on Surface

M. Aonuma, G. Akashi and Y. Tamai; U.S. Patent 3,966,510; June 29, 1976; assigned to Fuji Photo Film Co., Ltd., Japan have developed a ferromagnetic powder which contains at least Fe-B or Fe-Co-B, with the surface of the powder having a layer of Cr or a layer mainly consisting of Cr. The ferromagnetic powder is prepared by reducing a metal salt capable of forming a ferromagnetic powder containing at least Fe or Fe-Co in an aqueous solution thereof with one or more borohydride compounds or derivatives thereof and after the reduction is initiated adding an aqueous solution containing Cr ion or an anion containing Cr.

Example: In this example, all parts are by weight unless otherwise indicated.

		Quantity (mol/l)
Solution M	Ferrous chloride	0.45
	Cobalt chloride	0.05
Solution R	Sodium borohydride	1.0
Solution A	Potassium chromium sulfate	0.2

100 parts (by volume) of the above Solution M were put in a nonmagnetic container, a direct current magnetic field of 1,000 Oe was imparted to the entire container, and 100 parts of the above Solution R were added thereto over the course of 10 seconds while slowly stirring, to start the reaction. After 2 seconds, 25 parts of Solution A were added to the reaction system over the course of 5 seconds to further carry out the reaction, and the reaction was completed after 15 minutes.

After being fully washed with water, the ferromagnetic powder obtained was mixed with acetone to facilitate water removal and dried in an air drier at 60°C to obtain a ferromagnetic powder. Next, a composition of the following ingredients including the powder obtained was put in a ball mill and admixed and dispersed therein to obtain a magnetic coating composition.

	Parts
Ferromagnetic powder	300
Polyester-polyol	40
Polyisocyanate compound	20
Nitrocellulose	20
Aluminum oxide (particle size: 0.3 μ diameter)	5
Lecithin	3
Methyl ethyl ketone	450
Methyl isobutyl ketone	400

The coating composition was applied to one surface of a polyethylene terephthalate film (thickness: 25 μ) in a magnetic field and heated and dried to form a layer having a thickness of 5 μ. The thus obtained magnetic web of a broad width was treated with a supercalender and slit to form a video tape having a width of one-half inch. The surface character of the tape obtained was extremely good.

The ferromagnetic powders of this process have excellent oxidation resistance and are stable to variations of the coercive force, especially to a decrease thereof. In addition, when these substances are used in the formation of magnetic recording tapes, the dispersion and magnetic orientation thereof are excellent, and thus, tapes of a high squareness ratio (B_R/B_S) and an excellent surface character can be obtained. These substances are extremely excellent as ferromagnetic powders for high density recording.

Although these powders have particle sizes and ingredients almost similar to those of known Cr-containing ferromagnetic powders which are prepared by previously adding Cr to the reaction solution, the latter containing Cr uniformly distributed in the particles, only the process ferromagnetic powders have the abovedescribed advantages, being different from other conventional similar ferromagnetic powders. This is a characteristic feature of the process ferromagnetic powders containing the Cr-component principally in the surface of the particle.

Phosphate Ion Solution

In a process by *E. Tadokoro, M. Aonuma and Y. Tamai; U.S. Patent 4,069,073; January 17, 1978; assigned to Fuji Photo Film Co., Ltd., Japan* a ferromagnetic metal powder obtained by a reducing reaction by wet process or dry process is brought into contact with a solution containing phosphate ion. The ferromagnetic metal powder treated in this way is favorably used for magnetic recording media, in particular, high density recording.

More specifically, the ferromagnetic metal powders are produced by reducing a metal salt capable of forming a ferromagnetic metal powder with, as a reducing agent, an acid containing a phosphinate ion or a salt thereof; a boron hydride compound or the derivatives thereof; hydrazine and its derivatives; and metal powders such as Mg, Ca, Al, etc. If a phosphinic acid ion-containing acid or a salt thereof is used as the reducing agent, phosphorus, P, is incorporated into the ferromagnetic metal powder formed in a proportion of about 0.1 to 10% by weight during the course of the reaction. In the case of the boron hydride compounds or the derivatives thereof, boron, B, is incorporated into the ferromagnetic metal powder in a proportion of about 0.1 to 10% by weight during the course of the reaction.

At least about 75% by weight or more, preferably 80% by weight or more, of this ferromagnetic metal powder is a metal. At least about 80% by weight of the metal comprises a metal which is at least one member selected from Fe, Co, and Ni, i.e., Fe, Co, Ni, Fe-Co, Fe-Ni, Co-Ni and Fe-Co-Ni as the major component. In addition to these metals, about 20% by weight or less, preferably 0.5 to 5% by weight, of one or more of Al, Si, S, Sc, Ti, V, Cr, Mn, Zn, Y, Mo, Rh, Pd, Ag, Sn, Sb, Te, Ba, Ta, W, Re, Au, Hg, Pb, Bi, La, Ce, Pr, Nd, B, P and the like are present, if necessary in order to improve the properties of the ferromagnetic metal powder.

For example, where a boron hydride compound is used as the reducing agent, in a preferred example the metal comprises 80% by weight or more of the ferromagnetic metal powder, and the metal comprises about 68 to 89% by weight of iron, about 7 to 34% by weight of cobalt, and about 0.5 to 5% by weight of chromium. The remainder comprises water, hydroxides, oxides and the like.

Thus, a ferromagnetic metal powder is separated from the reaction solution, and the ferromagnetic metal powder so separated is, if the conductivity of the reaction solution is not more than 20 mV/cm, immediately, or if the conductivity is not less than 20 mV/cm after reducing the conductivity to not more than 20 mV/cm by washing the ferromagnetic metal powder with water, treated with an aqueous solution containing the anionic surface active agent, e.g., sodium oleate, to adsorb the surface active agent on the surface of the ferromagnetic metal powder.

In effecting the treatment, it is important to disperse the ferromagnetic metal powder uniformly in the aqueous solution. Then, any excess of the sodium oleate is removed directly or by washing with water. In carrying out the replacement, it has been found that the replacement can be carried out in a short period of time if an organic solvent is used. In this case, it is effective to incorporate a reducing agent into the organic solvent.

The thus treated ferromagnetic metal powder is, as is, dried in a nonoxidizing atmosphere such as nitrogen, an inert gas, or the like, or in vacuo, or in the atmosphere. Alternatively, the metal powder can be immersed in an organic solvent. Usually, the ferromagnetic metal is mixed with and dispersed in a binder, and the mixture is coated on a support and dried. This is used as a magnetic recording member in the form of a tape, a disc, a sheet, a card, or the like.

In a preferred embodiment, a ferromagnetic metal powder is prepared by the oxidation and reduction reaction in an aqueous solution using sodium borohydride as a reducing agent, separating the resulting ferromagnetic metal powder from the reaction solution, dispersing the powder in an aqueous solution containing phosphate ion followed by washing with water. The oxidation is thus prevented during the step of water washing. Instead of this water, a water-miscible organic solvent or a mixture of water and a water-miscible organic solvent can be used. Furthermore, after the powder is treated with water or a mixture of water and an organic solvent containing phosphate ion, the solution can be replaced by an organic solvent so as to remove the water from the powder and to increase the dispersibility in a binder.

The ferromagnetic metal powder treated in this way is dried as it is in an inert gas such as nitrogen, helium, neon, argon, krypton, xenon or radon gas or in the air, or immersed in an organic solvent. Then the powder is ordinarily mixed with a binder, dispersed, coated onto a support, dried and practically used as recording media in the form of a tape, disk, sheet, card, etc.

Iron-Cobalt-Chromium-Boron Powder on Support

Y. Tamai, M. Aonuma, M. Nakamura, H. Agawa and G. Akashi; U.S. Patent 3,943,012; March 9, 1976; assigned to Fuji Photo Film Co., Ltd., Japan have developed a magnetic recording medium for high density recording. It comprises a support having thereon a magnetic recording layer comprising a ferromagnetic metal powder of the Fe-Co-Cr-B system and having a coercive force of about 700 to 1500 Oe in the oriented direction of the magnetic substance, a squareness ratio of 0.70-0.93, and a residual magnetic flux density of about 1,500 gauss or more.

The residual magnetic flux density after storage for 160 hours of a temperature of 60°C and 90% RH is 85% or more of the initial value before storage under these conditions in a binder. The magnetic recording medium is especially suitable for high density recording, having sufficient sensitivity and excellent resistance.

The ferromagnetic alloy powder comprises the reaction product of reducing a mixed solution containing Fe, Co and Cr ions with a borohydride compound reducing agent having a hydroxide ion concentration of about 0.001 N to 0.6 N. The mixed solution contains metal ions ranging from 0.01 to 0.5 mol/liter in concentration such that the ferromagnetic alloy powder comprises about 65 to 89% Fe, about 7 to 34% Co, and about 0.5 to 5% Cr by weight.

Example: A direct current magnetic field of 500 Oe was applied to 100 liters of an aqueous solution of ferrous chloride, cobalt chloride and chromium alum, these components being dissolved in a predetermined proportion, and the solution was kept at 20°C. 20 liters of sodium borohydride solution were added to the above metal salt solution at a rate of 1.2 liters/sec. The concentration of the metal salt was 0.2 mol/liter total, and the concentration of the sodium borohydride solution was 2 mols/liter. This solution had a sodium hydroxide concentration of 0.06 N. The reaction started simultaneously with the addition of the reducing agent, and vigorous foaming was noticed. After the reaction, the black precipitate obtained was fully washed with water and then with acetone. Afterwards, the precipitate was dried at 25°C in an atmosphere of a humidity of 30% or less.

The analysis results (in %) of the metal components and the boron in the respective samples prepared are shown in the following Table 1.

Table 1

Sample	Fe	Co	Cr	B
A	74.0	17.3	2.3	6.4
B	65.7	26.3	1.8	6.2
C	83.9	7.5	1.9	6.7
D	80.3	13.9	0.0	5.8

The following Table 2 shows the saturation magnetization (magnetic field: 2,000 Oe) of these samples before and after weather-resistance testing.

Table 2

Sample	Initial Value	After Weather-Resistance Testing*
A	8650	7800
B	8400	7950
C	9290	4240
D	8700	4080

*Stored for one week at 45°C, 85% RH

Both Samples A and B of the process have an original saturation magnetization value (before weather-resistance testing) somewhat lower than that of the other Samples C and D, while after the weather-resistance testing the Samples A and B show a saturation magnetization value far higher than that of the Samples C and D.

Magnetic coating compositions of the following ingredients and these four samples were prepared.

	Parts by Weight
Ferromagnetic powder (average particle size of about 0.25 μ in length and 0.04 μ in width)	300
Polyester-polyol	40
Cellulose acetate butyrate	25
Polyisocyanate compound (Desdomur)	20
Chromium sesquioxide (average particle size: about 0.1 μ)	4
Soybean lecithin	6
Methyl ethyl ketone	400
Butyl acetate	500

Each of the resulting coating compositions was applied to one surface of a polyethylene terephthalate film (thickness: 25 μ) in a magnetic field and heated and dried to form a layer having a thickness of 4.5 μ. The thus obtained magnetic web was treated with a supercalender and slit to form a video tape having a width of ½". The magnetic characteristics of the tapes obtained are shown in the following Table 3.

Table 3

Sample	Magnetic Substance	Hc (Oe)	Squareness Ratio	Residual Magnetic Flux Density* (G) Initial Value	After Weather-Resistance Testing	Sensitivity** (dB)
No. 1	A	980	0.85	1900	1630	+1.6
No. 2	B	1070	0.86	1820	1660	+2.0
No. 3	C	720	0.82	2010	1305	−1.3
No. 4	D	970	0.85	1970	1070	0.0

*Stored for 160 hours at 60°C, 90% RH

**After weather-resistance testing, nonbias recording of 5 MHz, with No. 4 as standard.

Figure 4.1 shows a preferred composition range of the Fe-Co-Cr system of this process. The preferred range is the shaded portion in the figure. (This ternary figure only shows the three components of Fe, Co and Cr contained in the samples, and the boron component therein is not shown.) The respective Samples **A, B, C** and **D** are plotted in Figure 4.1 with respect to their Fe-Co-Cr composition.

Figure 4.1: Preferred Composition Range of Fe-Co-Cr System

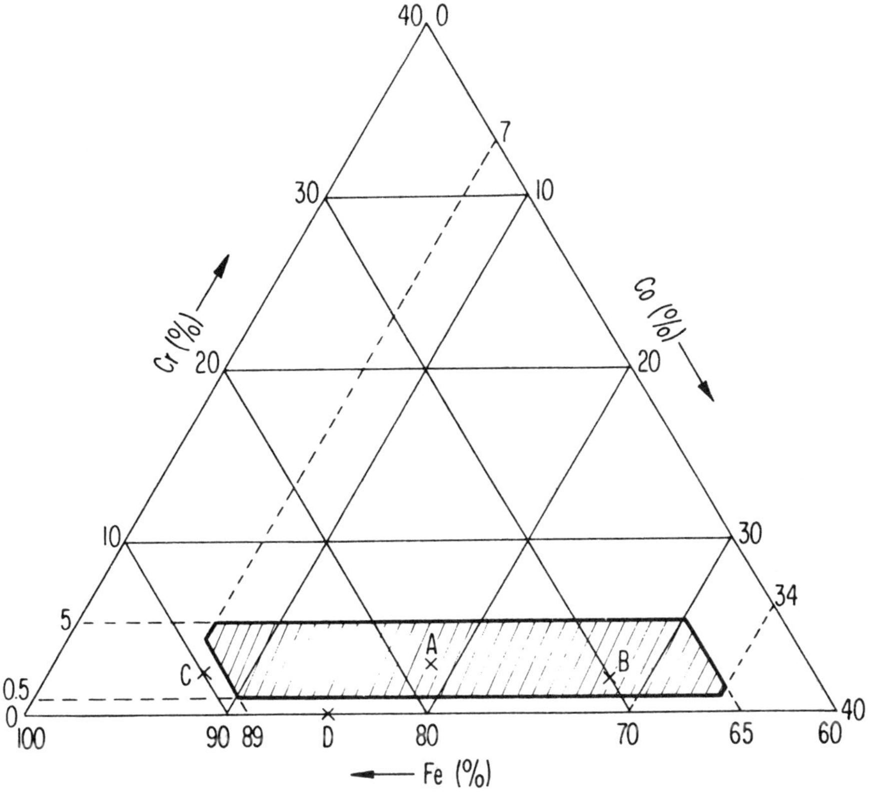

Source: U.S. Patent 3,943,012

The magnetic recording tapes **A** and **B** of this process have more excellent weather-resistance than other conventional tapes.

Hypophosphite Reduction of Cobalt and/or Nickel Salts

M. Aonuma, T. Kitamoto and G. Akashi; U.S. Patent 4,059,463; November 22, 1977; assigned to Fuji Photo Film Co., Ltd., Japan describe a process for producing ferromagnetic metal powder having a square hysteresis loop ratio whereby the coercive force may be easily controlled as desired.

The process is characterized in reducing nickel and/or cobalt salts in an aqueous solution which contains hypophosphite ions and an organic solvent while applying a magnetic field to the aqueous solution. While nickel and/or cobalt salts are the main components which form the ferromagnetic product, small proportions of other materials to be incorporated into the ferromagnetic material may be present, such as La, Ce, Nd, Sm, Al, S, Cr, Mn, Fe, Cu and Zn in an amount of 0-10% by weight of the total salts used; the concentration of total salts in the aqueous solution is in the range of 0.001 to 1 mol per liter.

The ferromagnetic metal powder obtained by this process has a coercive force of more than 300 Oe and a saturated magnetism of more than 8,000 G/cc, and comprises as a main component Co and/or Ni and several percent of P, e.g., usually 0.5-15% by weight based on powder weight. The desired grain size of the ferromagnetic metal or alloy powder is obtained by changing the reacting conditions, e.g., at higher temperatures larger grains are obtained. The grain size is generally 100 A to 2 μ at the temperature ranges provided above.

According to the process the squareness ratio of the hysteresis loop of the ferromagnetic metal or alloy powder obtained is greatly increased, the coercive force can be controlled within the range of 300-1,200 Oe by selecting the kind of the organic solvent, the grain size of the powder can be kept uniform, the ratio of the length to the width of the powder grains can be increased, and the strength and toughness of rod shaped powder grain is increased, whereby the dispersion of the powder in a binder and the magnetic orientation of the powder after coating are much improved, and a magnetic recording material made from the ferromagnetic metal powder has an extremely high squareness ratio.

The magnetic properties of the ferromagnetic metal or alloy powder obtained by this process are further improved by heating in a nonoxidizing environment, e.g., in a vacuum or in a flowing N_2, CO_2, CO or H_2 gas stream and the oxidation resistance of the ferromagnetic powder can be improved by heating the latter in an atmosphere containing very small amounts of moisture and oxygen. The conditions used are heating temperature at greater than 100°C, preferably more than 200°C, for 30 minutes—24 hours at a pressure less than 100 torrs.

These ferromagnetic powders preferably have a grain size distribution with the range of 100 A-2 μ.

Acicular Fine Powder in Presence of Hypophosphite Ion

M. Aonuma, T. Kitamoto and G. Akashi; U.S. Patent 3,902,888; September 2, 1975; assigned to Fuji Photo Film, Co., Ltd., Japan have developed a process for preparing a ferromagnetic alloy powder by reducing salts of ferromagnetic metals in an aqueous solution containing an acicular fine powder in the presence of hypophosphite ion. The aqueous solution can be subjected to the action of a magnetic field or ultrasonic waves during the reaction if desired, and the rate of the reaction can be increased by the addition to the aqueous solution of a noble metal ion.

The acicular fine powder acting as precipitation nucleus incorporated into the reaction bath, should have a particle size of less than 0.3 μ in length and less than 0.03 μ in width, since the final size of the alloy powder desired is 0.1 to 1 μ in length and a length/width ratio of 3:1. Examples of these fine powders include: those acicular fine powders of diatomaceous earth goethite (α-FeOOH), lepidocrocite (γ-FeOOH), acicular fine powder of γ-Fe_2O_3, magnetite, CrO_2 fine powder, attapulgus clay (Attagel), copper phthalocyanine, acicular ZnO, and acicular particles of V_2O_5.

The acicular particles used as the precipitation nucleus are preferably activated by a noble metal ion such as Pd, Au and Ag. In other words, the acicular particles used in the process are not the same substance as the alloy itself, therefore, there are many cases where growth of alloy particles is secured when the surface of the acicular particles is activated.

The preferred reaction condition employed in the process is a pH of higher than 5, but the reaction temperature is not critical. However, it is preferred to conduct the reaction at a pH ranging from 8 to 12 and at a temperature ranging from 70° to 95°C.

The ferromagnetic alloy powder obtained by this process has a coercive force of greater than 500 Oe, a saturation magnetization of greater than 8,000 G/cc, and the main component of the powder is Co, Co-Ni, and contains about 10% by weight or less of P depending on the reaction conditions employed. Further, the size of the particles obtained may be adjusted and it is possible to obtain any desired particle size by selecting the kind and amount of acicular particles added, and other reaction conditions.

The following example illustrates a preferred embodiment of the process. All parts and percents are by weight unless otherwise indicated.

Example:

Liquid A	
Cobalt sulfate	10 g
Sodium tartrate	50 g
Boric acid	15 g
Water in an amount to give 300 cc of solution	
Liquid B	
50% hypophosphorous acid	25 cc
Acicular goethite of about 0.2 μ in length and 0.05 μ in width	0.2 g
Liquid C	
Water in an amount to give 10 cc suspension	

Liquid A and Liquid B were prepared and mixed together. Into the mixture, an additional amount of water was added to give 500 cc of solution. The resulting solution was adjusted to a pH of 9.0 by adding a sodium hydroxide solution and was heated to 90°C. Into the solution, Liquid C activated previously with a

0.3% palladium chloride solution was added with stirring. The reaction was conducted while maintaining the temperature at 90°C and the pH in the range of 8.5-9.0 by adding a sodium hydroxide solution. The precipitate so formed was washed with water and dried to give an alloy powder. The yield of the alloy powder was 2.6 g. The average size of the rod shaped particles obtained was 0.8 μ in length, and a length/width ratio of about 5:1. The coercive force of the particles was 800 Oe, and the saturation magnetization was 14,000 G/cc.

For the purpose of comparison, this procedure was repeated but using Liquid C containing only palladium chloride as metal precipitation nucleus with no acicular particles and globular particles were obtained. The globular particles obtained had a remarkably low ratio of residual magnetic flux to saturated magnetic flux, i.e., a squareness ratio (B_R/B_S) of about 0.5, and had a coercive force of 500 Oe.

When the reaction was conducted using these acicular particles as a precipitation nucleus and applying a direct current magnetic field of 2,000 G during the reaction, rod-shaped particles of an average size of 1.2μ in length and a length/width ratio of 8:1 and having a coercive force of 1,000 Oe were obtained. The thus obtained alloy powder was dispersed thoroughly in a binding agent and the dispersion was coated on the surface of a tape. The B-H property of the tape was excellent and, more especially, the squareness ratio (B_R/B_S) was 0.90.

As described above, by using Liquid C, that is to say, using acicular particles as a precipitation nucleus, a rod shaped alloy powder was obtained, and it was possible to use the powder in such a way by utilizing its dimensional anisotropic property, just like the known and conventionally used acicular iron oxide powder.

Copper and Aluminum

COPPER

Cementation with Iron

C.H. Elges, III; U.S. Patent 3,877,932; April 15, 1975; assigned to U.S. Secretary of the Interior has developed a process for recovering copper from acidic aqueous solutions containing cupric, ferric, ferrous and other ions by a cementation reaction with elemental iron. The cementation step is performed under hydrogen pressure which minimizes iron consumption and allows the cementation treatment of highly acidic solutions.

The process is applicable to the treatment of those leach solutions now subjected to conventional cementation technology. Greatest benefits, however, will be obtained in the treatment of much more highly acidic leach solutions having a free acid in the general range of about 2 to 10 weight percent. This acid concentration corresponds to a pH substantially below 1.

Hydrogen pressures required for the efficient operation of the process range between about 50 and 500 psi, preferably in the range of about 100 to 300 psi. It is preferred to operate the process at the ambient temperature of the influent solution, i.e., in the range of about 35° to 95°F. Temperature, however, is not an important consideration and variations of temperature will not adversely affect the process.

It is preferable to operate with an excess of metallic iron as the precipitating metal. The reaction itself is carried out in a closed vessel capable of containing the hydrogen pressures required. Suitable means for pumping solutions through the precipitating vessel, introducing metallic iron, and recovering cement copper are those conventionally used in the art of hydrometallurgy. The following examples more fully illustrate the process.

Example 1: Samples of technical grade iron powder were mixed with a substantial excess of 5% sulfuric acid, placed in an autoclave, and pressurized with hydrogen gas. The mixture was then reacted with agitation for 1 hour at ambient temperatures. Pressures were recorded before and after each run. At the end of the 1 hour reaction time the autoclave was depressurized and the mixture remaining was filtered and the metal recovered was dried with acetone and weighed. Tabulated results are as follows:

Table 1

Run No.	. .Pressure, psig H_2 . . Before	After	Weight Loss, %
1	0	0	18.0
2	50	64	14.6
3	100	112	9.1
4	150	157	4.6
5	200	203	1.2
6	250	250	1.2

These tests illustrate that the kinetics of iron reacting with sulfuric acid are substantially affected by operating pressure. As is evident from the data, iron consumption at 200 psig hydrogen pressure was less than 10% of that consumed at ambient pressure.

Example 2: Two samples of technical grade iron powder were placed in reaction vessels with a 5% solution of sulfuric acid containing 2 g/l of copper as the sulfate. Iron was present in each sample in a weight ratio 10 times that of the copper. One vessel was pressurized to 200 psig hydrogen pressure; the other was left at ambient pressure. Temperature was ambient and the reactions were stirred intermittently. Reaction times were two hours in each case. At the end of the runs the contents of the reaction vessels were filtered and the metal was acetone dried and weighed. Results are as follows:

Table 2

Hydrogen Pressure (psig)	Cu Precipitated (%)	Fe Consumed (%)	Can Factor*
0	~100	74.9	7.49
200	~100	34.2	3.42

*A ratio of pounds of iron consumed per pound of copper precipitated.

A sulfuric acid concentration of 5% corresponds to a pH value of approximately 0.3. Iron was present in these experiments in a substantially greater excess than would be used in commercial practice and the reaction time was longer than that necessary to carry the precipitation reaction to substantial completion. Comparison of the two runs demonstrates that precipitation under substantial hydrogen pressure is not only feasible but also results in a much lower iron consumption.

Example 3: Technical grade iron powder was placed in an autoclave and mixed with a 5% solution of sulfuric acid containing 2 g/l of copper as the sulfate. In this experiment, the weight ratio of iron to copper was 1:2. The reaction vessel was pressurized with hydrogen to 200 psig, agitated, and allowed to react for 1 hour at ambient temperature. At the end of the run the pressure was released, the solution was filtered and the metal which was recovered was acetone dried, weighed and analyzed. The cement copper product assayed 81% metallic copper and the can factor for the precipitation reaction was 1.57. This experiment demonstrates that it is possible to produce cement copper of adequate grade from solutions of high acidity at economically realistic can factors.

Hydrometallurgical Leaching

C.W. Robinson and K.E. Merklin; U.S. Patent 3,958,947; May 25, 1976; assigned to Marcona Corporation describe a process for hydrometallurgical leaching of unseparated, crushed mineral ores using a large upright vessel having side walls and a bottom wall. Solids discharge outlets are formed in the bottom wall. Means are provided for supplying a leaching liquid in a stream or streams directed generally parallel to and above the bottom wall of the vessel. The stream or streams are progressively traversed as by rotation or oscillation to sweep over the bottom of the vessel and create a highly agitated reaction zone in the lower region thereof.

Ore solids are crushed to a top size of about ⅜" and are introduced without further grinding or separation at the top of the vessel to move under gravity, and pass countercurrent to a generally upward flow of leaching liquid. As the solids enter the lowermost region of the vessel, they are impacted by the streams and are brought into a highly agitated state. Solids are withdrawn from the lower region of the vessel after having passed therethrough and either may be recycled in the same stage or taken to subsequent stages if multistage operation is utilized. If the mineral recovery is satisfactory, the solids can be passed immediately to a washing stage.

Leaching liquid, now pregnant, is removed from the top of the treatment vessel and is either recycled, taken to the preceding stage in interstage countercurrent flow with respect to the solids if multistage operation is used, or processed to recover the contained values. In one preferred form of the process, the vessel is provided with a discharge trap which facilitates the accumulation of solids with a minimum amount of liquid for periodic discharge from the vessel. The frequency of discharge of solids determines the residence time of the solids in passing through the vessel. In addition, provision is also made for the introduction of quantity of gas, usually air, as a bubble which passes upwardly through the bed of solids and liquid to promote further agitation of the bed when in the vessel.

Example: A typical oxide copper ore bearing 0.7% copper oxides was crushed in a cone crusher to ⅜". A quantity of this ore was introduced into the reactor vessel by pumping the same as an aqueous slurry and draining off the entraining water. Lixivium of 6% H_2SO_4 was jetted into the lower region of the vessel. This acid rose upwardly through the solids forming an overflow which was passed through the thickener and the liquid phase returned through the jet nozzle. These operations were continued for about 3 hours, at the end of which copper recovery by leaching was about 90% complete and the pregnant liquor was suitable for further processing.

The pregnant liquor was then drained off, the solids washed, as with water, to remove any residual values. Thereafter, the solids were discharged as a slurry by the operation of the jet mechanism using water as the liquid entraining phase.

In the foregoing example, it will be noticed that the solids, after leaching is completed, were removed in a water suspension as a slurry. Like the leaching operation itself, this procedure is carried out under flooded conditions in which the high velocity jet stream works into the solids in the lowermost region of the vessel.

The solids impacted by the stream are repulped by it and drain back to the sump as a dischargeable slurry. By progressively moving the jet, new solids are continually being impacted and removed until the material in the lowermost region is reduced to less than that which will support the weight of the solids above. The entire load then shifts downwardly to be removed in turn.

Precipitation from Liquid Organic Medium

A.R. Burkin; U.S. Patent 3,844,763; October 29, 1974; assigned to National Research Development Corporation, England has developed a process for the production of a copper powder or coating which comprises contacting with a reducing gas, e.g., hydrogen or carbon-monoxide, a liquid organic medium containing a copper species in a compound of or in association with at least one tertiary carboxylic acid having at least five carbon atoms, to cause precipitation of copper from the organic phase, there also being present during precipitation a base in solution in the organic medium, e.g., ammonia.

The tertiary carboxylic acid having at least five carbon atoms is generally used in the form of a mixture of carboxylic acids containing mainly the required tertiary acids. A preferred such mixture is the C_{9-11} tertiary acid, for example, Versatic 911, although other acids or mixtures of acids may be employed, for example those known commercially as Versatic 9, Versatic 10, Versatic 11 and Versatic 16 which comprise respectively mainly C_9, C_{10}, C_{11} and C_{16} tertiary carboxylic acids.

The preferred base is ammonia although any other base which is soluble in the organic medium, for example an amine, can also be used. When ammonia is used, the solution preferably contains prior to reduction, about 2 gram mols of ammonia for each gram ion of copper in solution. The use of too great a quantity of ammonia may result undesirably in the formation of white crystals of the ammonium salt of the tertiary carboxylic acid which are usually removed after the reaction is completed.

As the yield of copper powder obtained is high, the addition of copper "seed" material is not normally necessary but such "seed" material may be used if so desired.

The reaction can be carried out in an autoclave at elevated temperature, preferably above 115°C, for example 140°C and is preferably carried out at elevated pressures above 100 psi, e.g., 300 psi.

Some tertiary acids, e.g., Versatic 911, are liquid at the temperatures at which reduction is accomplished and may be used in excess to provide the liquid medium in which the process is carried out.

The process is illustrated by the following example: A copper-containing organic liquid is prepared as follows. 1200 ml of a mixture of 2 vol of Shell Sol T and 1 vol of Versatic 911 is added to an aqueous solution of 250 g of cupric sulfate, $CuSO_4 \cdot 5H_2O$, made up to 1 liter in water and the mixture is stirred vigorously. A 1 molar solution of sodium hydroxide is added slowly so as to produce only a small amount of basic cupric sulfate or cupric hydroxide, which redissolves in the organic acid solution. Addition of the alkali is continued until the precipitate will not dissolve after stirring for 15 minutes without adding more alkali, and the stirring is then discontinued.

The organic liquid is separated from the aqueous phase and stored. Fresh Versatic 911 Shell Sol T is added to the residual aqueous cupric sulfate solution, the mixture is stirred, and alkali added until practically all of the copper has been removed from the aqueous phase, which is then removed and replaced by fresh cupric sulfate solution. The sequence of extraction is continued until 6 liters of Shell Sol T-Versatic 911 solution loaded with copper is obtained. The solution contains 25.2 g of copper per liter.

1200 ml of the copper-containing liquid is taken and ammonia gas is passed into the liquid until the desired increase in weight (8.0 g) has been obtained.

The liquid is put into a 2 liter autoclave, copper powder is added as seed where appropriate and the vessel is sealed. The autoclave is flushed with nitrogen and the pressure released through a valve which is then resealed. The temperature is taken to 140°C, the stirrer is turned on at 1,000 rpm and when the temperature has stabilized at 140°C hydrogen is admitted to about 360 psi. 0.452 g mol hydrogen are taken up.

A recorder is used to give a trace of the initial pressure, the pressure drop during reaction and the final value. When the reacted solution is cold it is removed from the autoclave and filtered under vacuum. The clear solution is removed and the copper powder washed with acetone, dried and weighed.

29.49 g copper is deposited from a solution which initially contains 0.475 g mol copper as the cupric salt of Versatic 911.

Copper Sulfate Solution Containing Catalyst

In order to improve the known process wherein copper values present in an aqueous copper sulfate solution are reduced with sulfur dioxide at elevated temperature and pressure and precipitated as copper powder, *N.R. Tipman; U.S. Patent 3,880,647; April 29, 1975* provides a catalyst, either dimethyl sulfoxide or dimethyl formamide, in the copper sulfate solution in an amount between 0.5 and 2.0% by weight of the solution.

Preferably, it is provided in an amount of about 1% by weight of the solution so as to provide about 1 mol of catalyst per mol of copper in solution.

The use of the catalyst results in three improvements. These are:

1. An increase in the yield of copper powder;
2. A lowering of the temperature needed to start the reduction; and
3. A change in the physical characteristics of the product powder.

More particularly, the copper recovery is increased from about 50 to about 75% in most cases. The temperature at which reduction commences is lowered to about 90°C. This minimizes contamination by iron and oxygen, which occurs at higher temperatures.

Copper sulfate ore is conventionally concentrated in a fluidized bed at 650°C. The roasted concentrate is then leached and the leach-end solution reduced as illustrated in the examples below.

The product powder particles are mainly hexagonal platelets having six side edges and two main surfaces. These platelets appear to be single crystals which have a smooth surface, thereby indicating a low surface area. Powder particles having a size of +120 mesh can be produced by this process. The powder is easily compacted and can be fabricated into rolled or extruded forms which are competitive in properties to those made from other types of copper powder.

According to another feature of the process, the reduction is preferably terminated at a temperature less than about 130°C. If the reduction reaction, which is exothermic, is allowed to go to higher temperatures, the product powder is found to be excessively fine and is discolored with impurities such as iron and copper oxide.

According to still another feature, sulfur dioxide gas is introduced into the reduction autoclave, preferably before the temperature of its contents reaches 90°C. If the sulfur dioxide is not added until the temperature of the autoclave contents is above 90°C, a fine product powder which is discolored with impurities is obtained. Also, the pH of the leach end solution fed to the autoclave preferably should be lower than 4.0. If the pH is higher, some of the copper sulfate precipitates as copper hydroxide.

Example 1: *Effect of Dimethylformamide*—143.5 pounds of leach liquor containing 21.4 g/l cupric ion, 20.6 g/l ferric iron and 5 g/l sulfuric acid were placed in a 20 gallon autoclave. 2.5 pounds of dimethylformamide were added to the liquor. The autoclave was closed and the trapped air purged with sulfur dioxide gas. The autoclave contents were then heated to 70°C with agitation, at which time sulfur dioxide gas was admitted to raise the pressure to 30 psi.

The temperature was increased to 90°C, at which time consumption of sulfur dioxide began. At this time the sulfur dioxide pressure was immediately increased to 80 psi and maintained there throughout the reduction. The temperature slowly rose over a period of 95 minutes to a final temperature of 125°C, at which time sulfur dioxide consumption ceased.

The copper product was recovered from the autoclave and washed with water to remove residual sulfuric acid. It was then washed in water containing 0.1% by weight ammonia to neutralize residual sulfuric acid. As a result of this operation, a protective copper-ammonia complex was formed, which reduced surface oxidation and preserved the bright metallic appearance of the product.

After neutralizing, the powder was washed with n-propyl alcohol and vacuum dried.

Analysis of the autoclave solution showed that 4.5 g/l copper remained in the solution. 2.43 pounds of metallic copper was removed for a recovery of 79%. The iron content of the solution remained at 20.6 g/l and the sulfuric acid content had risen to 41 g/l.

The product powder had the following composition:

Constituent	Wt %	Constituent	Wt %
Mn	0.01	Al	0.1
Sb	0.01	V	0.01
Mg	0.01	Ag	0.006
Mo	0.05	Zn	0.01
W	0.01	Ti	0.01
Pb	0.2	Ni	0.01
Sn	0.01	Zr	0.01
Cr	0.01	Co	0.01
Si	0.04	Sr	0.01
Fe	0.2	O	0.45
Bi	0.01	Cu	*

*Balance.

The product powder had the following typical screen analysis:

Tyler Sieve No.	Percent Retained
80	–
120	–
170	1
230	1
325	5
Pan	93

Further work has shown that particle size is governed by the rate of stirring. Larger particles can be produced by reducing the rate of stirring.

Example 2: *Effect of Dimethyl Sulfoxide*—130 pounds of leach liquor containing 18 g/l cupric ion, 18 g/l ferric ion and 5 g/l sulfuric acid were placed in a 20 gallon autoclave. 2.5 pounds of dimethyl sulfoxide were added to the liquor. The autoclave was closed and the trapped air purged with sulfur dioxide gas. The autoclave contents were then heated to 70°C with agitation, at which time sulfur dioxide gas was admitted to raise the pressure to 30 psi. The temperature was increased to 90°C, at which time consumption of sulfur dioxide began. At this time the sulfur dioxide pressure was gradually increased to 80 psi and maintained there throughout the reduction. The temperature slowly rose over a period of 95 minutes to a final temperature of 125°C, at which time sulfur dioxide consumption ceased.

The copper product was recovered from the autoclave and washed with water to remove residual sulfuric acid. It was then washed in water containing 0.1% by weight ammonia to neutralize residual sulfuric acid. As a result of this operation, a protective copper-ammonia complex was formed, which reduced surface oxidation and preserved the bright metallic appearance of the product.

After neutralizing, the powder was washed with n-propyl alcohol and vacuum dried.

Analysis of the autoclave solution showed that 5.0 g/l copper remained in the solution. 1.7 pounds of metallic copper was removed for a recovery of 73%. The iron content of the solution remained at 20 g/l and the sulfuric acid content had risen to 38 g/l.

The product powder had a composition and screen analysis similar to the product of Example 1.

Polyethyleneimine Polymer Added to Reducing Gas

R.M. Genik-Sas-Berezowsky; U.S. Patent 4,018,595; April 19, 1977; assigned to Sherritt Gordon Mines Ltd., Canada has developed a process for producing substantially oxide free copper from copper salts by reacting an acid copper sulfate solution which contains ammonium sulfate and the copper salts with a reducing gas at elevated temperature and pressure to precipitate metallic copper.

In the aqueous solution is provided a finite amount of an addition agent selected from the group consisting of polymeric polyamine compounds having amino groups linked by alkyl groups having 2 to 4 carbon atoms, e.g., a highly branched polyethyleneimine polymer having a molecular weight in the range of 40,000 to about 100,000, in an amount between about 100 and 200 ppm.

Example 1: A number of batch reductions were carried out in a one-gallon titanium-lined autoclave equipped with titanium internal fittings including a marine-type impeller-agitator. The feed solution was a synthetic acid sulfate leach solution prepared from reagent grade chemicals and which analyzed (g/l) 69.8 Cu, 21.2 Zn, 7.45 Mg, 1.23 Fe, 1.34 Ni, 60.0 H_2SO_4. For each reduction, 2.5 liters of this solution were charged into the autoclave. In one test, no additive was used, in another test, Acrysol A-3 was added in accordance with the prior art and in two further tests, 200 ppm of a different additive compound was added to each charge according to the process. Each batch was heated under 350 psi hydrogen overpressure to 163°C and held at temperature for 10 minutes, cooled and discharged. The metal powder products were given two displacement washes with 10 g/l H_2SO_4 solution, two with water and then one with acetone.

The results are reported in Table 1. It can be seen that in the absence of any additives, the product was in the form of coarse balls and heavy plastering on the walls and other internals of the reduction autoclave. Addition of Acrysol A-3 decreased plastering but produced a nonflowing powder of low apparent density. It can be observed also that the Acrysol addition resulted in a much higher level of impurities than is obtained with no additive or with PEI.

Example 2: The procedure of Example 1 was repeated except that for each reduction, 200 ppm of a different additive compound of the process was used. The results which are shown in Table 2 show that each of the additives is effective to produce free flowing powders of relatively high apparent density with the higher molecular weight additives producing powders with a smaller particle size than the lower molecular weight additives.

Other experimental procedures indicate that at levels as low as 5 ppm, the additive is quite effective, although the preferred range of additive for producing fine, free-flowing powder is 100-200 ppm; little additional benefit from levels of additive above 200 ppm is shown.

Hypophosphorous Acid Treatment for Electronic Grade Copper

E.X. Heidelberg; U.S. Patent 3,881,914; May 6, 1975; assigned to Owens-Illinois, Inc. has developed a process for producing electronic grade, elemental copper suitable for use in forming thick film conductive circuits.

Table 1

Additive	Final Copper Distribution: Level (ppm)	Filtrate (gpl)	Powder (g)	Impurities in Copper Products (ppm): Fe	Zn	Ni	Mg	S	C	Apparent Density (g/cc)	Flow Rate (sec/ 50g)	Remarks
None	—	9.6	—	19	37	17	15	88	38	N.M.**	No F.R.	Heavy plastering and coarse balls
Acrysol A-3	200	19.2	—	84	40	8	18	200	355	0.8	No F.R.	Light plastering on autoclave walls and internals
PEI 1000	200	13.9	129	10	5	3	3	26	42	3.10	32.2	Less than 5% plastering*, 75.4% minus 200 mesh powder
PEI 600	200	18.5	127	12	4	9	2	16	47	2.93	22.9	Less than 5% plastering*, 73.8% minus 200 mesh powder

(Columns Fe through Remarks: Assays and Physical Characteristics of Copper Products)

*In the PEI tests, the autoclave walls were clean with the only plastering occurring on agitator and thermowell

**Not measurable

Table 2

Additive	Additive Molecular Weight	Final Copper: Filtrate (g.p.l.)	Powder (g)	Impurities in Copper Powder (ppm): Zn	S	C	Apparent Density (g/cc)	Flow Rate (sec/ 50 g)	Tyler Screen (mesh) Cumulative % Passing: 100	200	325
PEI-1000	50–100,000	13.9	129	5	26	42	3.10	37.0	99.5	75.4	26.1
PEI-600	40–60,000	18.5	127	4	16	47	2.93	22.9	100	73.9	29.3
PEI-400	30–40,000	16.3	135	4	17	37	2.81	24.8	99.9	81.4	31.3
PEI-18	1800	15.6	134	7	30	26	2.80	28.8	45.4	11.8	1.5
PEI-12	1200	20.5	117	6	21	24	2.62	32.5	39.2	8.7	0.7
PEI-6	600	17.4	125	5	11	29	2.73	31.0	40.0	8.5	0.5
K&K PPI*	n.a.	15.1	120	5	12	25	2.63	33.0	22.8	3.7	0.3

(Columns Zn through 325: Chemical Assays and Physical Properties of Copper Powder)

*polypropylenimine of unspecified molecular weight from K&K Laboratories

The first step in the process involves the reaction of copper acetate or copper sulfate and hypophosphorous acid in water to form elemental copper. The amount of hypophosphorous acid will be at least about 2 mols per mol of copper sulfate, or copper acetate. When using less than that amount there has been a notable tendency of the elemental copper to be severely contaminated with large flakes and/or chunks which, obviously, is quite undesirable.

Excellent results will be obtained using between about 2 to about 3 mols of hypophosphorous acid per mol of copper sulfate or copper acetate, with the preferred amount being about 2 mols per mol. Any suitable and convenient time and temperature may be employed for the reaction but it is usually preferred to react the materials by heating at a temperature between about 80°C to a temperature of about 99°C and preferably to about 80° to about 85°C. Any pressure may be employed but the reaction proceeds quite well at ambient pressure.

The completion of the reaction may be monitored by the subsiding of foaming which results during the reaction because of effervescence. Heating times will, of course, vary with the size of the batch being employed but generally about 10-15 minutes is quite convenient. The amount of water employed may be varied with excellent results being obtained, in terms of product quality and reaction control, by employing about 50 to about 60 mols of water per mol of copper sulfate or copper acetate.

After the formation of the elemental copper by the reaction, the copper appears as a solid and it is separated in any convenient manner from the reaction medium. Filtration is generally preferred and will be found to be quite convenient. The resulting separated elemental copper is then washed with a water miscible, inert organic solvent so as to displace the water from, e.g., the filter cake.

It is gnerally preferred to employ a water miscible inert organic solvent which is at least as volatile as water. Exemplary solvents include the lower alkyl alcohols, e.g., the C_1 to C_3 alkyl alcohols as well as other solvents such as, e.g., lower boiling glycols or Cellosolves. The preferred solvents are methanol or ethanol. Subsequent to the washing, the copper is vacuum dried in any conventional vacuum drier. The time, temperature, and degree of vacuum does not appear to be critical and can be routinely selected by those skilled in the art.

Conveniently the vacuum drying will be done under a vacuum of about 20 to about 100 mm of mercury, preferably about 20 to about 40 mm of mercury with suitable temperatures being about 50° to about 100°C. Additionally, when the vacuum is released on the vacuum drier employed, this release will be effected by introducing an inert gas such as helium, neon, argon, or the like into the vacuum drier. Argon is preferred. The reason for the washing with the water miscible organic solvent and the vacuum release with an inert glass is to reproducibly insure the formation of the high quality product and that this is done in a safe manner. Without employing these procedures, ignition of the copper powder has resulted upon release of the vacuum, which ignition not only, obviously, is a safety hazard but which also significantly and detrimentally affects the quality of the product.

Generally, the copper powders produced in accordance with this process have surface areas on the order of 1-2 m^2/g, as measured by conventional absorption techniques, and have particle sizes less than about 1 or 2 μ. When employed in the manner indicated above for forming thick film conductive circuits they produce an excellent high quality product.

Reduction with Anthraquinols

J.N. Gerlach; U.S. Patent 4,033,765; July 5, 1977; assigned to Kennecott Copper Corporation has found that the reduction of cupric copper to cuprous copper by quinolic compounds is extremely fast (on the order of 5 seconds) when the cupric ions are contained in an aqueous ammoniacal solution, and further, that an excess of quinolic compound is not required for this reduction. Since the cuprous ions produced in such a process are stable in the presence of ammonia, a two stage process may be provided wherein the cupric to cuprous reduction may be effected rapidly, with minimal amounts of quinolic compound, and in a relatively small reactor, and thereafter the cuprous ions may be transferred to a second reactor containing a higher concentration of quinolic compound suitable for reducing cuprous ions to copper metal. Advantageously, this two step procedure, overall, requires only the stoichiometric quantity of quinolic compound, or a slight excess thereof, to be circulated in the recovery system.

In such a two stage process, the rate of reduction and precipitation of copper increases as the pH of the copper containing solution is increased. When the percentage of aromatic (or nonpolar) solvent in a mixed organic solvent system carrying the organic extractants is increased, the rate of reduction of copper to metal is increased. These discoveries permit a process to be designed which significantly increases the rate of copper production and has important economic advantages.

In a similar process by *J.N. Gerlach; U.S. Patent 4,032,332; June 28, 1977; assigned to Kennecott Copper Corporation* the rate of copper metal produced is significantly increased by the presence of chloride, bromide or thiocyanate (SCN^-) ions in the copper bearing pregnant liquor.

Even trace amounts of these ions have a beneficial, rate-increasing effect, and high concentrations may be used if desired. To take advantage of this behavior, these ions should be added in the form of a soluble salt in a concentration determined by economic factors such as the cost of the salt and the effect, if any, of various concentrations of the ions on the components of the recovery system.

In the case of acidic copper solutions, high concentrations of chloride have been observed to slow or even stop copper precipitation. On the other hand, low concentrations of chloride, i.e., on the order of 0.05 M or less, have a beneficial effect similar to that observed in basic solutions.

In order to conserve the quinolic compounds, *J.N. Gerlach; U.S. Patent 4,032,331; June 28, 1977; assigned to Kennecott Copper Corporation* deals with the solubility of the quinolic compounds, i.e., the reduced quinolic compound (quinol), as a general phenomenon, is much more soluble in water than is the oxidized quinonic compound (quinone). Because of this high solubility, there are significant losses of the quinolic reductant through the copper depleted raffinate. To prevent these losses, it has been discovered that water soluble quinolic compounds may be recovered by oxidizing them, in situ, to the corresponding quinonic compound. In this regard, the aqueous phase left over after the copper reduction is complete has been found to contain upwards of 600 ppm quinolic compounds before oxidation. However, after oxidation, the organic extractant concentration in the aqueous phase is on the order of 1 to 10 ppm.

The following examples illustrate the major aspects of the three processes.

Example 1: An aqueous, cupric ion containing solution was prepared by adding 19.65 g of $CuSO_4 \cdot 5H_2O$, 17.0 g of NH_3, and 36.0 g of $(NH_4)_2SO_4$ in enough water to make 1 liter of solution. 21.3 of 2-ethylanthraquinol were dissolved in enough mixed solvent comprising 70% xylene (nonpolar) and 30% 2-ethylhexanol (polar) to produce 1 liter of solution. At 50°C and under an inert atmosphere, 100 ml of the aqueous solution were placed in a 500 ml reactor vessel fitted with baffles and a 5 cm marine propeller rotatable at 2,000 rpm.

An amount of the mixed solvent containing the quinol prepared as disclosed above containing the stoichiometric amount of 2-ethylanthraquinol necessary to reduce all Cu^{++} to Cu^+ was then added to the vessel and the propeller was energized. When the Cu^{++} was reduced to Cu^+ the mixing was stopped and the spent organic phase removed. A twofold excess of quinol was added to the reaction vessel and the propeller energized. The results of two runs utilizing this procedure are set forth below:

Time (min)	.Cu^{++} in Aqueous Phase (ppm). No. 1	No. 2
0	5,000	5,000
1	–	292
2	23.7	4.83

As can be seen from these data, essentially quantitative reduction of Cu^+ to Cu° was achieved in a very short time. The partially spent organic phase can now be used to reduce Cu^{++} to Cu^+.

At this point it should be noted that the foregoing two stage reduction scheme is a preferred embodiment. However, the second process involves the use of ions such as chloride, bromide, thiocyanate, and mixtures thereof to increase the rate of reduction of soluble copper to copper metal.

In connection with the foregoing, effective amounts of these ions can be utilized to great advantage. The process does not have to be limited to the two stage reduction system discussed above.

According to another aspect of this process, it has been discovered that the pH of the aqueous phase, i.e., the pH of the phase containing the copper ions, influences the rate of copper reduction. In general, the higher the pH, the faster the reaction rate. A study of reaction rates versus pH in the alkaline range indicates that, in general, the rate increases as pH increases. Similar behavior has been observed in the acid range. However, extrapolation of pH-reaction rate curves over the entire pH range is not feasible. This latter behavior is believed to be a direct result of the different copper species present in ammoniacal and acid solutions.

During the reduction of copper, hydrogen ions are liberated when the quinonic compounds are formed and hence, the pH is lowered as the reduction proceeds. In view of this phenomenon, in accordance with the teaching of this process, the ideal process should employ copper solutions having a pH on the order of 9.5 or higher, e.g., fairly concentrated ammoniacal solutions, thereby decreasing the time required to achieve a suitable amount of copper reduction. The preferred pH value range is from about 9.5 to about 11.0. Obviously, this discovery may be utilized to advantage in conventional anthraquinol copper extraction processes.

The preferred mode for practicing this process utilizes chloride ions to accelerate the reduction of copper ions. Depending on the concentration and the pH of the copper solution, the presence of chloride ion in the copper containing aqueous solution can accelerate, decelerate, or even prevent the reduction of copper to metal by quinols. In the acid range, low chloride concentrations have been found to accelerate the reduction process; whereas, high chloride concentrations slow or even stop the process.

In the basic range, i.e., pH above about 7, any concentration of chloride in the aqueous copper bearing liquor increases the reaction rate. Low concentrations of bromide ion (Br^-) and thiocyanate ion (SCN^-) also increase the reaction rate and may be used in this modification of the process. Accordingly, it is apparent that a significant impact on the economics of the process can be made by the inclusion of appropriate amounts of chloride ions, bromide ions, thiocyanate ions and mixtures thereof. The following examples illustrate the principle.

Example 2: Four ammoniacal copper solutions were subjected to batch treatment with 2-ethylanthraquinol dissolved in a mixed solvent comprising 62.5% xylene and 37.5% 2-ethylhexanol at 40°C. At regular intervals, aliquots of the solutions were extracted and analyzed for the presence of soluble copper (Cu^{++} and Cu^{+}). The concentration of Cu^{+} in the solutions and the concentration of Cl^{-} are indicated below.

Table 1

	Test 1	Test 2	Test 3	Test 4
	M (Cl^{-})			
	0	0	0.015	1
	M (Cu^{+})			
Time	0.079	0.077	0.077	0.079
(min)	Cu^{++} Remaining in Solution (ppm)			
5	–	3,750	340	–
10	1,472	1,880	20	2
20	417	185	1	–
30	50	4	–	–

As can be seen from Table 1 by comparison of the results of the tests on solutions 3 and 4 with those of solutions 1 and 2, the inclusion of Cl^{-} significantly increases the rate at which copper is extracted from the solution. Further, the acceleration is substantial at chloride concentrations as low as 0.015 M and does not decrease as the molarity is raised.

Example 3: A 10% ammonia solution of copper sulfate was contacted with an excess of 2-tert-butylanthraquinol in an organic solvent consisting of 40% 2-ethylhexanol and 60% xylene. After mixing 5 minutes, copper metal was removed by filtration and the organic and aqueous phases were separated. Before bubbling air through the aqueous raffinate, 600 ppm of 2-tert-butylanthraquinol was present. After bubbling air through the raffinate to form insoluble 2-tert-butylanthraquinone, only 3 ppm of 2 tert butylanthraquinol remained.

It should be noted that the oxidation of the quinol which would otherwise be lost in the aqueous raffinate may take place either before or after the organic and aqueous phases are separated, but in any case, of course, should not be effected prior to separation of copper from the liquid phases. While bubbling oxygen through the raffinate is a preferred method of oxidizing the quinol, other oxidizing gases, such as chlorine gas, may be substituted for oxygen, and indeed, that solutions of oxidizing agents such as ferric or hypochlorite ions may be added to recover the selected organic extractant in the oxidized form.

This quinol recovery may be used in a conventional extraction copper recovery process. The excesses, typically 10 to 15%, of quinolic extractant used in such processes render the advantage achievable by this procedure particularly significant.

Anthraquinone Treatment in Organic Solvent

J. Manassen; U.S. Patent 3,820,979; June 28, 1974; assigned to Yeda Research

& Development Co., Ltd., Israel describes a process for the production of metals selected from the group consisting of copper, silver and mercury. It comprises contacting an aqueous solution of the metal compound with a solution of a reduced quinonic compound, preferably an anthraquinone compound at ambient temperature and pressure, in an organic solvent which is substantially insoluble in and immiscible with water, resulting in the precipitation of the metallic compound which is recovered. The organic phase may be separated, the quinonic compound may be reduced and used again for treating subsequent batches of metal salts.

An important consideration in the choice of the anthraquinones is its solubility in the organic solvent system used. If concentrated solutions are preferred, it is advisable to make use of compounds such as tert-butylanthraquinone, which is of higher solubility than, for example, methylanthraquinone. This applies both to the reduced as well as to the oxidized state of these compounds.

Solvent systems suitable both for the quinonic and for the reduced form are combinations of aromatic and polar solvents. In order to minimize losses due to the evaporation of the organic solvent, solvents such as alkyl toluenes, alkyl naphthalenes or diphenyls are advantageously resorted to. As polar solvents, there are advantageously used compounds such as octanol, ethyl hexanol or heptanol. Esters such as dialkyl phthalates, diaryl phthalates, alkyl benzoates, benzyl acetates, ethyl heptanoates, cyclohexanol acetates or propionates can be used as such or in combination with one of the solvents set out hereinbefore.

The quinonic form of the compounds used in the process can be reduced by means of conventional hydrogenation catalysts such as Raney nickel, or by catalysts of the platinum or palladium type, supported on conventional supports, such as activated carbon, barium sulfate, calcium carbonate, silica gel or the like.

It is possible to reduce and precipitate a large part of the metal contained in the solution, which is reducible by the reducing agent used, in one stage, employing an excess of the reducing agent. It is advantageous to effect the process in two or more stages, using in each stage a suitable quantity of fresh organic solution. It is also possible to effect the process in countercurrent.

Example 1: A quantity of 6 g tert-butylanthraquinone was dissolved in a mixture of 50 ml xylene and 50 ml 2-ethylhexanol. The anthraquinone was reduced at atmospheric pressure and ambient temperature by admixing a quantity of 140 mg of 10% palladium on calcium carbonate as catalyst and by introducing hydrogen until the absorption of hydrogen ceased. The catalyst was filtered off, and the thus obtained solution was divided into two equal portions, A and B.

Portion A was added to a solution of 10 mmol of copper sulfate in 150 ml water and the reaction system was stirred during one hour. Copper powder was precipitated, filtered off and the organic phase was separated from the aqueous one. The aqueous phase was admixed with the portion B and the process was repeated for a further hour. At the end of this period of time, the precipitate was filtered off, the organic phases were united and again subjected to catalytic reduction. The process was repeated four times, each time with an identical quantity of aqueous copper sulfate solution. The following table summarizes the results obtained:

Run No.	Copper from 1st Stage, mg	Copper from 2nd Stage, mg	Total Copper, mg
1	447	176	623
2	400	218	618
3	363	250	613
4	376	238	614

The copper powder was analyzed and found to be 100% pure copper. The theoretical value is 635 mg. The difference is due to mechanical losses and due to the work with small quantities. The organic layer did not contain any copper and the residual aqueous solutions contained only a few mg.

Example 2: A solution of 2.22 g 2-methylanthraquinone in 75 ml 2-ethylhexanol and 75 ml xylene was reduced as set out in the previous example. It was stirred with an aqueous solution of 1.5 g of silver acetate in 150 ml water for 45 minutes. There was formed a precipitate of metallic silver, which was removed by filtration. There was obtained a quantity of 969 mg of silver. The purity was 100%.

Example 3: A reduced organic solution as in Example 2 was stirred during 45 minutes with a solution of 2.5 g mercuric acetate in water. A precipitate of the metallic mercury was obtained. It was 100% pure.

Reduction of Ammoniacal Solutions

A.S. Rappas and J.P. Pemsler; U.S. Patent 4,047,940; September 13, 1977; assigned to Kennecott Copper Corporation describe a process for recovering copper from solutions containing cupric ammoniacal complexes by reducing copper in the complexes to the cuprous state and thereafter precipitating the copper as cuprous acetylide by passing acetylene through the solution. After the cuprous acetylide is separated from the solution and washed thoroughly, it is reacted with acetonitrile and an acid to form acetylene and a cuprous-acetonitrile complex: $Cu(CH_3CN)_2^+$. The solution containing this complex is then flash distilled to remove the acetonitrile and to disproportionate unstable cuprous ions into copper metal and cupric ions. Following the disproportionation, the components are separated and copper metal is collected.

Example: Approximately 13.4 g of basic cupric carbonate (55.9% assay as Cu) and 10 g of ammonium carbonate are together dissolved in 300 ml of ammonium hydroxide (30% NH_3), and water is added to bring the total volume to

1 liter. The dark blue ammoniacal solution is introduced into a flask for reduction with carbon monoxide. The temperature within the flask is kept constant at 45°±2°C, and the pH of the solution is adjusted to 10.5 by controlling the relative amounts of ammonia and/or carbon dioxide introduced. The air initially in the flask is displaced with carbon monoxide and the off-gases are passed through a gas flow indicator.

Stirring commences when the air is completely displaced from the flask. A piece of copper wire, which has previously been cleaned by dipping in nitric acid solution, is introduced into the flask to initiate the reduction of cupric to cuprous. With a small positive flow of carbon monoxide maintained through the flask, the progress of the reduction is monitored by withdrawing samples of the solution with a gas tight syringe.

The test samples removed are assayed for cuprous ion by being subjected to oxidation with a ferric solution, and thereafter titrating the resulting ferrous ions produced with a standard ferric solution.

When the reduction is complete, 600 ml of the colorless cuprous solution (containing 4.7416 g of copper) is pumped to the flask, while maintaining a low argon flow rate to prevent oxidation. Prior to introducing the cuprous solution into the flask, the air therein is displaced with carbon monoxide. With the 600 ml sample of cuprous solution in the flask, prepurified acetylene gas is introduced and the solution is stirred moderately. The reaction between the cuprous ions and acetylene is almost instantaneous at the temperature employed, i.e., 25° to 30°C.

In a few minutes, precipitation of cuprous acetylide is complete employing only a small positive flow of acetylene gas and moderate stirring. The slurry is then filtered under an argon atmosphere, the filtrate being collected in a collecting chamber and analyzed by atomic absorption to determine the amount of copper left in the solution. The filter cake is washed with 200 ml of a dilute ammonia solution (pH 9.0-9.5) and the wash solution is analyzed by atomic absorption to determine percent copper in the wash solution.

The amount of solid in the filter cake is determined by dissolving a weighed amount of the wet acetylide in HCl and analyzing the solution for copper. The amount of solid is calculated assuming that all copper in the solid is present as Cu_2C_2.

28.80 g of the filter cake obtained as disclosed above is then transferred into a four necked, round bottom flask. The handling of the wet cuprous acetylide is conducted such that contact with air is avoided or minimized in order to prevent oxidation. A solution comprising 44.8 ml H_2O, 6 ml H_2SO_4 (96.4%), and 23.0 ml CH_3CN, at approximately 40°C, is added to the wet acetylide in the flask.

Prior to this step, argon is introduced at a low flow rate to remove all oxygen.

The acetylide dissolves completely within 25 minutes at 40°±2°C to give a light, yellowish clear solution. The acetylene produced from the dissolution of the acetylide in the flask is carried from the flask in the argon stream and may be scrubbed as desired in either of two acetylene scrubbing systems.

When the dissolution is complete the outflow of gas is switched from the refluxing condenser to the spiral condenser. The temperature in the water bath is increased in order to distil off the acetonitrile from the cuprous solution, and the cuprous ions are thus disproportionated into copper metal and cupric ions.

An azeotrope, $CH_3CN:H_2O$ = 83.7:16.3, boils at 76.5°C. The heating of the flask continues until the temperature reaches 89°C at which point the condensate is collected (32 ml). The solution remaining in the flask (66 ml) contains 2.051 g of Cu^{++}. The copper metal is collected by filtration, washed with water, acetone, and finally benzene, and thereafter dried under argon.

To determine the total amount of copper metal produced, the empty flask is rinsed with water and then a solution of HNO_3 is used to dissolve any copper that adheres to the walls. The concentration of copper in this solution is determined by atomic absorption, and the total weight of copper recovered is calculated by adding the weight of the copper in this solution to the copper powder collected. Total copper metal produced under the above conditions is 1.899 g.

Recovery from Acidic Leaching Liquors

This process by *A.S. Rappas and J.P. Pemsler; U.S. Patent 4,038,070; July 26, 1977; assigned to Kennecott Copper Corporation* is useful as a recycle step for the previous process. It is a continuous process for recovering copper from acidic leaching liquors containing cupric ions in admixture with other metal ions. Briefly, copper values are recovered from an acidic solution of solubilized cupric ions by the reduction of cupric copper to cuprous copper with hydrogen gas in the presence of a solid hydrogenation catalyst and in the presence of a cuprous ion stabilizing ligand.

Catalyst poisoning by precipitation of copper metal on the surface of the catalyst during the reduction is avoided by controlling the ratio of Cu^+/ligand/Cu^{++} and by terminating the hydrogenation before copper metal precipitates on the catalyst. The solution containing the cuprous ions is separated from the solid catalyst and the cuprous ions are then disproportionated to produce copper metal and cupric ions.

The removal of the last amounts of cupric ions from solutions, after several stages of reduction and disproportionation is achieved by allowing the cupric ions to precipitate on the catalyst as copper metal. The catalyst is then separated from the copper barren raffinate and regenerated by exposing it to a fresh pregnant cupric ion solution in the presence of a cuprous ion stabilizing ligand.

Example 1: *Process of Acidic Liquors*—Dissolve 165.16 g of $CuSO_4 \cdot 5H_2O$ in 800 ml H_2O and introduce solution in the catalytic reactor under an argon atmosphere. Heat to 47.5°±1.5°C. Adjust pH to approximately 2 with concentrated H_2SO_4. Bring total volume of solution to 850 ml. Make a final pH adjustment to pH = 2.0. Add 0.60 g of catalyst (5% Pd on Al_2O_3), and let it settle (no stirring). Introduce H_2 at low flow rate to displace the argon. Add 150 ml of acetonitrile.

The solution contains 43.05 g of copper ions. Keep a low positive flow of hydrogen through the reactor. Commence stirring (time zero for reduction). Take periodic samples of solution and determine Cu^+. In 38 minutes stop stirring: 41.01 g Cu^+/l, 2.04 g Cu^{++}/l; [(Cu^+,final/Cu^{++},inital) x 100 = 95.2%]. Filter under inert atmosphere to separate catalyst from cuprous solution. Transfer 800 ml of filtered cuprous solution to the distillation flask. Distill to disproportionate the cuprous complex to a final temperature of 90°±1°C (condensate: 145 ml of acetonitrile and water).

Filter to separate copper powder from cupric solution. (Cu°, 15.077 g; [Cu^{++}], 28.49 g/l.) Take 500 ml of this filtrate (28.49 g Cu^{++}/l) and transfer to the catalytic hydrogenation reactor. Repeat above procedure for the reduction to cuprous acetonitrile complex. T° = 47.5°±1.5°C, S (wt of catalyst) = 0.5 g of previously used catalyst after washing and drying in argon; 50 ml acetonitrile added. The [Cu^{++}] in the mixed solution is 25.9 g/l. In 13 minutes stop stirring: 25.00 g Cu^+/l. Filter to remove catalyst from solution. Distill to disproportionate the cuprous-acetonitrile solution. Final temperature 91°C, condensate 60 ml (acetonitrile and water). Filter to separate copper metal (6.48 g) from cupric solution (15.816 g Cu^{++}/l).

Take 300 ml of the above filtrate and introduce in the catalytic reduction reactor. Adjust pH to 1.0 with solution of NaOH or H_2SO_4 as needed. Raise the temperature to 72°±1°C and add 0.617 g catalyst Pd 5% on Al_2O_3. Introduce H_2 gas and start stirring as described before. Notice that no acetonitrile is added this time. Copper metal precipitates. Take periodic samples and determine total copper ions left in solution. Stop stirring after 150 minutes. Filter to separate Cu°/Pd catalyst. Determine copper ions in raffinate. Raffinate contains 15 ppm Cu.

The solid on the filter is a mixture of Cu° and Cu°/Pd. To regenerate the Pd catalyst, take the solids on the filter, introduce into a flask with fresh (initial) cupric sulfate solution add 10% acetonitrile and stir under inert atmosphere (argon) at 30°-45°C until all copper dissolves. Filter under inert atmosphere. The solid on the filter (regenerated catalyst) is washed with acidified water (H_2SO_4, pH 2) and then with water and dried in an argon atmosphere. The catalyst can be reused.

Example 2: *Recycle Stop*—Dissolve 31.45 g of $CuSO_4 \cdot 5H_2O$ in 260 ml H_2O. Adjust pH to approximately 0.7 with H_2SO_4. Bring total volume of solution to 280 ml and make a final pH adjustment (pH = 0.7). Heat solution to 47.5°±1.5°C and add 40 ml acetonitrile. Add 0.4 g of catalyst (5% Pd on Al_2O_3) and follow the procedure of Example 1, to reduce cupric to cuprous acetonitrile complexes. Within 30 minutes the cuprous ion concentration is 23.99 g Cu^+/l; total copper in solution is 7.9983 g. Filter, under inert atmosphere, to separate catalyst from solution. Transfer 200 ml of the cuprous solution to an addition flask. Mix 26 ml of acetonitrile into solution in the addition flask.

Cuprous acetylide is precipitated, filtered, repulped, refiltered and thoroughly washed on the filter. The amount of acetylide in the wet filter cake (16.29%) is determined as before. 16.17 g of the acetylide cake is introduced into another flask under inert atmosphere (argon). 113 ml of the solution is transferred from the first flask into the second flask and the slurry is magnetically stirred. Dissolution of acetylide and scrubbing of acetylene evolved occur. The temperature is maintained at 60°C. Samples are taken to determine the concentration of cuprous ions and total copper in solution. Within 25 minutes the concentration of cuprous ions is 37.208 g Cu^+/l and the total copper in solution is 38.016 g Cu/l. The acetylene produced during dissolution of acetylide is scrubbed and determined quantitatively. It is found that 369.6 ml (standard temperature and pressure) of acetylene are recovered.

The cuprous solution is then distilled to disproportionate the cuprous ions into metallic copper and cupric ions in solution. The copper powder is separated from the cupric solution. The amounts of metallic copper and the cupric ions in solution are determined as described in the example of the previous patent.

The following quantities are obtained at the end of the distillation:

Copper powder, 2.2863 g Cu°
CH_3CN/H_2O condensate, 33 ml
Cupric sulfate solution, 25.967 g Cu/l

Reduction of Cupric Sulfate Solutions Using Acetonitrile

A.J. Parker, W.E. Waghorne, D.E. Giles, J.H. Sharp, R. Alexander and D.M. Muir; U.S. Patent 3,966,890; June 29, 1976; assigned to Anumin Pty. Limited, Australia describe a process for providing a composition of matter suitable for processing copper via cuprous ions. It comprises at least 15% by volume water, an acid selected from the group consisting of sulfuric, sulfurous and nitric acid, at least 4% by volume of an organic nitrile and a copper salt of the corresponding acid.

The particular nitrile used is acetonitrile and the reductant is sulfur dioxide. The method can be applied commercially to leaching ores containing oxidized copper to obtain cuprous ion solutions. Such solutions can be treated by electrochemical or thermal disproportionation methods to obtain very pure copper.

Example 1: 20 g of copper concentrate (analyzing 75% copper) prepared by the Torco process (Nchanga Consolidated Copper Mines, Copper Refineries Pty. Ltd.), was −60 mesh and was stirred with 60 g of $CuSo_4 \cdot 5H_2O$ in 100 ml of acetonitrile, 250 ml of water and 10 ml of sulfuric acid at 50°C for 1 hour under nitrogen. The solution was filtered and distilled slowly under nitrogen to give 11.4 g of copper. The residue was 56 g containing 0.36 g of copper. Thus the extraction was 97.4% of the available copper.

Example 2: 25 g of −60 mesh freshly prepared cupric sulfide was stirred in a reaction vessel under reflux for 1 hr with a composition of 110 g $CuSO_4 \cdot 5H_2O$, 1,100 ml acetonitrile, 1,000 ml water, and 110 ml sulfuric acid. The blue color rapidly decreased in intensity and coagulated sulfur was observed.

The solution was filtered to leave 5 g of a gray material, containing 0.15 g of copper, soluble in carbon disulfide and which gave yellow sulfur on evaporation of the carbon disulfide. The filtrate contained 44 g of copper, corresponding to the theoretical value.

It was thermally disproportioned to give 9 g of copper powder, contaminated with a little cuprous sulfide.

Leaching of Cuprous Sulfate and Bisulfate Solutions

In a related process, *A.J. Parker, W.E. Waghorne, D.E. Giles, J.H. Sharp, R. Alexander and D.M. Muir; U.S. Patent 3,961,028; June 1, 1976; assigned to Anumin Pty. Limited, Australia* describe a method of preparing cuprous sulfate and bisulfate solutions from materials containing copper and impurities by a leaching process. The leaching composition contains sulfuric acid, cupric ions, water and an organic nitrile. The leaching composition may be used with copper sulfide ores in particulate form. The cuprous solutions resulting from the leaching can be treated by disproportionation methods to recover pure copper.

Recovery by Continuous Gaseous Reduction

D.R. Neskora, C.M. Sliepcevich and R.L. Brown; U.S. Patent 3,877,931; April 15, 1975 have developed a continuous gaseous reduction process for obtaining relatively pure copper, nickel, cobalt, silver or gold from a solution of the salt or salts of one or more of these metals by reduction with a reducing gas.

They use a tubular reactor and pass the reducing gas in countercurrent flow with respect to the treated solution containing the metal values to be recovered. Thus, less of the reducing gas is needed, and both the downstream gas/liquid-solid separator and the heat exchanger of the reducing gas as required in the cocurrent flow procedure can be eliminated. Improved product quality has also been observed.

In the process, an aqueous solution containing reducible copper, nickel, cobalt, silver or gold ions is charged to one end of an elongated tubular reactor and is passed through the reactor in countercurrent flow with respect to a reducing gas. Metal powder produced by the reduction reaction is withdrawn from the tubular reactor at a point downstream from the end of the reactor to which the raw solution containing the unreduced metal values is initially introduced. In the course of the reduction occurring in the tubular reactor, steady state conditions are maintained over the length of the reaction path, and localized back mixing, with modified plug flow, is engendered in the reaction mixture, either by the reducing gas being bubbled therethrough in countercurrent flow, or by this action in combination with suitable static mixing devices.

A reducing gas partial pressure of at least 50 psig is maintained in the reactor in conjunction with a total pressure sufficient to prevent boiling of the solution at the temperature prevailing in the reactor. The reactor is operated so that a minimum temperature of at least 200°F is attained by the reaction mixture in traversing the reaction path. The amount of reducing gas employed is in excess of that stoichiometrically required to reduce all of the reducible metal values present in the solution and sought to be recovered as metal powder.

The flow of materials in the process, and the major components of the apparatus utilized, are schematically illustrated in Figure 5.1a. In the illustrated arrangement, a strongly acidic leach solution is being utilized as the source of metal values to be recovered, and in view of the acidity, no pretreatment is required and no pretreating equipment or components are illustrated in the system. The feed solution enters the top of an elongated tubular reactor **10** which extends substantially vertically, and flows downwardly by gravity through the reactor.

Reducing gas is charged to the lower end of the tubular reactor via a charging conduit **12** which extends concentrically inside an effluent discharge conduit **14**. Both reducing gas and depleted solution pass through a product cooler and heat exchanger device **16** which facilitates heat exchange between a relatively cool heat exchange medium, used in a manner hereinafter described for heating the tubular reactor, and the product solids and spent solution discharged from the reactor.

It should also be noted that in charging the gas to the reactor through a conduit which is concentrically located with respect to the effluent discharge conduit, the reducing gas is warmed before it enters the reactor by the relatively warm effluent. The spent or depleted solution and the particles of product metal discharged from the reactor are passed to a separations unit for recovering the metal, then to a washing step and ultimately to a drying zone.

The tubular reactor **10** is heated in the manner hereinbefore described to provide a temperature at some location in the reactor which is at least 200°F. In the illustrated embodiment, single stage heating is provided for establishing a

Figure 5.1: Recovery of Metal Powders by Continuous Gaseous Reduction

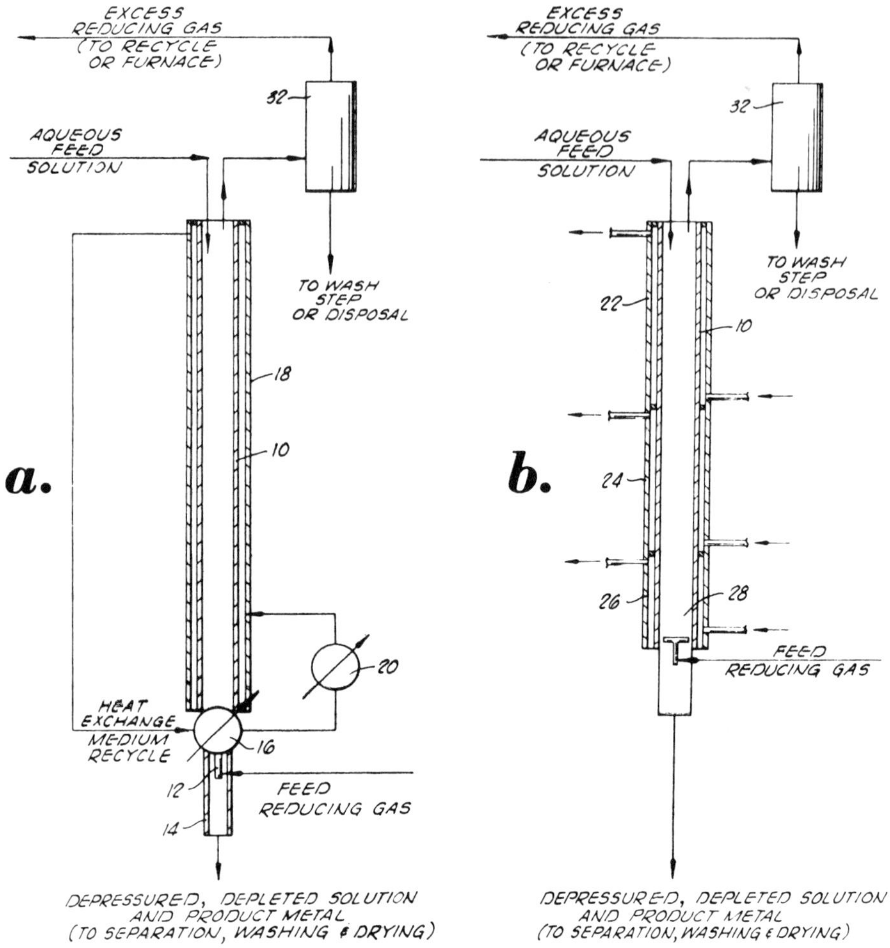

Source: U.S. Patent 3,877,931

uniform temperature gradient ranging from ambient temperature at the upper end of the reactor to about 600°F at the lower end of the reactor. The heating system includes a jacket **18** surrounding the tubular reactor over a major portion of its length and a conduit for recycling the heat exchange medium via the heat exchanger **16** and a make-up heater **20** to the lower end of the reactor **10**.

Alternatively, and preferably, as illustrated in Figure 5.1b, two stage heating may be used in conjunction with the tubular reactor **10**, with a first stage jacket **22** providing a temperature gradient of from ambient temperature at the upper end of the reactor to a temperature of from about 400°F to about 600°F at about the mid point of the reactor, followed by a second stage provided by a jacket **24** which surrounds the portion of the reactor below the jacket **22**.

A heat exchange medium is circulated in the jacket **24** to provide either a constant temperature of around 400°F or a decreasing temperature gradient of from about 400° to 600°F at the mid point of the reactor to a temperature of about 300°F at that portion of the reactor adjacent the lower end of the jacket **24**.

It has been determined that when the system depicted in Figure 5.1a is utilized for cooling the spent solution and solid product discharged from the reactor **10**, constrictions in the cooling assembly and valving in the cooling assembly, coupled with the inherent rapid cooling encountered upon discharge, have apparently caused some hang-up or retention of solid product in the valving and cooling assembly. The multistage system depicted in Figure 5.1b alleviates the problem to a significant degree. As illustrated, a cooling jacket **26** is provided around the lower end of the tubular reactor **10**, and continues the cooling of the spent solution and solid product before the mixture is discharged from the reactor.

The effluent is then passed directly to the separation, washing and drying stages of the process. In this arrangement, the reducing gas is introduced substantially directly into the lower end of the reactor **10**, and is preferably admitted through an annular, perforated manifold **28** which charges the gas to the reactor at a number of spaced points in the lower end thereof.

In either structural arrangement utilized for carrying out the process, the tail gas, consisting essentially of unused reducing gas, is passed from the top of the tubular reactor and into a liquid knock-out drum **32** where entrained liquid from the reactor is removed from the reducing gas. The liquid is disposed of in any suitable manner, and the relatively dry reducing gas can then be recycled, or, if combustible, supplied to a furnace to provide a part of the heat required in the process.

In the equipment provided for handling and conveying the spent solution product, an inert atmosphere is preferably employed over the solid product to prevent oxidation of the product and to enhance and extend the effective shelf life of the pure metallic powder.

A series of continuous reduction runs using the described countercurrent flow were carried out on acidic solutions of cupric sulfate. These solutions included both actual leach solutions and synthetic solutions. The reductions were effected in two tubular reactors having differing diametric and length dimensions. The conditions utilized in the runs, and the results obtained are set forth in the following table:

Continuous Reduction of Cupric Sulfate Solutions in the Countercurrent-Flow Tubular Reactor

	Reactor No. 1 (¾″ x 60″)									Reactor No. 2 (1″ x 72″)			
	1	2	3	4	5	6	7	8	9	10	11	12	13
Cu^{++} (g/l)	43.2	43.2	52.5	16.5	36.8	45.9	45.9	20.8	16.5	55.7	54.8	9.1	9.3
Liquid feed rate (g/min)	69.0	69.0	67.0	58.9	58.4	65.9	45.9	43.3	43.3	81.1	73.9	108.8	54.6
H_2 feed rate (scfm)	0.076	0.076	0.054	0.054	0.054	0.045	0.047	0.022	0.023	0.056	0.044	0.020	0.020
Reactor pressure (psig)	601	642	703	699	701	706	704	711	714	703	704	711	707
Reactor temp.* (°F)													
(a) 4″ from top	495	602	571	517	524	83	233	154	331	222	445	323	255
(b) 29″ from top**	525	532	541	518	531	428	475	471	483	420	534	449	476
(c) 59″ from top	523	532	548	515	532	355	367	373	376	445	492	454	474
(d) average	517	550	550	517	530	324	388	367	418	377	501	419	421
Residence time (min)	1.78	1.78	1.92	1.89	2.04	2.27	2.87	3.51	3.51	3.89	4.25	3.11	3.85
Conversion of Cu^{++} (%)	73.7	74.3	52.5	53.2	66.1	35.4	70	54.0	68	41.5	54.1	51.3	64.5
pH of liquid feed	1.93	1.96	–	0.31	0.3	0.3	0.3	0.3	0.31	0.21	0.26	2.08	–

*Reactor temperatures (a), (b) and (c) are reactor skin temperatures.

**In the runs carried out in Reactor No. 2, the (b) thermocouple was placed to measure the skin temperature 33 inches from the top of the reactor.

In earlier work by *D.R. Neskora, R.L. Brown and C.M. Sliepcevich; U.S. Patent 3,833,351; September 3, 1974; assigned to University Engineers, Inc.*, the raw or feed aqueous solution is pretreated with ammonia to attain a selected free ammonia to metal mol ratio in the general range of about 0.0:1 to 6:1, depending upon the metal to be reduced. Thus, where copper solutions are pretreated, the free ammonia to copper ion ratio is adjusted by ammonia or acid addition (in the case of ammoniacal leach solution) to from about 0.1:1 to about 1.6:1, and the pH is maintained at less than about 7.0.

In nickel and cobalt reductions, the ammonia to metal ion mol ratio in the feed solution is adjusted to from about 1.25:1 to about 6:1, and a catalytic or heteronucleating agent is added to the solution in order to achieve a satisfactory reaction rate. Silver and gold can be recovered from aqueous solutions having a very low pH, and silver can also be recovered at pH values up to 11.5. Silver solutions are preferably pretreated with ammonia to provide a free ammonia to silver mol ratio of about 0.5:4.5.

Improving Green Strength

S.V. Wilson and P.E. Matthews; U.S. Patent 3,832,156; August 27, 1974; assigned to United States Bronze Powders, Inc. describe a process for forming high green strength particles suitable for powder metallurgical applications, which particles possess irregular shapes and highly desirable strength and hardness characteristics. According to the process, malleable spherical elemental powders, blends of spherical elemental powders at least one of which is malleable, and malleable prealloyed powders may be converted to a particle form which exhibits high green strength on cold consolidation.

The process comprises mechanically working the powders to flake form, such as in a ball mill, annealing the flakes above the recrystallization temperature thereof in a nonoxidizing atmosphere to form a sinter cake, and mechanically disintegrating the cake into irregularly shaped particles having high green strength. These high green strength particles may then be compacted and sintered according to conventional techniques to form the desired article.

Example 1: Mono-metallic spherical, ductile copper powder manufactured by particulation from a molten stream with air was converted to irregular particles according to the present process. Spherical copper, U.S. Bronze Powders, Inc. Grade #2 (percent Cu 99.2 min) having a bulk density of 4.56 g/cc was used as the starting material powder. The copper powder exhibited a green strength on cold consolidation to 7.7 g/cc of less than 10 psi and had a particle size distribution according to the following screen analysis:

	Wt. % Retained
+ 80 mesh	0.1
+100 mesh	0.2
+140 mesh	6.1
+200 mesh	11.0
+325 mesh	22.3
Pan	60.3

The starting material copper powder was ball milled in a continuous type mill using 7⁄32" diameter balls, a ball mill speed of about 54 rpm and a recirculating air atmosphere. About 0.36% by weight of stearic acid was added with the feed metal powder. The mill was externally cooled with water. The resulting flakes which had a bulk density of 0.6 g/cc, were annealed in a Drever furnace in a dissociated ammonia atmosphere for 15 minutes at 1650°F, after which the resulting cake was passed through a Weber Brothers pulverizing mill and the pulverized powder hand screened through an 80 mesh U.S. standard sieve. The resulting irregular particles had a bulk density of 2.08 g/cc, a green strength at 7.77 g/cc of 5,840 psi and a screen analysis as follows:

	Wt. % Retained
+ 80 mesh	–
+100 mesh	9.2
+140 mesh	17.2
+200 mesh	18.0
+325 mesh	21.7
Pan	33.9

Example 2: A bi-elemental powder blend of copper and nickel, in the proportions of 75% by weight copper to 25% by weight nickel, was converted to an alloy powder according to this process. Starting materials were U.S. Bronze Powders, Inc. Grace C-132 copper powder and Alcan Metal Powders MD-101 nickel powder. The bulk densities of the copper and nickel powders were 2.45 g/cc and 3.70 g/cc respectively. The powders had a particle distribution as shown by the following screen analysis:

	. . . .Wt. % Retained. . . .	
	Copper Powder	Nickel Powder
+ 80 mesh	Trace	0
+100 mesh	Trace	0.1
+140 mesh	1.1	12.9
+200 mesh	5.7	17.0
+325 mesh	29.2	25.0
Pan	64.0	45.0

The elemental powders were blended in the proportion 75% by weight copper/25% by weight nickel for approximately 15 minutes to serve as the ball mill feed powder. The powder was ball milled in a batch type mill utilizing 5⁄16" diameter ball and a mill speed of about 46 rpm in a carbon dioxide purged atmosphere. The mill was not externally cooled. No lubricant was added for the first two hours of mill operation to maximize cold welding of copper to nickel. After two hours of milling 0.5% by weight stearic acid was added to the mill based on the weight of metal powder in the mill.

The mill was repurged with carbon dioxide and restarted. Following eight hours total milling time the bi-metal flake product exhibited a bulk density of 1.34 g/cc. The flakes were annealed in a Drever furnace for 30 minutes at 1800°F in a dissociated ammonia annealing atmosphere. Following annealing the flakes were precrushed using a mortar and pestle and pulverized in a Weber Brothers pulverizing mill, after which they were hand screened through a U.S. standard sieve 60 mesh. The resulting particles had a bulk density of 2.39 g/cc and a particle distribution as evidenced by the following screen analysis:

	Wt. % Retained
+ 60 mesh	0.4
+ 80 mesh	12.1
+100 mesh	12.9
+140 mesh	23.3
+200 mesh	24.2
+325 mesh	17.0
Pan	10.1

The irregular particle shaped alloy product was comparatively tested against the 75% copper/25% nickel elemental powder blend starting material and against the powder blend starting material annealed for 30 minutes at 1800°F in a dissociated ammonia annealing atmosphere. By blending each of the starting material powder blends, the annealed powder blend and the particles prepared by the present process with ¾%, by weight, lithium stearate and subjecting each blend to a compacting pressure of 30 tons per square inch to achieve a compacted density of 7.6 g/cc the green strength, transverse strength and Rockwell Hardness could be tested by standard test procedures. The transverse strength and Rockwell Hardness tests were conducted on Metal Powder Industries Federation standard 1.25" x 0.50" x 0.25" compacted transverse bars sintered for 30 minutes at 2000°F in a dissociated ammonia atmosphere.

75% Cu, 25% Ni	Blend	Blend-Annealed	This Process
Green strength (lb/in^2)	1,480	1,137	1,938
Transverse strength (lb/in^2)	26,100	28,700	58,100
Rockwell hardness	H 36	H 37	H 74

Carbon Monoxide in Winning Process

M. Dubeck and S.J. Im; U.S. Patent 3,918,962; November 11, 1975; assigned to Ethyl Corporation describe a process for winning copper from chloride containing solutions by treating with carbon monoxide.

It comprises contacting the solution with carbon monoxide, to form a complex of copper chloride and carbon monoxide, removing the complex of copper chloride and carbon monoxide from the solution, decomposing the complex to copper chloride, and reducing the copper chloride to copper metal.

This process can be applied to separation of iron from copper values in a solution prepared by leaching a copper ore of the sulfide type with HCl, the leaching being conducted in the presence of oxygen-containing gas.

For the process, better results are achieved when it is conducted using metal values in the ionic state and in which an anionic portion of the metal salts consists of chloride ion. It is not necessary for all metal salts in the starting material to be chlorides. Better results are achieved, however, when there is at least 1 gram ion of chloride ion present per each gram ion or gram atom of copper. The ionic state of the copper in the starting material is not critical. Thus, cupric or cuprous copper can be used to initiate the process. Of these, cuprous copper is preferred.

Example: To illustrate the feasibility of this process, a test was carried out to produce the carbon monoxide complex with cuprous chloride utilizing carbon monoxide and cupric chloride solution. The solution contained 58.9 g/l copper and 50.6 g/l iron. This test solution simulates copper-iron containing solutions available by treatment with sulfide copper concentrates. Treatment of the above-identified solution at 170°C using a pressure of carbon monoxide of 445 psia for two hours yielded 60% of the copper as the complex of carbon monoxide with cuprous chloride. When washed, this material contained about 0.05% Fe.

Similar results are obtained when the process is conducted at temperatures of from about 100° to about 250°C and at carbon monoxide pressure of from about 100 to about 1,500 psia. Similar results are obtained when the concentration of copper is from about 20 to about 80 g/l and the concentration of iron is from about 40 to about 80 g/l. Similar results are obtained when the process is conducted at the temperatures and pressures noted from a period of about ½ to about 24 hours. Similar results are obtained at the times, temperatures and pressures noted above when substantially pure carbon monoxide is employed, or the carbon monoxide is added in producer gas, synthesis gas, or water gas.

The complex of carbon monoxide and cuprous chloride produced by this process can be heated to drive off the carbon monoxide. Then the cuprous chloride remaining can be reduced to copper as, for example, with hydrogen.

Copper Powder for Infiltrating Iron Compacts

W.M. Shafer; U.S. Patent 4,168,162; September 18, 1979; assigned to SCM Corporation has developed an infiltrating powder composition for infiltrating porous iron compacts. It comprises a copper powder mixture containing a minor amount of iron powder, copper-manganese alloy powder, stainless steel powder, refractory metal oxide, and aluminum powder.

The preferred amounts are at least 90% copper powder having a particle size less than about 177 μ; between about 2% and 8% iron powder having a particle size less than 20 μ; between about 1% and 5% copper-manganese alloy powder; between about 0.4% and 1% refractory metal oxide powder; between about 0.3% and 0.8% stainless steel powder having a particle size less than about 150 μ; and between about 0.05% and 0.2% aluminum powder.

In practice, the infiltrating composition of this process can be contacted with the porous ferrous work-piece and thereafter heated to sufficiently raise the temperature of the infiltrating composition to cause the infiltrant material to infiltrate the porous iron compact and fill the voids in the iron compact. Infiltration can be carried out at temperatures slightly above the melting point of copper, that is about 1980°F, and preferably between about 2000° to 2350°F. Temperatures greater than 2350°F may cause molten copper to rapidly increase the dissolving of iron.

The tendency of molten copper to dissolve iron will cause excessive erosion of the powder iron compact. A one-step infiltration and sintering process, e.g., can be achieved in about 15 minutes at 2050°F. Desirably, infiltration takes place in an inert or reducing atmosphere. Reducing atmospheres, e.g., include hydrogen, cracked ammonia and endothermic atmosphere, whereas suitable inert gases are nitrogen, argon, and the like.

The infiltrating composition of this process produces a nonsticking residue over a wide range of dew points from about 30° to 50°F when used in an endothermic atmosphere whereby the residue no longer has to be chipped or ground off after infiltration. It can be utilized for simultaneous infiltration and sintering at temperatures of about 2000° to 2100°F in an endothermic atmosphere at a dew point between about 0° and 65°F.

The following examples illustrate the process. All parts and percentages are by weight and all temperatures are in degrees Fahrenheit unless otherwise specifically stated.

Example 1: Infiltrating compositions produced in accordance with this process are as follows. Percentages are weight percent.

Material	Sample C	Sample H
Copper powder*	90.7%	90.7%
Iron powder (−325 mesh)	4.0%	4.0%
Copper-manganese (40/60)	2.8%	
Copper-manganese-silicon (33/65.6/1.4)		3.2%
Titanium dioxide (−80 mesh)	0.7%	0.7%
Stainless steel (−100 mesh)	0.5%	0.5%
Zinc stearate	0.5%	0.5%
Graphite	0.3%	0.3%
Aluminum powder	0.1%	0.1%
Rubrex light blending oil (SG = 0.854 gm/cc)	6 ml/100 lb.	6 ml/100 lb.

*Sieve Analysis	Tyler %
−80 +100	trace
−100 +150	0.5
−150 +200	15.4
−200 +325	34.8
−325	49.3

The foregoing infiltrating composition samples had the following characteristics.

	Sample C	Sample H
Apparent density	3.24 gm/cc	3.26 gm/cc
Flow rate	33 sec/50g	31 sec/50g
Green density, compressed		
at 12 tsi	6.19 g/cc	6.21 g/cc
at 20 tsi	6.89 g/cc	6.97 g/cc
Green strength, compressed		
at 12 tsi	277 psi	305 psi
at 20 tsi	641 psi	832 psi

The foregoing samples were tested on a 15 g porous iron test bar comprising 99 parts iron powder, 1 part graphite, and 1 part zinc stearate lubricant compacted to 5.85 g/cc. These test bars were then infiltrated with 5 g slugs produced from infiltrating composition samples C and H above. The slugs were compacted to a density of about 6.80 g/cc. The tests were conducted in a best furnace preheated to 1100°F and then to a high heat of 2060°F. The test parts were held for 40 minutes in the high heat zone at a dew point of 44°F in an endothermic atmosphere.

TEST RESULTS

	Sample C	Sample H
Number of samples	5	8
Adhesion	none	none
Erosion	very slight	slight
Efficiency	87.6%	90.4%
Dimensional change	−.12%	+.06%
Infiltrated density	7.47 g/cc	7.53 g/cc
MOR (modulus of rupture)	167 ksi	170 ksi

Example 2: An infiltrating composition tested under conditions comparable to Example 1 and containing copper-manganese silicon powder but containing higher amounts of silicon, over 2.1%, caused severe adhesion and unacceptable growth believed to be due to the high oxygen content of the Cu-Mn-Si component (32 Cu-Mn-2.1 Si). In contrast, a minor amount of silicon in the Cu-Mn-Si component such as 1.4% silicon as used in sample H improves the efficiency of the infiltrating mixture.

Example 3: Infiltrating material sample H was infiltrated at different dew points of 32° and 49°F respectively. The compacted test bars were infiltrated as in Example 1 at 2060°F and endothermic atmosphere with the following results. The results are based on 3 samples compacted to 6.6 g/cc.

Dew point	32° F.	49° F.
Adhesion	none	none
Erosion	very slight	very slight
Efficiency	88.4%	87.3%
Dimensional change	−.49%	−.17%
Infiltrated density	7.53 g/cc	7.52 g/cc
MOR (modulus of rupture)	180 ksi	177 ksi

ALUMINUM

Treatment of Raw Ores

M.F. Gautreaux, J.H. McCarthy, W.E. Foster, D.O. Hutchinson and F.W. Frey; U.S. Patent 3,860,415; January 14, 1975; assigned to Ethyl Corporation provide a process for producing substantially pure aluminum from mined aluminum silicate ore, especially kyanite ore. A kyanite ore is defined herein as any aluminum silicate ore which contains 20% or more of a mineral having equal parts Al_2O_3 and SiO_2. Under some circumstances, a raw ore containing as little as 15% of the desired minerals might be suitable. A kyanite concentrate is defined as a kyanite ore which has been beneficiated to remove substantial amounts of impurities or materials other than kyanite. It comprises the following basic steps:

> The mined kyanite ore is crushed and ground in suitable equipment to a particle size of about 600 μ to about 44 μ and preferably less than about 500 μ or about −35 mesh.
>
> The ground kyanite ore is beneficiated to form a kyanite concentrate. A kyanite-quartz flotation process is preferred. The ground kyanite ore is subjected first to a scrubbing and desliming process in which various micas are primarily removed. The ground kyanite ore is then subjected to a combination magnetic separation and flotation treatment and tabling if desired. The kyanite is nonmagnetic. In some instances, the ground kyanite ore is then subjected to a reductive roast and further magnetic separation to remove residual iron species. The particular beneficiation selected will vary with the type of ore.
>
> The kyanite concentrate is then compacted into agglomerates. Briquettes of from about 1 inch x 1½ inches x ¾ inch to about 2 inches x 2 inches x 1 inch formed in suitable briquetting or like equipment utilizing suitable binders have been found to be particularly satisfactory. In some instances it may be desired to compact ore-carbon pellets, briquettes, or other suitable agglomerates.
>
> The ore briquettes are then fed to an electric arc furnace along with a predetermined amount of carbon in the form of coal, coke, charcoal and/or wood chips wherein the ore is carbothermically reduced to form an aluminum-silicon alloy. The carbon may also be introduced in the form of compactions with the ore. Additional carbon in a suitable form may be added as necessary.
>
> The aluminum-silicon alloy is then comminuted to a particle size of from about 150 μ to about 10 μ. A powder with a medium particle size of about 100 μ is particularly desirable. The alloy may be cast and ground to the desired size, blown from the melt, or comminuted by water spray.

The aluminum-silicon alloy particles are transferred to a suitable reactor and treated with propylene and hydrogen and a sodium catalyst under desired temperatures and pressures to form tripropylaluminum (TNPA) and dipropylaluminum hydride (DNPAH). Some TNPA may be used to initiate the hydroalumination reaction.

The TNPA and DNPAH are pyrolyzed or decomposed in an inert diluent such as a hydrocarbon oil in a suitable reactor to form aluminum powder, propylene and hydrogen. Some propane is produced in the pyrolysis and most of it is vented off and recovered. The recycle gas stream is compressed back to the hydroalumination reactor and propane is separated out during compression. Propylene and hydrogen purged from the process after they are reacted are transferred to the hydroalumination reactor.

The aluminum powder is filtered and washed with a light hydrocarbon or oil such as hexane and then dried and compacted into a desired form or shape. The diluent oil is separated from the wash and preferably recycled to the decomposition step. The process can also be ended here.

The aluminum powder compacts are then melted and fluxed with chlorine gas or metal chlorides and fluorides, and then cast into pigs, sows or other desired shapes. A flux comprised of sodium chloride, potassium chloride and cryolite is especially beneficial, but other suitable fluxes may be used. In gas fluxing, chlorine gas is bubbled into the molten metal. Preferaby an inert gas carrier, such as nitrogen is used. A most preferred fluxing gas is chlorine plus carbon monoxide and nitrogen.

In the preferred form of the process, silicon residue, which is filtered from the hydroalumination product, is heated in a suitable furnace with calcium oxide, silicon dioxide and if needed, additional iron to produce a ferro-silicon alloy of a desired ratio of silicon and iron. The alloy is separated from slag and cast into chills, pigs, sows, or other desired shapes.

Figure 5.2 is a flow diagram of the complete process for producing pure aluminum and ferro-silicon alloy.

In related work by *M.F. Gautreaux, J.H. McCarthy, W.E. Foster, D.O. Hutchinson and F.W. Frey, Jr.; U.S. Patent 3,860,416; January 14, 1975; assigned to Ethyl Corporation* instead of kyanite, a diaspore clay is used, that is any alumina ore which contains 20% or more by weight of a hydrated alumina mineral, such as diaspore, plus alumina-silicate minerals. Under some circumstances, a raw clay containing as little as 15% hydrated alumina might be suitable.

Figure 5.2: Flow Diagram for Producing Pure Aluminum and Ferro-Silicon Alloy

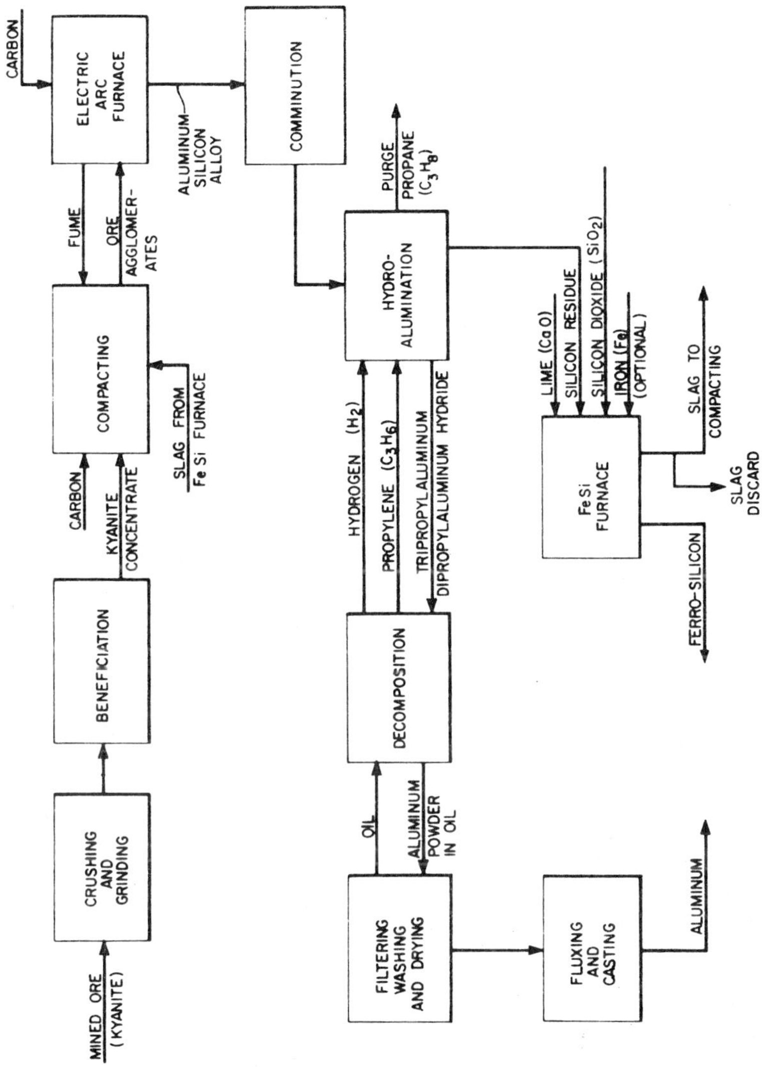

Source: U.S. Patent 3,860,415

In a preferred embodiment, slag from the aluminum purification process and fume produced from the carbothermic reduction of the beneficiated diaspore product in the electric arc furnace are transferred to the compaction operation, hydrogen and propylene produced in the decomposition phase are routed to the hydroalumination reaction, and oil from the washing and drying of the aluminum powder is circulated to the decomposition step.

Most preferably, the silicon rich residue from the hydroalumination reaction is conducted to a furnace wherein lime, silicon-dioxide and iron, if necessary, are added to produce ferro-silicon alloy.

Decreasing Content of Intermetallic Compounds

A.R. Valdo and F.M. Sanderford; U.S. Patent 3,910,787; October 7, 1975; assigned to Ethyl Corporation describe a process for reducing the content of one or more undesirable intermetallic compounds in a carbothermically produced alloy by transferring molten alloy directly from a carbothermic reduction furnace to a ladle or other suitable holding container and maintaining the molten alloy at a high enough temperature for a sufficiently long time to inhibit the formation of the intermetallic compounds and then rapidly cooling the alloy to minimize formation of the intermetallic compounds.

The heat retained prevents formation of the intermetallic compound or compounds to the constituent elements and the subsequent quenching retards formation of these undesirable compounds by effecting a rapid temperature drop through the formation temperature of the compound.

Substantially any alloy may be easily purified according to the techniques of this process, magnesium alloys, zinc alloys and the like being exemplary. Alloys containing aluminum are particularly well suited to upgrading by the process. Aluminum alloys which are readily susceptible to purification may have an aluminum content varying from a fraction of a percent up to 99% and higher.

Aluminum alloys which are particularly susceptible to purification by this process are those containing from about 40% to about 90% aluminum by weight. The aluminum may be alloyed with substantially any element so long as one or more intermetallic compounds are present, the removal of which is desired.

Intermetallic compounds which are frequently found to be undesirable constituents of aluminum alloys, are at least two of the elements, aluminum, silicon, titanium, iron, copper, magnesium and manganese. They are particularly significant because they are frequently found in those alloys obtained by the carbothermic reduction of clay. Common intermetallic compounds contained in carbothermically produced aluminum alloys are $FeSi_2Al_4$ and $TiAl_3$.

Example 1: Approximately 500 g of an alloy containing by weight 68% aluminum, 27% silicon, 3% iron and 2% titanium and having a $FeSi_2Al_4$ content of about 20 weight percent were heated to 1000°C in a large crucible and held at this temperature for 30 minutes. The molten metal was then rapidly quenched in ice water by pouring the alloy directly into a water-ice mixture having a temperature of 2°C. After the quenching operation, the alloy was analyzed and found to contain 5.5% by weight of the $FeSi_2Al_4$ intermetallic compound.

Example 2: Twenty-five grams of an alloy containing by weight 73% silicon, 5% aluminum, 22% $FeSi_2Al_4$ and a trace of $TiAl_3$ were heated in an argon atmosphere to 1000°C, held for 15 minutes and subsequently quenched in ice water. Upon x-ray analysis, the $FeSi_2Al_4$ content was found to be 2.2% by weight and there appeared to be no trace of $TiAl_3$ left in the alloy.

Example 3: Three 10 g samples of aluminum alloy containing by weight 27.4% aluminum, 50.2% silicon, 8% iron ($FeSi_2Al_4$ was present in an amount equal to 19% by weight) and 0.4% titanium were heated to temperatures of 800°, 900° and 1000°C, respectively. The alloy samples were held at these respective temperatures for 15 minutes and were then rapidly quenched in ice water. X-ray analysis of the three samples showed the $FeSi_2Al_4$ content of the samples by weight to be: 800°C, 14.9%; 900°C, 14.9%; 1000°C, 1.9%.

Example 3 clearly illustrates that for best results the alloy to be purified should be heated to a temperature significantly above the melting point of the intermetallic compound to be decomposed. The melting point of $FeSi_2Al_4$, as previously noted, is about 870°C, and the results of Example 3 show that much more $FeSi_2Al_4$ impurity is removed by quenching from a temperature of 1000°C (well above 870°C) than from the temperatures below and even slightly above the intermetallic compound melting point.

From an analysis of the preceding examples and embodiments it is apparent that the process provides an easy and economical method for decomposing a quantity of undesirable intermetallic compound in virtually any alloy capable of undergoing peritectic decomposition. Operation procedures are simple and necessary equipment is not expensive and is readily available. Accordingly, the process aids in alloy purification.

Aluminum-Titanium-Boron Mother Alloy

M. Alliot, J.-C. Beguin, M. Moutach, and J.-C. Percheron; U.S. Patent 3,961,995; June 8, 1976; assigned to Aluminum Pechiney, France have developed an aluminum-titanium-boron mother alloy having a boron content of 0.2 to 0.8% by weight and a titanium content such that Ti − 2.2 B ≥ 3.9%. The matrix has a preponderant proportion of grains of less than 30 μ in size, and contains fine TiB_2 crystals having an average size of about 1 μ primarily dispersed along the grain boundaries.

The method of preparation is by the formation of titanium diboride by the action of liquid aluminum on titanium oxide and boron oxide in solution in molten cryolite, mixing the reactants in a manner to utilize the starting materials, and then quenching the formed alloy rapidly to cool and solidify the mother alloy, preferably by pouring the liquid alloy in water to produce the alloy in the form of granules or fine powder.

Microscopic examination of the mother alloy, thus obtained, shows a matrix consisting of aluminum crystallized into grains of 50 to 300 μ and more in size, which is traversed by acicular crystals of titanium aluminide, and in which most of the TiB_2 grains, with a size of around 1 μ, are distributed along the grain boundaries where they form accumulations.

If, in accordance with the process, the mother alloy is solidified by cooling in such a way that the size of the grains of the matrix does not exceed 30 μ, the TiB_2 crystals are present in highly dispersed form, on the one hand because the grain boundaries are more numerous and, on the other hand because some of the TiB_2 crystals are actually dispersed inside the grains.

Aluminum Oxide Cartridge Case Coating

In order to provide a high strength aluminum alloy for small arms cartridge cases having excellent burn-through resistance and having about one-third the weight of a similar conventional brass case, *F.R. Gruner; U.S. Patent 3,924,534; December 9, 1975; assigned to United States Secretary of the Army* has discovered that conventional high strength aluminum alloys can be rendered resistant to burn-through by providing an Al_2O_3 coating around each aluminum alloy particle and/or an Al_2O_3 dispersoid throughout the aluminum alloy matrix.

About 6 to 13% by weight of Al_2O_3, preferably 10-12%, when incorporated into high strength aluminum alloys results in a material which by subsequent compacting and/or sintering, hot and cold working will yield a lightweight high strength product suited for small arms cartridge cases.

Any of the high strength aluminum alloy powders, e.g., the 2000, 6000 or 7000 series, when processed to obtain proper coating of the alloy particles with Al_2O_3 and dispersion of Al_2O_3 throughout the matrix, have been found suitable and satisfactory for this process.

The process steps are:

1. The aluminum alloy powder of about 1 to 20 μ in diameter contains an oxide or hydrate film of about 0.01 μ thickness around each particle. This powder is ball milled using stainless steel balls (or corundum balls if freedom from iron is required). Ball milling is continued until the desired Al_2O_3 content is obtained. The extent of powder grinding is based on the desired oxide content, which is proportional to the generated surface area. The preferred oxide content is about 10 to 12%. A lubricant

such as isopropyl alcohol should be used in the grinding operation. The resulting powder should then be vacuum dried at about 300°F to complete alcohol removal.

2. The dried powder is then cold pressed isostatically and sintered at approximately 540°C in any suitable neutral atmosphere, such as argon, to produce a billet.
3. The billet is then hot-pressed at about 700°F at about 90,000 psi into a rod.
4. The rod may next be hot extruded at a reduction ratio in a range of about 10:1, to 30:1, into a rod of smaller diameter or configuration to partially break down the oxide coating and provide aluminum alloy to aluminum alloy bonding of the particles to increase the mechanical strength of the alloy and to also provide very uniform dispersion of Al_2O_3.
5. The extruded rod is now hot and cold rolled to obtain the desired thickness of sheet for subsequent production processess of conventional stamping, heat treating, and drawing operations.

Highly Reflective Flake

J.T. Casey and F.A. Badia; U.S. Patent 3,901,688; August 26, 1975; assigned to The International Nickel Company, Inc. produce aluminum flake powders having high specular reflectivity (R_S) by wet ball milling aluminum powder, e.g., substantially spherical powder, such as that produced by atomization, in the presence of a lubricant using a volume ratio of grinding balls to powder of at least about 15:1, e.g., at least about 20:1 and up to about 75:1.

Good results have been obtained at ball to liquid volume ratios of 2:1 to 1:1.25, and preferably the ratio of grinding balls to liquid employed will be substantially equal. The time of milling to achieve high R_S will be relatively short, e.g., on the order of about 2 hours or about 1 hour or even less. It has been found that milling for an excessive amount of time can result in thinning the flake product to the point at which it starts to transmit light with an accompanying decrease in specular reflectivity. The product powder will have good water coverage.

Aluminum powder suitable for treatment is preferably of a type having a substantially spherical shape. The average particle size of the starting material may range from about 4 to about 300 μ. It is preferred in instances wherein flake product of small dimension is desired, e.g., a flake product which will substantially pass a 325 mesh Tyler screen (44 μ opening) to employ a fine starting powder, e.g., about 6 μ average size. When larger product flakes are desired, larger starting powders can be employed. Desirably, the milling process principally represents a flattening or flaking action exerted upon the starting powder.

The milling liquid employed preferably is a light liquid hydrocarbon called "mineral spirits," usually having a specific gravity of about 0.7 to about 0.8 at 60°F. Other liquid hydrocarbons may be employed. A lubricant such as stearic acid or other fatty acids, e.g., oleic, erucic, etc., are employed in amounts of about 0.5% to about 5%, e.g., about 1%, by weight of the powder being processed. It is found that when stearic acid is used as the lubricant, the product powder is coated with aluminum stearate apparently generated in situ during processing. The powder product is generally shipped in admixture with a minor amount of mineral spirits in the form of paste. If, desired, the excess mineral spirits can be removed and the flake product can be subjected to a polishing operation further to improve specular reflectivity.

Precious Metals

GOLD

Reducing Silica-Containing Catalyst Solution

D.S. Harrington; U.S. Patent 4,113,467; September 12, 1978 has developed a method for recovering gold from solution by adsorbing the solution onto powdered gypsum-bound sodium silicate, which serves as a catalyst for the reduction reaction of the ionic gold in solution. The grade of the elemental gold product which is obtained is directly proportional to the grade of sodium silicate gel used, thereby allowing control of the resulting elemental gold particle size. When the feed solution has been adsorbed onto the sodium silicate and a gel formed, this gel is treated with oxalic acid or other suitable reducing agent, whereupon an oxidation-reduction reaction occurs and reduced gold may be observed suspended in the gel.

The oxalic acid is oxidized to form carbon dioxide gas and water. The system is then agitated to allow the metal to separate from the catalyst. The elemental gold, due to the consistency and electrostatic adherence of the hydrated catalyst, will remain suspended above the catalyst, and may now be easily washed from the catalyst. The catalyst may then be dried to remove the water formed in the oxidation-reduction reaction and recycled to the adsorption step without further treatment. Isolation of the system from ultraviolet radiation is preferable in order to prevent the formation of colloidal forms within the system, which will irretrievably suspend the elemental gold.

The size of the gold particle produced is directly proportional to the grade of the catalyst used; recovery is greater than 99% effective.

The process may be performed in any suitable container since the use of toxic and/or corrosive materials is avoided. Suitable containers include steel, polyethylene, polypropylene, polyvinyl chloride, and glass. Furthermore, the gold or silver may be in the presence of other chemicals in the adsorbed solution (e.g., iron, zinc, copper, lead, etc.) and be reduced from this mixture by this process.

Example 1: A solution of 5% gold was prepared by dissolving gold chloride ($HAuCl_4 \cdot 4H_2O$) in a suitable amount of water. Dry sodium silicate gel powder was added in an amount of solution sufficient to hydrate all of the dry gel. The excess solution was poured off so that the level for the solution equalled the height of the gel. Saturated oxalic acid solution was added in an amount equal to approximately one-half the volume of the gel and the test tube covered with film to prevent evaporation. Upon completion of the reduction reaction, the suspension turned purplish-brown in color, and the solution was swirled and decanted. The decanted solution was then filtered to remove the elemental gold dust. Purity of the recovered elemental gold was 99.99%; 100% of the gold in the feed solution was recovered.

The gold solution may have any concentration of gold attainable and it may be a gold solution of water or alcohol. The amount of silicon-containing compound used is not critical but must be in an amount sufficient to act as a suitable catalyst for the reducing agent to reduce all of the gold present. Although the oxalic acid used was a saturated solution, a more dilute solution may be used so long as enough oxalic acid is used to reduce all of the gold present; however, a more dilute solution will take more time for the complete reduction to occur. Various mechanical means may be used to separate the gold from the solution or gel.

Examples 2-10: The following additional tests were made using the identical procedure as that used in Example 1.

Example	Reductant	Metal	Catalyst
2	Cinnamic acid	Gold	Gypsum-bound silicate (Na^+)
3	Cinnamic acid	Gold	Silicic acid
4	Tartaric acid	Gold	Gypsum-bound silicate (Na^+)
5	Tartaric acid	Gold	Silicic acid
6	Citric acid	Gold	Gypsum-bound silicate (Na^+)
7	Citric acid	Gold	Silicic acid
8	Oxalic acid	Gold	Silicic acid
9	Oxalic acid	Gold	Grade 7, chromatographic silicate
10	Oxalic acid	Silver	Gypsum-bound silicate

Examples 2 through 9 gave good yields of gold, the best yields being obtained using oxalic acid as the reductant and gypsum-bound (Na^+) silicate as the catalyst. A good yield of silver was obtained in Example 10.

Metal Coated Heat Resistant Substrate

O.N. Collier and S.J. Hackett; U.S. Patent 4,130,506; December 19, 1978; as-

signed to Johnson, Matthey & Co., Ltd., England provide a metal which, at temperatures approaching, equal to or exceeding the melting point of the metal, does not melt or tend to sinter.

Thus, a metal powder suitable for use at high temperatures is comprised of particles made from a heat resistant substrate, individually coated or encapsulated with a layer of platinum, palladium, rhodium, ruthenium, iridum, osmium, gold and silver or an alloy containing one of the metals. According to another feature of the process, the metal constitutes between 50 and 95 weight percent, e.g., 50 to 65 weight percent, of the total weight of the powder.

The heat resistant substrate may be made from any material from the range of naturally-occurring and synthetic refractories. Examples of naturally-occurring refractories are clays, silica, alumina, titania, zirconia and/or mixtures of these and examples of synthetic refractories are silicon nitride, silicon carbide and/or mixtures of these.

The method of making a metal powder includes the steps of nucleating the metal, for example gold, or alloy onto the surface of the substrate particles and then growing further metal or alloy in bulk onto the resulting nuclei. The resultant metal or alloy film or coating forms an encapsulation around each particle of the substrate. It is generally necessary, in order to produce and to reproduce consistently a metal powder in which the particles are of the desired size and the encapsulation is of the desired thickness, to control closely the steps of nucleation and growth.

Example 1: 75.0 g of gold ammonium chloride was dissolved in 2.4 liters of distilled water in a 5 liter beaker. 40 ml of a 10% gum acacia solution was added and the mixture stirred to ensure complete dissolution of the gold salt. Meanwhile, 25 g of china clay was activated by boiling in 100 ml of distilled water containing 10 drops (= 0.45 ml) of a 6% hydrazine hydrate aqueous solution. The china clay/hydrazine hydrate suspension was then added with vigorous stirring to the gold solution. Upon addition, the color of the mixture changed from yellow to yellow-green. After stirring for 10 minutes, 400 ml of "40 volume" hydrogen peroxide was added, as a result of which the color changed from green to brown and foam was formed from reaction gases generated.

The foam was suppressed using the minimum quantity of isopropanol from a laboratory spray. After 10 minutes, a further 100 ml of "40 volume" hydrogen peroxide was added, which caused further foaming, and the final 100 ml of hydrogen peroxide was added after a further 10 minutes. The reaction mixture was then stirred for 5 hours to complete the reaction, after which the clear supernatant liquor was decanted off and the powder was filtered off, washed and dried until constant weight was achieved.

Example 2: 80 g of gold as gold ammonium chloride was dissolved in 2 liters of distilled water in a 5 liter beaker. 40 ml of a 10% aqueous gum acacia solution was added and the mixture stirred to ensure complete dissolution of the gold salt. Meanwhile, 20 g of bentonite was activated by boiling in 200 ml of distilled water containing 0.45 ml of an aqueous 6% hydrazine hydrate solution. The bentonite/hydrazine hydrate mixture was then added with vigorous stirring to the gold solution. After stirring for 10 minutes, a 6% hydrazine hydrate aqueous solution was added to the mixture until the reaction was complete.

The foam formed from the reaction gases generated was suppressed using the minimum quantity of isopropanol spray. When the reaction was complete the gold/bentonite composite was allowed to settle, the clear supernatant liquor was decanted off and the powder was filtered, washed and dried to constant weight.

Example 3: 80 g of gold as gold ammonium chloride was dissolved in 2 liters of distilled water in a 5 liter beaker. 40 ml of a 10% aqueous gum acacia solution was added and the mixture stirred to ensure complete dissolution of the gold salt. Meanwhile, 20 g of zirconium dioxide powder was dispersed in 100 ml of water containing 0.45 ml of 6% hydrazine hydrate aqueous solution.

The zirconia/hydrazine hydrate suspension was boiled and added with vigorous stirring to the gold solution. After stirring for 10 minutes, a solution of 200 g of sodium sulfite dissolved in 1 liter of water was added. The reaction was complete within 10 minutes. The resulting powder was allowed to settle and the supernatant liquid was decanted off. The powder was filtered, washed free from dissolved salts and was dried to constant weight.

One advantage of a metal powder according to the process is that economy of the metal is achieved, as the particulate substrate can be considered as an "extender" which does not "dilute" or weaken in any way the physical properties of the metal, which is the case with the usual extenders which are used in the form of an intimate mixture or dispersion in the untreated metal powder.

One example of a disadvantage of the platinum group metals and silver at high temperatures, which could be overcome by using the metal in the form of a metal powder according to the process, is that alloys of platinum and rhodium, near their softening point, lose their cohesive strength with the result that faults such as sagging develop. Therefore, according to the process, thinner than normal sections of metal may be prepared by powder metallurgy.

Reduction of Precipitate from Chloroauric (III) Acid

K. Lutz and M. Golla; U.S. Patent 3,885,955; May 27, 1975; assigned to Demetron Gesellschaft fur Elektronik-/Werkstoffe mbH, Germany have found that gold powder of different predeterminable particle size, particle shapes and specific surface area predeterminable by the choice of reaction conditions is obtained by precipitation of chloroauric (III) acid with ammonia from an aqueous hydrochloric acid solution and reducing the precipitate as a suspension with a reducing agent.

In the precipitation of acidic gold chloride solutions with ammonia there is formed a precipitate known as fulminating gold. This fulminating gold explodes in the dry condition upon heating. However, it is harmless to handle in the wet condition.

As the starting solution for the production of fulminating gold there is used a hydrochloric acid solution of gold chloride which contains the gold essentially as chloroauric (III) acid. Since the composition of the precipitate formed in the addition of ammonia is dependent on the hydrochloric concentration of the starting solution, it has proven advantageous to use acidic gold chloride solutions containing 2-15% of free hydrochloric acid. Especially approved are gold chloride solutions with 5-10% free hydrochloric acid. The concentration of chloroauric (III) acid can vary between 1 and 35% of the solution.

The fulminating gold formed by the precipitation with ammonia need not be isolated but after the decanting off of the solvent and several washings by decantation with water can be directly worked up further.

The washed fulminating gold is suspended in a suspension medium and reduced to gold by passing into the suspension or adding dropwise a chemical reducing agent. According to the choice of the suspension medium, the reducing agent and the reaction temperature there is obtained gold powder of different particle size and particle form as well as different specific surface area.

As suspension media, water, glycol ethers and water-glycol ether mixtures have proven satisfactory with ethylene glycol monoethyl ether being the preferred glycol ether. Other suitable glycol ethers are for instance ethylene glycol monomethyl ether, ethylene glycol monopropyl ether, ethylene glycol monobutyl ether, diethylene glycol monobutyl ether, ethylene glycol dibutyl ether, diethylene glycol monoethyl ether or ethylene glycol diethyl ether. The ratio of water to glycol can be varied widely but usually when such a mixture is used 30 to 70% of the total of water and glycol ether by weight is water and the balance glycol ether.

While there can be used any of the reducing agents mentioned above or other reducing agents known to reduce acidic gold chloride solutions to gold it is preferred to use sulfur dioxide, hydroxylamine, hydrazine or triethanolamine whereby there can be obtained powders of different particle sizes and forms as well as of various specific surface areas by using specific combinations of suspension agent and reducing agent in conjunction with the reaction temperature and, in a given case, the pH value.

The following examples further explain the process. Unless otherwise indicated all parts and percentages are by weight.

Example 1: 3 kg of a chloroauric (III) acid solution containing 33% Au and a pH value of 0.5 were diluted with 3 liters of water and a solution of 1 liter of concentrated ammonia in 2 liters of water was added dropwise with stirring in the course of about 30 minutes. The temperature thereupon rose to 50°C, the pH value to 8 to 9. It was allowed to react further for 60 minutes and after settling of the precipitate the supernatant solution filtered off with suction. With powerful stirring up it was decanted several times with water and ethylene glycol monoethyl ether. The fulminating gold thus formed was worked up further directly.

Example 2: 100 g of fulminating gold in the form produced in Example 1 was suspended in 500 ml of water. A weak stream of sulfur dioxide was led in with stirring while the temperature was not permitted to go above 40°C. After the end of the reaction the product was washed several times with water and ethanol, filtered with suction and the gold powder dried at 70°C. 99.7% of the amount of gold added was obtained in the form of gold balls having a size of 3-10 mμ and having a specific surface of about 1.0 m^2/g.

Example 3: 100 g of gold in the form of fulminating gold produced according to Example 1 was suspended in a mixture of water and ethylene glycol monoethyl ether (1:1 by volume) and heated under reflux to 100°C. A solution of 70 g of hydroxylamine hydrochloride in 150 ml of water was added slowly dropwise to the solution with powerful stirring, whereupon there was noted a strong development of gas. After addition of the reducing agent it was allowed to react further for one hour. The gold powder formed was separated off as in Example 2, washed and dried. There were obtained about 99.7% of the gold added as crystallites having a granule size of 1-10 mμ with a specific surface area of 0.6 m^2/g.

Example 4: 100 g of gold in the form of fulminating gold were suspended in 400 ml of water/ethylene glycol monoethyl ether mixture (1:1 by volume) and acidified to pH 2 with hydrochloric acid. 60 g of triethanolamine in 50 ml of water/ethanol (1:1 by volume) were added dropwise at 100°C. There were obtained in 99.5% yield small gold fibers having a length of 3-15 mμ, a thickness of 1-2 mμ, and a specific surface area of about 0.6 m^2/g.

Example 5: 1,000 g of gold in the form of fulminating gold were suspended in 4 liters of water and treated dropwise at 55°-60°C with 1 kg of 25% hydrazine hydrate solution. There were obtained about 99.5% of the added gold as crystallites having a particle size of 1-20 mμ with a specific surface area of 3.5 m^2/g.

Aqua Regia Solutions

V.R. Daiga; U.S. Patent 3,843,379; October 22, 1974; assigned to Owens-Illinois, Inc. has developed a method for producing gold powder useful in the formation of thick film microelectronic circuitry, that is, having a bulk density greater than about 5.0 g/cc and having an average particle size of less than about 20 μ.

It is formed by dissolving a gold bearing material in aqua regia, thereafter, adding to the solution so formed a sufficient amount of an emulsifying agent or a particle size inhibitor such that upon precipitation of the gold from the solution the average particle size of the precipitate will be less than about 20 μ, and thereafter adding, in a rapid fashion, a precipitating agent to the solution in an amount sufficient to precipitate the gold from the solution.

The freshly formed gold particles are encapsulated with the emulsifier and a powder formed. By using the described technique, the powder so formed, after washing and drying, generally has a bulk density of greater than about 5.0 g/cc and each of the fine particles of gold have a coating of the emulsifying agent that acts as a lubricant when used with an organic binder in a thick film printing formulation.

Example 1: 20.2 g of sponge gold are admixed with 100 ml HCl (37-38% HCl concentration) and 20 ml HNO_3 (70-71% HNO_3 concentration). The admixture is then heated for a period of 1 hour at a temperature just below the boiling point of the admixture. After 1 hour, all sponge gold has been dissolved in the acid medium. The heat is then increased to the extent that the solution boils vigorously. As the solution boils, an additional 200 ml HCl is added over a period of about 1½ hours.

The distillate coming from the boiling solution is constantly removed during this time and after this 1½ hour period, about 130 ml of solution remain. This boiling process removes substantially all traces of nitric acid from the solution as evidenced by a lack of brown gases emitted at the end of the boiling period. The solution is then cooled to room temperature and filtered to remove any solids contained therein. The filtered solution is then diluted with distilled water so as to form a 200 ml solution.

A sodium sulfite solution is formulated by dissolving 30 g of Na_2SO_3 in 200 ml distilled water. 2 ml of butyl stearate are then measured out. The temperature of all ingredients is then adjusted to about 25°C.

The 200 ml solution of gold is then placed in a 1 liter glass-baffled reaction vessel and 200 ml of distilled water are added thereto to bring the total volume of the solution to 400 ml. Agitation is begun using a marine type propeller at about 200 rpm. The 2 ml of butyl stearate are then added and agitation is allowed to proceed for about 3 minutes to insure adequate dispersion of the particle size inhibitor substantially homogeneously throughout the solution.

The 200 ml solution of Na_2SO_3 is then added over a period of 10 seconds to the solution and agitation is allowed to proceed for about 1 minute after the addition of the Na_2SO_3 solution is completed. Agitation is then discontinued and a precipitate is found to exist in the vessel which is substantially pure gold. The mixture is then filtered to obtain the precipitate using a standard Buchner funnel. The precipitate is then repeatedly washed with distilled water and air-dried at 100°C for about 48 hours. The product so formed is found to be a dense gold powder having a bulk density within the range of 6.8-7.5 g/cc and having a particle size distribution of:

	Percent
$<$0.6 microns	0
$<$2 but $>$0.6 microns	20
$<$2.8 but $>$0.6 microns	50
$<$5.0 but $>$0.6 microns	98.8
$>$5.8 microns	0

A printing paste was formulated from the powder previously made by admixing 96 parts by weight of the gold powder with 4 parts by weight of a lead barium borosilicate glass consisting of: (by weight) 15% SiO_2, 10% B_2O_3, 40% PbO, 20% BaO and 15% ZnO. The average particle size of this glass binder admixed with the gold powder was about 1 μ. This glass powder-glass binder admixture was then added to a liquid organic vehicle consisting of 15% by weight N-4 ethyl cellulose and 85% by weight of a mixture of 2 parts by weight butyl Carbitol acetate (di-ethylene glycol monobutyl ether acetate) and 1 part by weight isoamyl salicylate.

The vehicle was mixed with the glass powder-glass binder admixture in an amount of 10% by weight vehicle to 90% by weight gold and glass binder. The paste so formulated was printed by a conventional screen printing technique onto a known alumina substrate and dried at 100°C for 15 minutes, thereafter fired at 875°C peak for five minutes with an 8-minute heat-up and cool-down period. The resultant film exhibited high film density and was in every respect an excellent thick film microelectronic conductor. The above example through further experimentation is found to be highly reproducible.

Example 2: Essentially the same procedure is followed as in Example 1 except a stoichiometric amount of Na_2SO_3 (20.9 g in a 200 ml solution of water) was employed to precipitate the gold. The resulting powder had the requisite particle size and bulk density to form an excellent thick film microelectronic conductor.

Example 3: Essentially the same procedure is followed as in Example 1 except that the solution was adjusted to 20 g of gold per 600 ml of solution by adding 400 ml of distilled water prior to precipitation. 1.2 ml of oleic acid is employed as the particle size inhibitor and the addition rate of the Na_2SO_3 solution is 200 ml over a 60 second period. The resulting gold powder had the requisite particle size and bulk density to form a high quality thick film microelectronic conductor.

In a related process for recovering gold from solution in aqua regia, *J.E. Brug and E.X. Heidelberg; U.S. Patent 3,856,507; December 24, 1974; assigned to Owens-Illinois, Inc.* add base to the solution in an amount sufficient to raise the pH of the resulting reaction mixture to in the range of about 2 to about 4, and preferably in the range of about 2 to about 3, while maintaining the temperature of the reaction mixture below about 50°C, and preferably in the range of about 20° to about 40°C, and then adding reducing agent to the reaction mixture in an amount sufficient to precipitate metallic gold from the reaction mixture. The metallic gold can then be recovered from the reaction mixture by filtration or decantation.

This technique obviates the necessity for removal of the nitric acid from aqua regia solution of gold by destructive distillation. In the pH range of about 2 to about 4, the effective oxidizing power of the nitric acid is decreased to the point where reduction and precipitation of metallic gold is no longer inhibited.

Any conventional and well-known gold-bearing material can be processed according to this process, e.g., sponge gold, gold bars, gold granules, gold pellets, and previously prepared gold powder of low bulk density and relatively high particle size.

Practically any base can be employed to adjust the pH, although bases such as the alkali, and alkaline earth hydroxides, and carbonates are usually employed for efficiency and economy. These bases can be employed in solid form or as a concentrated aqueous solution. They include sodium hydroxide, potassium hydroxide, lithium hydroxide, sodium carbonate, potassium carbonate, calcium hydroxide, magnesium hydroxide, etc. Solid sodium carbonate and concentrated aqueous solution (e.g., 25 to 60% by weight) of sodium hydroxide and potassium hydroxide are commercially expedient to minimize the volume of water that must be handled in the process.

Practically any reducing agent such as sulfites (e.g., sodium sulfite, and sodium acid sulfite), hydrazine, stannous chloride, sulfur dioxide, hydroquinone, sugar, sulfurous acid, formaldehyde, metal dust such as zinc or aluminum dust, ferrous sulfate, hydrazine sulfate, or hypophosphoric acid can be used to reduce and precipitate metallic gold from the aqua regia solution having the pH adjusted to in the range of about 2 to 4. Concentrated aqueous solutions of sodium sulfite are preferred for efficiency and economy.

The reducing agent is usually added in amounts slightly in excess of the stoichiometric amount necessary to precipitate substantially 100% of the gold from the aqua regia solution.

Emulsifying agents that can be used for this process include the organic compounds known as lubricants, dispersants, and emulsifiers. The preferred emulsifying agent is an ester of fatty acid such as an alkyl stearate with the best results being obtained with a butyl stearate. The alkyl esters of fatty acids are those in which the alkyl group is about 1 to 8 carbon atoms and preferably about 3 to 5 carbon atoms and a fatty acid generally represented by the formula $C_nH_{2n+1}COOH$ in which n is generally greater than 10 and preferably greater than 17 up to about 24.

Examples of the suitable fatty acid component of the alkyl ester of fatty acid are lauric acid, palmitic acid, and stearic acid. Hence, the suitable fatty acids are those in which n is 11 such as lauric acid, where n is 13 such as myristic acid, where n is 15 such as palmitic acid, and where n is 17 such as stearic acid. Also suitable are unsaturated aliphatic acids such as oleic (having 18 carbon atoms including the carboxyl group), erucic acid (having 22 carbon atoms), linoleic acid (having 18 carbon atoms) and linolenic acid (having 18 carbon atoms).

These emulsifiers can be added at any time prior to the reduction and precipitation of the gold from the aqua regia solution. In the interest of efficiency and economy of commercial practice, the emulsifying agent is usually added after the pH has been adjusted in the range of 2 to 4 and prior to the addition of the reducing agent.

The proportion of an emulsifying agent employed in producing gold powder is usually within the range of about 0.01 parts by weight up to about 10 parts by weight per hundred parts of gold in solution.

SILVER

Hydrogen Treatment of Suspension of Oxide

F. Montino and L. Colombo; U.S. Patent 4,039,317; August 2, 1977; assigned to Montedison Fibre SpA, Italy have developed a process for preparing silver powders with a controlled apparent density, wherein an aqueous suspension of silver oxide is treated with hydrogen at a temperature between 60° and 115°C, and at a partial hydrogen pressure between 5 and 25 atmospheres.

The starting silver oxide may be prepared according to any known method, for instance by precipitation with soda or caustic soda from an aqueous solution of silver nitrate. Preferably an aqueous suspension of silver oxide is prepared having a solids content between 10 and 200 g/l of suspension.

The reducing treatment with hydrogen of the suspension thus prepared is conducted in a stainless steel autoclave fitted with a stirrer and a heating system. The autoclave is loaded with the suspension and is brought up to the desired reduction temperature, that is, up to a temperature between 60° and 115°C. The hydrogen preheated to the reduction temperature is fed into the autoclave at a pressure between 5 and 25 atm, and preferably between 10 and 20 atm. The temperature in the autoclave is maintained until there is no more drop in pressure.

In general, the reduction times vary from 10 to 40 minutes, depending on the degree of turbulence of the motion of the fluid in the autoclave and on the temperature.

In the event it is desired to obtain a silver powder of low apparent density, the aqueous silver oxide suspension, before being loaded into the autoclave, is additioned with one or more alkali metal, alkaline earth metal or silver sulfides. Preferably silver sulfide, sodium sulfide or potassium sulfide is additioned. For this purpose a colloidal solution of silver sulfide, for instance one obtained by the addition of sodium sulfide to diluted ammoniacal silver sulfate $Ag_2(NH_3)_4SO_4$, is especially desirable. The sulfide is added to the aqueous silver oxide suspension in quantities corresponding to 15-50 mg of S per 100 g of Ag as oxide. In this way, silver powders are obtained having an apparent density between 0.3 and 1 g/cc.

If it is wished to obtain a silver powder of high apparent density, before being loaded into the autoclave the aqueous silver oxide suspension is added with one or more silver salts, either soluble or made soluble by complexing with ammonia. Preferably ammoniacal silver sulfate, silver nitrate or ammoniacal silver nitrate is added. If a soluble silver salt is used, the addition must be made at a ratio of 0.5-5 g of soluble Ag/100 g of Ag as oxide. If a soluble ammoniacal salt is used, the addition must be made at a ratio of 100-1,000 g of soluble Ag/100 g of Ag as oxide. In this way, silver powders are obtained having an apparent density between 1 and 3 g/cc.

If desired, the process may also be carried out with an aqueous silver oxide suspension in which is present contemporaneously a soluble silver salt and a sulfide.

At the end of the reduction with the hydrogen, and after cooling down, the autoclave is discharged. The silver powder thus obtained is filtered, washed with water until the ions present have disappeared, and finally dried.

The most significant advantages offered by the process are: short reaction times; the possibility of obtaining silver powders having a controlled apparent density over the whole interval of 0.3-3 g/cc, by varying the operating conditions; and a good homogeneity of shape and dimensions of the silver powder particles.

Electrically Conductive Surfaces

C.C. Dannelly and J.M. McIntire; U.S. Patent 3,940,261; February 24, 1976; assigned to Eastman Kodak Company describe a process for the reduction of water soluble silver salts, e.g., silver nitrate, to prepare crystalline silver particles ranging in size from about 10-100 μ. The silver particles have exceptionally clean, bright electrically conductive surfaces. The process comprises reducing a water soluble silver salt in a redox reaction at a temperature of about 0° to about 50°C for about 2 to 24 hours, the silver salt being present as an aqueous solution in a concentration of about 0.25-1.0 mol/l of water, and dilute nitric acid being present in the amount of 0.1 to 3 mol/l of silver salt solution. A reducing agent characterized by an oxidation potential within −0.150 volts of the silver salt under the conditions of the reaction solution, solubility of about 0.025 or more mol/l of water at 25°C, is also present in a concentration of 1 to 1.25 molar equivalents, based on the silver ion to be reduced, dissolved in about 0.5 to 2 liters of water.

A preferred process comprises (a) preparing an aqueous solution of a silver salt, the concentration of the silver salt being in the range of about 0.25-1.0 mol/l of water; (b) adding dilute nitric acid in the ratio of about 0.1-3.0 ml/l of the aqueous solution; (c) preparing an aqueous solution of reducing agent characterized by an oxidation of −0.150 volts of the silver salt under the conditions of the reaction solution, solubility of at least about 0.25 mol/l at 25°C, and being present in a concentration of 1 to 1.25 molar equivalents based on the silver ion to be reacted, dissolved in about 0.5 to 2 liters of water; (d) blending the aqueous solution of reducing agent and the aqueous solution of silver salt to thereby reduce the silver salt in a redox reaction, the reaction being at a temperature of about 0°-50°C for about 2 to 24 hours; (e) filtering and washing the silver particles until the filtrate is essentially clear; and (f) drying the silver particles.

Example: A 3 liter, three-necked, round-bottom flask is equipped with a thermometer, a mechanical stirrer, and inlet ports from a Polystaltic pump which drains two reservoirs. One mol of silver nitrate dissolved in 1 liter of water and 2 ml of dilute nitric acid are added to the flask and cooled to approximately 20°C. One liter of water containing 0.62 mol of hydroquinone is placed in one reservoir and 500 mols of water containing 1 mol of sodium acetate is placed in the other reservoir. The mechanical stirrer agitates the solution at approximately 75 rpm.

The addition of the hydroquinone solution is initiated and fed at such a rate as to complete addition in 30 minutes. Fifteen minutes after beginning the addition of the hydroquinone solution, the addition of sodium acetate solution is begun. It is also added at such a rate as to complete addition in 30 minutes. The reduction is allowed to proceed for 4 hours from the start of the hydroquinone addition. At this time, the larger crystalline silver particles are filtered and washed with water until the filtrate is clear. This requires about 10 liters of water. The silver powder is dried in a vacuum oven at 50°C. This yields a high quality silver powder in 94.2% yield.

COMBINATION SYSTEMS

Precipitation from Carbon Monoxide Treated Solutions

R. Ugo; U.S. Patent 4,163,664; August 7, 1979; assigned to PROTEC Processi e Tecnologie SpA, Italy describes a process for the practically total precipitation of precious metals (Pt, Pd, Au, Ag, Bu, Rh, Os, Ir) from solutions which contain them.

The process consists essentially of treating the solutions with carbon monoxide in the presence of quantities of water or primary or secondary alcohols equal or greater in weight than the total quantity of precious metals present. The process is carried out preferably at a pH from 0 to 7, at a temperature from 20° to 200°C and under carbon monoxide pressures from 1 to 100 atmospheres. At pH values greater than 7 there is a certain precipitation of the metals in the form of oxides or hydroxides which slow down the reaction with the carbon monoxide.

Temperatures greater than ambient temperature accelerate separation, but in the case of dilute solutions there is no economical convenience in providing such temperatures unless under special conditions, such as the availability of hot process effluents which can be worthwhile treating directly without passing through an intermediate cooling stage. There is no actual upper limit to the carbon monoxide pressure used, other than that imposed by economic convenience.

Example 1: One liter of an aqueous solution of ruthenium trichloride hydrate containing 2 ppm of ruthenium is brought to pH 6 and saturated with carbon monoxide. Precipitation is complete after 20 hours. The ruthenium is separated in the form of a microcrystalline, easily decantable powder, with a purity of 99.9%. The mother liquors when analyzed show a ruthenium content less than analytical limits, i.e., less than 0.01 ppm. The ruthenium recovery can therefore be considered total.

Example 2: 1% of water is added to one liter of acetone containing palladium nitrate and rhodium nitrate in quantities equivalent to 1 ppm of palladium and 2 ppm of rhodium, and the mixture saturated with CO at ambient temperature and under atmospheric pressure.

The reaction is complete after 25 hours. A microcrystalline, easily decantable powder separates, containing palladium and rhodium in a ratio by weight of 1:2 and 99.9% purity.

The residual acetone solution shows an analysis of palladium concentration of 0.03 ppm and a rhodium concentration of 0.09 ppm, and the recovery of the two metals can therefore be considered total.

Example 3: An industrial effluent consisting of an aqueous solution from which the precious metals have been separated by cementation, contains a total of 3 ppm of palladium, platinum, gold and rhodium, 100 g/l of $ZnSO_4$, 10 g/l of Na_2SO_4 and 5 g/l of $Al_2(SO_4)_3$.

This solution, which is initially acidified with H_2SO_4, is brought to a pH of 3 with NaOH and then saturated with CO at ambient temperature and under atmospheric pressure. The reaction is complete after 30 hours.

A microcrystalline solid is separated, containing palladium, platinum, gold and rhodium in approximately the same ratio as in the initial solution. The analysis of the residual liquid shows a total concentration of precious metals of the order of the analytical limit (0.01-0.05 ppm). All the precious metals are therefore recovered practically totally.

Recovery from Organophosphorus Solution

T. Onoda, Y. Tsunoda, T. Nomura, T. Nonaka and T. Masuyama; U.S. Patent 3,920,449; November 18, 1975; assigned to Mitsubishi Chemical Industries Ltd., Japan describe a process for recovering a Group VIII noble metal from an organic solvent solution of a soluble complex of the Group VIII noble metal and an organophosphorus compound. It comprises supplying the organic solvent solution and an oxygen-containing gas into a combustion zone, combusting the organic solvent solution, immediately introducing the combustion product into an aqueous absorbing solution to catch the particles of the Group VIII noble metal and phosphorus oxide in the aqueous solution, and recovering the noble metal particles from the aqueous solution.

The solution remaining is an aqueous solution of phosphoric acid and it may be discharged in a drain after neutralization or can, if desired, be used for various purposes.

Example 1: In a submerged combustion device having an inside capacity of 0.5 m^3 was charged 0.3 m^3 of water and a toluene solution containing tris(-triphenylphosphine) rhodium carbonylhydride, $RhH(CO)[P(C_6H_5)_3]_3$, in an amount of 0.3% by weight, as metallic rhodium. Triphenylphosphine and triphenylphosphine oxide in an amount of 3% and 2% by weight, respectively, and high boiling by-products resulting in the hydroformylation of propylene in an amount of 21.2% by weight were supplied at a rate of 5 kg/hr and air at a rate of 60 Nm^3/hr; they burned at about 1180°C.

The combustion product was, then, immediately introduced in water to be absorbed. The residence time of the combustion product was 0.5 second. Phosphorus oxide formed in the combustion was absorbed by water to provide phosphoric acid and the rhodium formed was dispersed in the water as solid particles.

While supplying water so that the volume of water in the absorbing drum was maintained always at 0.3 m^3, combustion was conducted continuously for 20 hours and then the combustion products in the combustion device were filtered, whereby 95% of the rhodium was recovered as solid particles and almost 100% of the phosphorus was recovered as phosphoric acid.

Example 2: A xylene solution containing rhodium in an amount of 0.3% by weight, triphenylphosphite and oxide of triphenylphosphite in an amount of 3.0% and 2.0% by weight, respectively, and high boiling by-products resulting in the hydroformylation of propylene in an amount of 18.3% by weight was burned as described in Example 1, using an aqueous 2% sodium hydroxide solution as the absorbing solution.

As a result of the combustion, 93% of the rhodium was recovered and 100% of the phosphorus was recovered as sodium phosphate.

Example 3: A toluene solution containing 0.5% by weight of a palladium iododitriphenylphosphine complex, $PdI_2[P(C_6H_5)_3]$, and 19.8% by weight of triphenylphosphine was treated as described in Example 1. As a result, 90% of the palladium was recovered and 100% of the phosphorus was recovered.

Example 4: A toluene solution containing 0.2% by weight of tris-(triphenylphosphine)iridium carbonylhydride, $IrH(CO)[P(C_6H_5)_3]_3$, and 2.7% by weight triphenylphosphine was treated as in Example 1. As a result, 91% of the iridium and 100% of the phosphorus were recovered, respectively.

Reacting Atomized Spray of Metal Solution

P. Douglas and T.I. Stewart; U.S. Patent 4,023,961; May 17, 1977; assigned to Plessey Incorporated have developed a method of producing powdered material which includes the steps of providing at least one metal or its oxide in the form of a solution; forming the solution into an atomized spray; and reacting the atomized solution to produce a metal or metal oxide powder having a particle size which is determined by the composition of the solution in relation to the size of the individual droplets of the atomized spray.

Prior to the formation of the atomized spray, at least one metal oxide is provided in the form of a solution; the solution of the one metal oxide is intimately mixed with the solution of the metal or its oxide, the material solution being a metal/metal oxide or a metal oxide/metal oxide composite in powder form.

The reaction can be effected either thermally and/or chemically. The particle size distribution of the atomized solution droplets is a function of the spray parameters. Control of the atomized solution droplet size and concentration and control of the decomposing solution concentration or decomposition temperature can be used to produce powders of the desired characteristics.

Typical powdered materials that can be produced by this method are silver, gold, silver-palladium, palladium, platinum, any combination of these metals, cadmium oxide, tin oxide, zinc oxide, ruthenium oxide, cobalt oxide, copper oxide; indium oxide, tantalum oxide, tungsten oxide, lanthanum oxide, strontium oxide, chromium oxide, iron oxide, magnesium oxide, manganese oxide, calcium oxide, barium oxide, tellurium oxide and lithium oxide. Alternatively, a composite powdered material can consist of at least one of the listed metals and at least one of the listed oxides or at least two of the listed oxides.

The formation of the individual droplets can be effected by forming an atomized spray of the solution using, for example, a rotating disc atomizer, an aerosol spray or an annular two fluid atomizer. The size of the individual droplets of the atomized spray can be in the range $<1\ \mu m$ to $<100\ \mu m$.

With a thermal reaction the atomized spray can be projected through a gas flame, for example, an oxy-propane flame, in order to effect decomposition of the individual droplets. The powder produced by this method is collected in a liquid medium, for example, water.

In the production of a gold powder by this method, the gold solution can be provided by (a) gold metal dissolved in aqua regia, or (b) chloroauric acid ($HAuCl_4$) dissolved in water, or (c) gold chloride or cyanoauric acid dissolved in aqueous or organic solvents.

In the production of silver powder the silver solution can be provided by silver nitrate and the reducing agent can be provided by a solution of ammonium formate containing an excess of formic acid. The size of the silver powder particles produced by this method is determined by the silver nitrate concentration and the droplet radius and there is a linear relationship between the log radius of the powder particles and ⅓ log silver concentration.

In the production of palladium powder the palladium solution can be provided by either (a) palladium metal dissolved in aqua regia, or (b) palladium chloride ($PdCl_2$) dissolved in water, and the reducing agent can be a solution of a reducing agent such as hydroquinone or formic acid. The size of the palladium powder particles is determined by the concentration of the sprayed solution and the droplet radius.

For platinum (Pt) powder, the platinum solution can be provided by platinum metal dissolved in aqua regia and the reducing agent can be hot formic acid. The chemical reaction which occurs as a result of this method is as follows:

$$H_2[PtCl_6] + 2HCOOH \rightarrow Pt + 6HCl + 2CO_2$$

In an alternative method of producing a metal or metal oxide powder, the solution of the metal or metal oxide can be simultaneously formed into individual droplets and decomposed to produce the powder by spraying the solution into a reducing agent such as ferrous sulfate.

In the production of a powdered material containing at least two metals, for example Pd/Au, Pt/Au, Pd/Ag, and Pd/Au/Pt, combined in any desired proportions, a solution containing the desired metals is sprayed into a suitable reducing agent.

Thus, in the production of a Pd/Au or a Pt/Au powder, the two metals are dissolved in the desired proportions in aqua regia and the solution is sprayed into a solution of either (a) hydroquinone maintained at a temperature of about 90°C, or (b) formic acid maintained at a temperature of about 90°C.

For example, in the production of a Pd/Au powder containing 70 weight percent of gold and 30 weight percent of palladium, a solution containing 70 g of Au/l of solution and 30 g of Pd/l of solution is prepared by dissolving the two metals in aqua regia and then spraying the solution into a hydroquinone solution maintained at a temperature of about 90°C. The resulting powder particles were spherical, had a mean diameter of 0.4 μm and contained Au and Pd in the ratio 70:30.

A Ag/Pd powder is produced by dissolving the desired amount of palladium metal in nitric acid, adding silver nitrate and distilled water to give a solution of the desired composition, and spraying the mixed solution into a solution of ammonium formate. For example, a solution of 50 g/l of Pd and 70 g/l of Ag, when sprayed into a solution of 240 cc of NH_4OH/176 cc of HCOOH/l which is being maintained at a temperature of 20°C resulted in the production of a powder containing 40 weight percent of palladium and 60 weight percent of silver.

A Pd/Pt/Au powder can be produced by the method outlined in preceding paragraphs for the Pd/Au and the Pt/Au powders.

Palladium-Silver-Gold

V.R. Daiga; U.S. Patent 3,816,097; June 11, 1974; assigned to Owens-Illinois, Inc. has developed a process for producing a substantially homogeneous, finely divided powder, substantially free of silver chloride or cyanide comprising at least one metal other than silver or gold, preferably palladium, and silver and gold, and an antiagglomerating agent such as particulate silica submicron in size and substantially homogeneously dispersed throughout the powder. The powders are particularly useful in the electronic arts.

This process includes two alternative techniques for forming homogeneous, finely divided powders of one or more metals in combination with silver and gold (or two other metals which give rise to similar problems of contaminated coprecipitation as silver and gold).

The first alternative technique generally comprises forming a soluble salt solution of a metal other than silver or gold, and silver and thereafter reducing the solution in accordance with prior art techniques similar to those described in U.S. Patents 3,390,981 and 3,385,799 and British Patent 1,004,652. Such a reduction reaction with agitation forms a metal-silver slurry. To this metal-silver slurry is then added gold in the form of a gold salt solution generally of the chloride type. The slurry solution now containing gold is then contacted with a reducing agent either already present or additionally added, to precipitate the gold into the metal-silver slurry and thereby form the trimembered powder of metal, silver and gold.

The second alternative method provides for the total elimination of the formation of any insoluble silver chloride or cyanide in the system and thus is preferred where noncontaminated powders are required to be used such as in the microelectronics art, particularly for making resistors and the like. This second alternative method generally comprises initially forming a soluble salt solution of the metal or metals other than silver or gold, and silver in a manner similar to the first alternative method.

To this silver-metal salt solution there is then added a finely divided gold powder usually having a particle size less than about 5 μ, preferably less than about 2 μ, and most preferably substantially submicron in size. The solution containing the gold powder, which is not soluble in the salt solution, is then thoroughly mixed as by agitation, to form a slurry of the gold and there is then added thereto a reducing agent for the metal and silver.

While some small amount of homogeneity is sacrificed because the gold powder is not precipitated, precipitation of the metal and silver into the slurried gold powder effects a substantial amount of homogeneity to the extent that excellent products can be made therefrom for the purposes of microelectronic circuitry, particularly in the resistor area.

Example 1: *First Alternative Technique*—A gold reactant solution ($HAuCl_4$) was prepared from pure gold sponge. This was accomplished by reacting 20 g of Au in an aqua regia solution and diluting the solution to 160 ml with H_2O. A palladium nitrate solution was prepared by dissolving 53.8 g of the metal in 150 ml (70%) HNO_3. The resultant $Pd(NO_3)_2$ solution had a volume of 160 ml. A silver nitrate solution was formed by dissolving 115.3 g of silver nitrate crystals in 150 ml of distilled water. The total volume of the resultant solution was 175 ml.

To start the process 80 ml of the $Pd(NO_3)_2$ solution, 87 ml of the $AgNO_3$ solution, and 100 ml of H_2O were charged into a reactor. The reactor employed was a one liter baffled glass kettle provided with a pitched marine-type propeller agitator capable of operating at up to 500 rpm. A cooling bath surrounded the kettle and reaction temperature was measured by a thermometer immersed in the reaction solution.

After mixing the initial charge thoroughly, an admixture containing 3.75 g of Cab-O-Sil (ultrafine silica) in 50 ml of H_2O was added and slurried. With continued agitation the reduction reaction was carried out by adding 170 ml of a solution containing 75 g of $NaHSO_3$ to the reactor. The $NaHSO_3$ addition took place over a 14 minute time interval and the reaction media was maintained at 38°C.

Agitation was continued to slurry the precipitated silver and palladium metals with the antiagglomerant and the slurry was prepared for $HAuCl_4$ addition and reduction by the addition thereto of 15 g of Na_2SO_3 in 100 ml of H_2O. This was followed by rapid addition of 80 ml of the $HAuCl_4$ solution. An additional 100 ml of the Na_2SO_3 solution was added to insure complete reduction of the gold. After 5 minutes of further mixing, the agitator was shut off.

After permitting the reactants to settle out for 30 minutes, the solids were separated from the liquid by filtration. The solids were washed by slurrying in

H_2O and filtering several times. The final wash was followed by drying at 110°C overnight. A finely divided, homogeneous powder was obtained. The powder analyzed as containing on a metal basis by weight, 47.4% Ag, 14.55% Au, and 38.05% Pd. The powder also contained 13.02% by weight chloride salt, assuming all chlorides to be silver chloride.

Example 2: *Second Alternative Technique*—250 g of a powder consisting of by weight 52% Ag, 35% Pd, and 13% Au were prepared by the following technique.

A $Pd(NO_3)_2$ solution was prepared by adding 87.5 g of palladium metal to 245 ml of a 70% HNO_3 solution. The addition of Pd to the acid was conducted slowly and the nitrous oxide gas formed was vented. An $AgNO_3$ solution was prepared by adding 204.3 g of $AgNO_3$ crystals to 500 ml of distilled water with stirring. The gold powder employed had a particle size of less than 5 μ and was produced in accordance with Example 1. 32.5 g of the powder were washed free of emulsifier in acetone and thoroughly dried for use hereinafter.

The $AgNO_3$ solution was charged into a reaction vessel which consisted of a 5 liter glass baffled flask with a propeller type agitator. The reaction vessel was provided with a heating and cooling bath to regulate the temperature. One liter of distilled water was added and the solution mixed rapidly by the propeller agitator. At this time, 12.5 g of Cab-O-Sil (ultrafine SiO_2) antiagglomerating agent was suspended in the reaction medium which was then stirred vigorously for five minutes.

Next, the $Pd(NO_3)_2$ solution was added and the 32.5 g of gold powder suspended in the combined solution with continuing agitation. An additional 375 ml of distilled water were also added to aid in the washing in of the gold powder.

The reduction reaction was initiated, with continued vigorous agitation, by adding to the reaction mass, dropwise, 390 g of a 50% hypophosphorous acid solution diluted in 400 ml of distilled water. The total time to add all of the H_3PO_2 was about ½ hour. During that time the resulting exothermic reaction was maintained at about 50°C by cooling water. After the reducing agent had been added the reaction media was held at the reaction temperature with mixing for 1 hour to insure completeness and optimum homogeneity of the metal slurry.

Thereafter, the metal slurry was filtered and washed with about two liters of distilled water and placed in a 110°C drying oven overnight.

The resulting dry powder was a microscopically homogeneous black mass having substantially no particles greater than about 5 μ. Analysis indicated the powder to consist on a metal basis of, by weight, 52% Ag, 35% Pd, and 13% Au and further indicated that no contaminating salts were present therein. This powder may be used as valuable starting material for microelectronic resistor production.

Ag and Pd/Cu for Conductor Patterns

Metallizations which are fired onto ceramic dielectric substrates to produce conductor patterns usually comprise finely divided noble metals and an inorganic binder, and usually are applied to the substrate as a dispersion of the inorganic powders in an inert liquid medium. The metallic component provides the functional (conductive) utility, while the binder (e.g., glass, Bi_2O_3, etc.) bonds metal particles to the substrate and to one another.

Mixtures of Pd, Ag and binder are often used to make adherent conductor patterns on dielectric substrates. However, due to the high cost of Pd, there is a distinct need for compositions which are less expensive by virtue of the elimination of at least some Pd, while at the same time retaining the high performance characteristics of fired Pd/Ag patterns, in terms of adhesion and solder leach resistance. Likewise, the composition should be air fireable for practical commercial utilization, even if some base (nonnoble) metals are introduced.

J.R. Larry; U.S. Patent 3,843,350; October 22, 1974; assigned to E.I. DuPont de Nemours and Company describes powder compositions useful for producing low-cost, air-fireable conductor patterns adherent to dielectric substrates. The compositions comprise finely divided silver particles and finely divided palladium/copper coprecipitated alloy particles; there is at least 35% by weight Pd in the alloy particles, based on total weight of Pd and Cu in the alloy particles. The weight ratio of palladium/copper alloy to silver in the compositions is in the range of 0.1:1 to 0.5:1, preferably 0.2:1 to 0.4:1. Preferred powder compositions are those wherein the alloy contains 35-75% Pd, based on the total weight of Pd and Cu therein.

The following examples and comparative showings are presented to illustrate the advantages of this process. In the examples all parts, percentages, proportions, etc., are by weight unless otherwise stated.

Example 1: A Pd/Cu alloy coprecipitate powder was formed from an aqueous solution containing palladium and copper ions, as follows.

A Pd/Cu coprecipitate (55 Pd/45 Cu based upon weight of Pd and Cu charged into the reactor) was prepared by charging to a 1 liter beaker 77.5 g of $PdCl_2$ solution (equal to 14.7 g Pd), 31.1 g of $CuCl_2 \cdot 2H_2O$ (equal to 11.6 g Cu) and 150 ml H_2O. The pH was adjusted to 7.2 with 42 ml of 50% NaOH solution. The temperature increased to 74°C. A reducing solution of 30 g dihydrazine sulfate dissolved in 100 ml water was added over a 6-minute period. The temperature and pH decreased to 70°C and 2.0, respectively. The reduction was incomplete at this stage; 15 ml of 50% NaOH was added to completely reduce the remaining $PdCl_2$ and $CuCl_2$.

The temperature and pH increased to 77°C and 6.9, respectively. Analyses of the washed and dried black powder showed 49.63% Pd, 39.04% Cu and 9.89% O. The surface area was 42.8 m^2/g. X-ray diffraction of the powder showed a broad, moderately intense peak at 2 theta equal to 40.5° (attributed to Pd/Cu coprecipitate) and small peaks at 2 theta equal to 32.5° and 38.8° (attributed to CuO). Pd normally occurs as a sharp intense peak at 2 theta equal to 40.1°. No peak was present at 2 theta equal to 43.3°, which would appear if free Cu was present.

Examples 2 and 3: These examples illustrate a practical lower limit of the Pd/Cu ratio. A Pd/Cu coprecipitate was prepared whereby the ratio of Pd/Cu as added to the precipitation vessel was 38.8/61.2. The precipitated material had a surface area of 46.4 m^2/g and was prepared as in Example 1 except that the initial pH was 4.3 and the reaction temperature was 50°C. Analysis of the precipitate showed 32.64% Pd, 49.95% Cu, 17.03% O. Compositions based on this powder were printed and fired once at either 850° or 760°C (8 minutes at peak temperature in both cycles). The table sets forth the composition and evaluation of these compositions. The solderability in Example 3 was poor, and significant solder leaching was evident. This ratio of Pd/Cu would be near the lower limit of Pd.

	. . Conductor Paste (wt %) . .				Ratio Pd/Cu			
Example	Pd/Cu	Ag	Glass A Binder*	Vehicle	Alloy to Ag	Solderability	Adhesion (lb) Initial	Aged
2	18	45	12	25	0.4	Good**	5.2	4.1
3	12.5	50	12	25.5	0.25	Poor***	-	-

*Glass A contained 10.87% PbO, 9.37% SiO_2, 2.45% CaO, 1.07% Al_2O_3, 1.22% B_2O_3, 75% Bi_2O_3.
**The sample fired at 850°C had good solderability.
***The sample fired at 760°C had poor solderability.

Recovery of Gold and Silver from Aqueous Cyanide Solutions

G. Dietz, Jr., R.M. Skomoroski and R.G. Zobbi; U.S. Patent 4,039,327; August 2, 1977; assigned to American Chemical & Refining Co., Inc. describe a process for recovering precious metals such as gold and silver from aqueous cyanide solutions by a stepwise process. The pH of the solution is, if necessary, initially adjusted to about 13 and thereafter is adjusted and maintained at 10 or above throughout the process. A "starter" carbonyl compound (e.g., formaldehyde or dextrose) which reacts with free cyanide ions is dissolved in the solution. Following this, an "accelerator" (e.g., hydrogen peroxide or a persulfate) is added and the solution is heated to at least about 85°C for at least one-half hour to aid in conversion of free cyanide ions to other chemical species.

Finally, a "clarifier" (e.g., hydrazine or a hydrosulfite) is added and the elevated temperature is maintained for at least a brief period after the addition. The solution is then cooled and the elemental precious metal particles formed are allowed to settle and are separated from the supernatant solution.

In a preferred embodiment, an interval of time for the chemical reactions to occur is provided after addition of the second material and before commencing heating of the solution, preferably, at least about 15 minutes. Heating of the solution to facilitate reaction of the cyanide ions is preferably for a period of between about one-half hour to about one and one-half hours.

Example 1: A 45 gallon (170.6 liter) batch of a gold solution has the following analysis at room temperature: Au, 6.06 g/l; pH, 12.7 and free CN^-, 7.17 g/l. It is treated as follows:

Time (min)	Steps and Data
0	Add 4108.3 g dextrose
5	Add 1000 cc H_2O_2 (35 wt %)
10	Add 1000 cc H_2O_2 (35 wt %)
55	Solution temperature = 39.4°C
57	Add 1000 cc H_2O_2 (35 wt %), strong NH_3 odor
63	Solution temperature = 43.3°C
64	Add 1000 cc H_2O_2 (35 wt %)
66	Solution temperature = 46.1°C
70	Solution temperature = 48.8°C
78	Solution temperature = 51.7°C
83	Add 3368.1 g dextrose
85	Solution temperature = 51.7°C
86	Add 1000 cc H_2O_2 (35 wt %)
90	Solution temperature = 55.6°C
100	Heat solution with steam
105	Solution temperature 79.4°C
109	Solution temperature 87.8°C. Decrease heating steam flow rate and maintain temperature at 82°-85°C
195	Solution temperature 83.3°C. Increase steam flow rate and raise solution temperature to 93.3°C
210	Solution temperature 93.3°C. Add 4994.7 g $Na_2S_2O_4$ to reactor. Gold metal became visible in solution in 1 or 2 minutes after $Na_2S_2O_4$ was added
240	Solution temperature about 93.3°C. Steam flow rate shut off. Allow solution to stand overnight
18 hr 45 min	Solution temperature 72.2°C Decanted solution: Au less than 1 ppm

Example 2: A 250 cc silver plating bath has the following analysis: Ag, 82.1 g/l and HCN, 82.1 g/l.

Steps and Data

Add 53 g of dextrose to solution.
Heat solution to 82.2°C.
Add 21 g of Na_2O_2 in small increments; some foam and spattering. Continue heating for 15 minutes.
Add 10 g $Na_2S_2O_4$. Allow solution to stand for 1 hour.
Filter off silver precipitate.
Filtered solution: Ag, 38 ppm.

As indicated by the above examples, the sequence of addition of the materials may be departed from somewhat after initial additions have been made in the prescribed sequence. For example, after addition of an incremental portion, say one-fifth or more, of the total of the first material, some of the second material may be added and then the balance of the first material. Other similar variations may be followed without departing from the scope of the process.

Oxidation and Sinter Resistant Powders

A.V. Fraioli and J.A. DeRosa; U.S. Patents 3,966,463; June 29, 1976; 4,010,025; March 1, 1977 and 4,036,634; July 19, 1977; all assigned to Plessey Inc. prepare a high density precious metal powder by conucleation and precipitation of the metal and titania or zirconia from an ammoniacal solution maintained at a pH in the range of 3-4 with sodium bisulfite. As starting materials, chloroplatinic acid, palladium chloride, chlorauric acid or similar solutions are used. For the refractory oxide component, titanium tetrachloride or disulfato zirconic acid are preferred.

The metal-oxide combination is precipitated with a bisulfite, and there is no requirement for firing to achieve both high density and increased sintering temperature. The specific procedure for preparing a high density gold powder in accordance with the process is set forth below in Example 1.

For preparing nickel powders according to the process, while the nickel powder must be fired, this is done in hydrogen or another reducing atmosphere to prevent nickel oxidation. Because of this firing, it is preferred to use zirconia rather than titania as the oxide component, because the former has greater resistance to reduction in the hydrogen environment. The reaction rates must be controlled, and are very important to achieve the desired results, but must be quite slow.

Generally, a nickel powder in accordance with the process is prepared as follows.

A first solution is prepared containing the metals in desired proportions. This can conveniently be nickel chloride and disulfato zirconic acid. A second solution is then prepared containing sodium hydroxide, hydrazine hydrate and water. The latter is heated to about 90°-95°C under conditions of agitation. Because the precipitate will be magnetic, a propeller rather than a magnetic stirrer should be used. The nickel-zirconia solution is then added thereto very slowly, essentially drop-by-drop, at a rate of no more than about 10 ml/min. Agitation is continued until all gas generation has ceased. The precipitate is filtered out, dried, pulverized, washed to remove residual chloride, dried again, screened and fired in hydrogen for about an hour at 400°-500°C. A specific example of nickel powder production is set forth below in Example 2.

Example 1: It was desired to produce a gold powder including 0.5% by weight of titania. A solution of the following composition was prepared: 166

ml of chlorauric acid (0.3 g Au/ml), 6 ml of titanium tetrachloride in 3 N HCl (0.25 g Ti/ml), and 338 ml of deionized (DI) water.

The gold solution is poured, over a span of one minute, into stirred, diluted ammonia, containing 110 ml of 28-30% NH_3, and 780 ml DI-water. A clean 2,000 ml pyrex beaker serves as the reaction vessel. After being stirred an additional 5 minutes, the ammoniated gold slurry is neutralized (pH 6-7) with 3 N HCl. For reduction to metallic gold, this slurry is heated to 50°C, preferably with a quartz or pyrex immersion heater, and then 10 g of anhydrous sodium acetate is added and dissolved. With the source of heat removed, 40 g of anhydrous sodium bisulfite is added at once to the stirred mixture.

The temperature rises to about 60°C, and the pH drops to and remains at 4 throughout the reduction. The product coagulates in about 15 minutes, but stirring is continued for 15 minutes more, while the reaction temperature is maintained at about 55°C, using the immersion heater if necessary. After being washed several times with DI-water by decanting, the product is further dispersed and densified in a one quart commercial Waring Blender.

The wet product is added to the blender containing 100 ml DI-water and 1.25 g cyanoguanidine. After blending at low speed twice for 30 seconds, the contents are poured into a 400 ml beaker for settling and decanting. The gold is then dried at 125°C and screened through a 140 mesh sieve. The yield is about 98%; tapped packing density is 6.7 g/cc; average particle size (Fisher Subsieve Sizer) is 4.1 μm; and surface area (BET) is 0.8 m^2/g.

Example 2: It was desired to produce a nickel powder containing 2% ZrO_2. A solution containing 162.5 ml of 6.25 N NaOH, 32.5 ml of 85% hydrazine hydrate, and 55 ml DI-water is heated at 90°-95°C in a 1,000 ml beaker, which is fitted with a polypropylene propeller mixer. A solution of nickel chloride, containing 24.5 g Ni, and disulfato zirconic acid, equivalent to 0.5 g ZrO_2 is added to the hot solution from a separatory funnel, at a rate of 10 ml/min.

After the addition, mixing is continued for about 15 minutes at 90°C. The mixture is filtered without washing through a 600M glass Buchner funnel, then dried on the funnel at 125°C. The mass is then pulverized in a mortar to pass a 100 mesh sieve. In this condition the product is washable by decanting with DI-water until free of chloride. After drying at 125°C and screening through 100 mesh, the product is placed in a shallow graphite boat and fired in a hydrogen furnace at 450°C and at 600°C. The resulting powder properties at these firing temperatures are set forth below:

	450°C	**600°C**
Loose packing density	1.5 g/cc	1.8 g/cc
Tapped packing density	1.9 g/cc	2.4 g/cc
Surface area	42.7 m^2/g	30.4 m^2/g
Average particle size	1.46 μm	2.38 μm

Example 3: The product of Example 2 was heated in air for 1 hour at 450°C and gained no weight, indicating significant resistance to oxidation.

Refractory Oxides and Other Metals

TANTALUM AND COLUMBIUM

Reclaiming by Fused Salt Treatments

H. Vartanian; U.S. Patent 4,141,720; February 27, 1979 assigned to NRC, Inc. has developed a process for reclaiming tantalum powders which have high oxygen, high sodium contents by treating them with selected fused salt compositions. This process converts normally unleachable sodium metatantalate, recognized as a significant source of the powder's content of oxygen and sodium, to leachable forms of oxygen and sodium bearing compounds.

The conversion is accomplished by placing the high oxygen, high sodium tantalum powder, containing such metatantalate, in a fused salt bath which has the major portion of its composition equivalent to the final salts composition which would result from a reduction of potassium fluotantalate with sodium metal, under conditions where 90-95% of required sodium stoichiometry was used and as a minor portion of composition contains small amounts of added potassium fluotantalate. The required amount of potassium fluotantalate is small in relation to the quantity of reclaimable tantalum powder treated by the process; i.e., the weight ratio of tantalum powder to be treated to tantalum contained in the added potassium fluotantalate (K_2TaF_7) is in the range of 9:1-40:1.

Several experimental batches of tantalum powder having oxygen contents of 2,101, 4,500 and 2,930 ppm respectively, were provided in 227 gram batches and combined with mixtures of potassium fluotantalate and mixed halide salts to provide, in relation to the 227 grams of tantalum supplied, 55.4 grams of potassium fluotantalate, 368 grams of sodium fluoride, 539 grams of sodium chloride and 187 grams potassium chloride. This composition corresponds to a fused salt mixture with the following weight and mol ratios:

	Weight Ratio	Mol Ratio
Tantalum Powder	1	1
K_2TaF_7	0.24	0.11
KCl	0.82	2
NaF	1.62	7
NaCl	2.37	7.37

Two melts were made with the 2,930 ppm powder, which had sodium and potassium levels of 197 and 435 ppm respectively and a nominal size distribution of −100 mesh +325 mesh but, in fact, containing some 50% −325 mesh components. The first melt was conducted according to the nominal amounts of salt given above in relation to the 227 grams of tantalum. The procedure was to dry the mixture of salts in a reactor at 540°C in a vacuum and then cool to room temperature, release vacuum, add the input tantalum powder to the dried salt, reevacuate and heat the tantalum powder and salts with mixing to 540°C under vacuum, backfill with argon at 540°C, purge with argon between 700°-800°C (at which point the salt mixture would be molten), and heat to 900°C and hold for three hours before cooling to room temperature.

The mixture would then be removed from the reactor, micropulverized, leached in a tank and then further leached in barrels and subsequently dried and screened to desired fractions. Through this processing, oxygen levels were reduced to 650 ppm and sodium and potassium levels were reduced to 16 and 289 ppm respectively.

A second melt, using similar processing, was repeated with double the amount of tantalum (454 grams) relative to the total weight of mixed salts. The 2,930 ppm original oxygen was reduced to 570 ppm and the sodium and potassium to 87 ppm and 112 ppm respectively.

The 4,500 ppm oxygen powder which had 1,399 ppm and 306 ppm sodium and potassium levels, respectively, and a −17 micron +3 micron nominal distribution, was similarly treated using the proportion as shown above and the oxygen level was reduced to 710 ppm, sodium to 10 ppm, and potassium to 125 ppm. Powder compacts were pressed from both the original high-oxygen powder and the treated powder (1.8 gram compacts at 8.0 g/cc green density).

These were sintered in vacuum at 1850°C for 30 minutes, and anodized to 200 V in accordance with conventional anodizing practice of the industry. A wet test for capacitance showed average values for anodes of the treated powder of 17.9 μF/g and 0.82 μA/g leakage whereas anodes made from the untreated powders could not be measured because leakage values were so high.

Similarly prepared powder compacts were anodized to only 100 volts. Wet test average values for anodes prepared from the treated powder were 35.3 μF/g and 0.61 μA/g leakage, versus 36.9 μF/g and 36.0 μA/g leakage for anodes prepared from the untreated powder.

A summary of chemical, physical and wet test electrical properties of the treated and original powders is given in Table 1.

Table 1

	O	Na	K	N	FAPD	PR	SBD	-325 Mesh	Capacitance . .(μF/g). .		Leakage* . (μA/g) . .	
									. . Formation Voltage . .			
	. . . . . (ppm)				(μ)	(%)	(g/in^3)	(%)	200	100	200	100
Source powder	4,500	1,399	306	20	6.7	72.8	51.7	100	**	36.9	**	36.0
Treated powder	710	10	125	34	8.0	75.0	49.5	85***	17.9	35.3	0.82	0.61

*200 V formed anodes tested at 140 V; 100 V formed anodes tested at 70 V.
**Would not form to 200 V.
***-100+325 mesh, 12.1%; -100+200 mesh, 2.5%; and -40+100 mesh, 0.5%.

Preparation of Tantalum Anodes Without Using Binder

V.T. Bates, S.S. Fry and J.B. Hakko; U.S. Patent 4,017,302; April 12, 1977; assigned to Fansteel Inc. have developed high-purity agglomerated tantalum powders characterized by exceptionally high green strength when pressed without using a carbonaceous binder into low density anodes and high electrical capacitance along with low direct current leakage and dissipation factor in the anodes after sintering. Such powder is produced by milling hydrided high-purity tantalum metal ingots or powder to a critical particle size powder, subjecting the powder to a two-step heat treatment to degas and preagglomerate it, and screening and milling the oversize portion to the degassed and preagglomerated powder to achieve intermediate mesh size powder.

This powder is then subjected to a higher temperature treatment to agglomerate it, the agglomerated powder is milled to an intermediate mesh fraction, the powder screened to remove the fine particle size portion, this fine particle material subjected to an additional heat treatment to reagglomerate it, and the reagglomerated fines blended with the coarser mesh size portions of the agglomerated powder.

Example:

(a) Tantalum hydride powder was prepared by crushing an electron-beam melted tantalum ingot that had been saturated with hydrogen gas at an elevated temperature. The hydride was milled and classified to a Fisher subsieve size of 4.15 μ. Apparent density, hereafter called "Scott density," was determined on this powder by the procedure of ASTM designation B212-48, "Determination of Apparent Density of Metal Powders," and was found to be 75.5 g/in^3.

(b) The tantalum hydride powder was heated under a high purity flowing argon atmosphere to 800°C and held for two hours at temperature to evolve the hydrogen. Then the furnace was evacuated to about 10^{-3} torr, and the tantalum powder (degassed tantalum hydride) was further heated to 1200°C and held for ½ hour at temperature. Then this degassed and preagglomerated powder was cooled under vacuum for 2 hours and then under helium during cooling to ambient temperature.

(c) The degassed and preagglomerated powder was screened using an 80 mesh U.S. standard screen. The oversize material was milled so that all powder was −80 mesh. The −80 mesh material was 89.5% −325 mesh by a screen test, had a Fisher subsieve size of 5.6 μ and a Scott density of 76.8 g/in^3.

(d) The −80 mesh powder was heated in vacuum (about 10^{-3} torr) and 1410°C and held for 1 hour at temperature, then cooled under vacuum for 2 hours and finally under helium to ambient temperature to obtain agglomerated tantalum powder.

(e) The agglomerated powder was milled and screened using a 35 mesh screen, with any oversize material remilled and rescreened so that all powder was −35 mesh.

(f) The −35 mesh agglomerated powder was further screened using a 250 mesh screen. The −35+250 mesh portion of the powder was set aside so that it could be blended later with the −250 mesh portion of the powder, which was further processed.

(g) The −250 mesh powder was subjected to an additional agglomeration treatment by heating it in vacuum (about 10^{-3} torr) to 1475°C and holding for 1 hour at temperature, followed by cooling in vacuum for 2 hours, then in helium to ambient temperature.

(h) The agglomerated −250 mesh material was milled and screened to −35 mesh.

(i) This −35 mesh powder from step (h) was blended with the −35+250 mesh powder from step (f).

(j) This final tantalum powder was sampled and tested. The Fisher subsieve size was 9.6 μ, Scott density was 65.7 g/in^3, and a screen test showed:

+80 mesh, 37.3%
−80+200 mesh, 13.1%
−300+325 mesh, 5.7%
−325 mesh, 43.9%

A Roller analysis of the −325 mesh fraction showed a particle size distribution of:

0-3 μ, 0.35%
3-5 μ, 0.86%
5-10 μ, 8.44%
10-15 μ, 17.40%
15-20 μ, 40.65%
20-25 μ, 21.43%
25-30 μ, 10.86%
30-35 μ, 0.01%

The final tantalum powder was found to have a Hall flow of 49 seconds when measured in accordance with "Standard Method of Test for Flow Rate of Metal Powders," ASTM designation B213-48 (reapproved 1965), except that the test unit was modified to vibrate the Hall flow cup, with the cup vibrated at a frequency of 3,600 cycles and an amplitude of 0.024 inch.

The powder was pressed into individually weighed 2.010±0.020 gram compacts in a 0.261 inch diameter die to a green density of 7.0±0.05 g/cm^3 for determination of green strength. The compacts individually were laid sideways under the anvil of a Chatillon Model LTCM Universal Tensile, Compression and Spring Tester provided with a flat anvil and base and crushed at a compression rate setting of 2.0.

The pressure in pounds required to crush the compact was recorded as the green strength. Four compacts were tested for green strength and the data averaged. In this test, if any compact shows a green strength value that is an outlier, as determined in accordance with ASTM designation E-178-61T, an additional anode is pressed and tested. (An outlying observation, or outlier, is one that appears to deviate markedly from other members of the set in which it occurs.)

Green strength was determined on the final powder and on the +325 and −325 mesh fractions of screened samples of the powder. Green strength results were in the range of 12 to 30 pounds as follows:

	Green Strength (lb)
Final tantalum powder	18.1-18.4
+325 Mesh fraction	27.2
-325 Mesh fraction	12.3

Chemical analyses on the final powder showed an oxygen content of 1,615 ppm (0.1615 weight percent), a nitrogen content of 18 ppm (0.0018 weight percent), a carbon content of 20 ppm (0.0020 weight percent), and an iron content of 15 ppm (0.0015 weight percent).

For electrical tests, the powder was pressed into individually weighed 2 g anodes with an embedded lead wire in a 0.261 inch diameter die to a green density of 7.2±0.05 g/cm^3. The anodes were sintered either for 30 minutes at

1800°C (optical) or for 30 minutes at 2000°C (optical) in a cold-wall, vacuum sintering furnace (10^{-5} torr absolute pressure), and then were tested for shrinkage in diameter during sintering and for electrical properties including direct current leakage, specific capacitance (CV/g) and dissipation factor.

The electrical testing procedure involved anodizing the sintered anodes in 0.01% phosphoric acid in water. Anodizing was carried out at a current density of 35 mA/g until 200 V was reached. The anodes at 1800°C were held at 200 V for 2 hours. The anodes sintered at 2000°C were anodized using a current density of 35 mA/g to 200 V, and then at 12 mA/g from 200 to 270 V; they were held at 270 V for 1 hour.

The anodes, after anodizing, rinsing and drying, were first tested for direct current leakage (DCL). A phosphoric acid solution was employed. The test conditions were as follows:

Anode Formation Voltage	Test Electrolyte Concentration	Test Voltage
200	H_3PO_4, 10.0%	140
270	H_3PO_4, 0.01%	240

The anodes were immersed in the test solution to the top of the anode and the proper voltage was applied for 2 minutes, after which the DCL was measured. After DCL measurements were completed, the anodes formed to 200 V were placed in a tray containing 10% phosphoric acid and permitted to soak 30 to 45 minutes.

The anodes formed to 270 V were washed for 3 to 5 minutes at 105°±5°C in air. They were then soaked in 10% phosphoric acid for 30 to 45 minutes. The capacitance was measured on the anode immersed in 10% phosphoric acid employing a type 1611B General Radio Capacitance Test Bridge with an ac signal of 0.5 V and a dc bias of 3 V. The dissipation factor also was determined from this bridge set. The results were:

	Anodes Sintered for . . . . 30 Minutes.	
	1800°C	2000°C
Shrinkage in diameter, %	4.3	8.3
Direct current leakage (DCL), μA/μF - V x 10^5	26.9	80.3
Capacitance, μF - V/g	3,936	3,011
Dissipation factor, %	14.8	11.0

Addition of Phosphorus

S.S. Fry; U.S. Patent 4,009,007; February 22, 1977; assigned to Fansteel Inc. has found that the electrical conductance of anodes produced from tantalum powder can be increased by the addition of phosphorus-containing materials in amounts from about 5 to about 400 ppm based on elemental phosphorus.

It is necessary, that the phosphorus-containing material be added to the tantalum powder, or to a tantalum hydride powder before reduction thereof to tantalum. When phosphorus is present in a tantalum powder as an incidental impurity, either carried over from the original ore or introduced as an impurity in the chemicals used in the normal preparation of the tantalum powder, the results of this process are not obtained.

Nor are the results of this process obtained when phosphorus-containing materials are added to tantalum powder which has been pressed and sintered into anode form. However, the results are obtained when a phosphorus-containing material is added to a tantalum powder which has been pressed into anode form but not sintered.

The preferred phosphorus-containing materials are the inorganic phosphate salts, such as ammonium, sodium, potassium, calcium, barium and lead orthophosphates, ammonium monohydrogen orthophosphate, ammonium dihydrogen orthophosphate, sodium monohydrogen orthophosphate, sodium dihydrogen orthophosphate, and potassium dihydrogen orthophosphate. Other suitable phosphorus-containing materials include elemental phosphorus, metallic phosphides, phosphorus oxides and acids, and organic phosphorus-containing materials, such as alkyl phosphates.

Phosphate materials containing no metallic cations, such as ammonium monohydrogen orthophosphate, ammonium dihydrogen orthophosphate and phosphoric acid, are particularly preferred because they do not introduce other metals into the tantalum powder with possible adverse effects on the dc leakage and breakdown voltage properties of the anodes produced therefrom.

When calcium orthophosphate is added as the phosphorus-containing material and when the material is added after agglomeration, an added advantage is obtained in that the flow characteristics of the final powder are improved.

Examples 1 and 2: An agglomerated sodium-reduced tantalum powder, designated Example 1, was used as a control.

Calcium orthophosphate was produced by adding orthophosphoric acid to calcium oxide and was washed free of acids before use. Methanol and the washed precipitate of calcium orthophosphate were added to an agglomerated sodium-reduced tantalum powder to form a slurry, and the slurry was dried in air at 90°-100°C. The dried mixture was then stirred in a V-shell blender for three minutes.

The amount of calcium orthophosphate was sufficient to make a master mixture containing 800 to 1,000 ppm of the orthophosphate ion, PO_4^{-3}, corresponding to about 261 to about 326 ppm of elemental phosphorus. The mixture was blended with an additional amount of the same lot of agglomerated sodium reduced tantalum powder as in Example 1 to produce a final concentration of 115 ppm of the orthophosphate ion, corresponding to 37.5 ppm of elemental phosphorus. This phosphorus-containing powder is designated Example 2.

The tantalum control powder of Example 1 was found to have a Hall flow of 46 seconds when measured in accordance with "Standard Method of Test for Flow Rate of Metal Powders," ASTM designation: B213-48 (reapproved 1965), except that the test unit was modified to vibrate the Hall flow cup, and the cup was vibrated at a frequency of 3,600 cycles and an amplitude of 0.024 inch. The phosphorus-containing powder of Example 2 had a Hall flow of 27 seconds when measured under the same conditions.

The control powder (Example 1) and the calcium orthophosphate-containing powder (Example 2) were formed into 2-gram anodes by pressing the powder to a density of 6.45 g/cm^3. The anodes were sintered either for 30 minutes at 1800°C or for 30 minutes at 2000°C in a cold-wall, vacuum sintering furnace (10^{-5} torr absolute pressure), and then were tested for density and specific capacitance (CV/g).

The testing procedure involved anodizing the sintered anodes in 0.01% phosphoric acid in water. Anodizing was carried out at a current density of 35 mA/g until 200 V was reached. The anodes sintered at 1800°C were held at 200 V for 2 hours. The anodes sintered at 2000°C were anodized to 270 V but at a current density of 12 mA/g. The latter anodes were held at 270 V for 1 hour.

The formed anodes were washed in deionized water and then dried in clean air at 105°C. They were then soaked in 10% phosphoric acid for 30 minutes. The capacitance was measured on the anode immersed in 10% phosphoric acid employing a type 1611B General Radio Capacitance Test Bridge with an ac signal of 0.5 V and a dc bias of 3 V. The results are summarized in the table below.

		. . . . Sintered at 1800°C			. . . . Sintered at 2000°C.		
Ex. No.	Remarks	Sintered Density, g/cm^3	Capacitance, μFV/g (CV/g)	CV/g Improvement, %	Sintered Density, g/cm^3	Capacitance, μFV/g (CV/g)	CV/g Improvement, %
1	Control	7.31	4,783	–	8.58	3,488	–
2	$Ca_3(PO_4)_2$*	7.08	5,036	5.3	8.54	3,550	1.8

*Added to 37.5 ppm P.

In additional work by *S.S. Fry; U.S. Patent 4,084,965; April 18, 1978; assigned to Fansteel Inc.,* the same procedure is applied to columbium, but the range of phosphorus added is about 5-600 ppm.

Reduction of Hydrided Tantalum

To improve upon the processes described in the two preceding sections, *J.B.*

Hakko; U.S. Patent 4,141,719; February 27, 1979; assigned to Fansteel Inc. describes a process of hydriding a tantalum metal ingot, reducing and classifying the resulting hydride tantalum to a powder having a predetermined particle size, degassing and preagglomerating the reduced tantalum powder, reducing the preagglomerated powder to a screen size of less than 80 mesh, reagglomerating the reduced powder, reducing the reagglomerated powder to an average intermediate particle size which is larger than the reduced preagglomerated powder, and blending the reduced powder to intermediate particle size.

The resulting tantalum powder has all of the desirable characteristics of the powder produced by the process of U.S. Patent 4,017,302 and also has improved green strength and a lower oxygen content which is desirable when making capacitor electrodes to minimize embrittlement of tantalum lead wires embedded in the compact before sintering. During sintering the embedded wires may become embrittled by absorption of sufficient oxygen from the tantalum powder.

Capacitor anodes made from this improved powder also have less shrinkage, improved reliability, longer useful life, and in some instances less direct current leakage and somewhat higher capacitance than anodes produced by the powder of U.S. Patent 4,017,302.

Thus, this process is simpler and less expensive to utilize than previously known processes, produces a powder having good flow characteristics which can be recycled in compacting operations, and requires no binders to produce a compact. When compacted it has high green strength and can be sintered with little shrinkage and minimum embrittlement of any tantalum lead wires embedded therein; and when the compact is sintered has low density and can be anodized to provide capacitor anodes having low direct current leakage, low dissipation factor, high capacitance, high working and breakdown voltages, improved reliability, and a longer useful life in service.

Example:

(a) Tantalum hydride powder was prepared by crushing an electron-beam melted tantalum ingot that had been saturated with hydrogen gas at an elevated temperature. The hydride was milled and classified to a Fisher subsieve size of 4.15 μ.

The particle size distribution of this precursor powder as measured by Roller analysis ["Metal Powder Size Distribution with the Roller Air Analyzer," by P.S. Roller, ASTM Special Publication No. 140 (1952)], referenced for performance of ASTM B293-60 (70) (Subsieve Analysis of Granular Metal Powders by Air Classification) and also recognized as the appropriate standard by the Metal Powder Association "12-51T (1951)," was found to be:

Size Micron (μ)	% by Weight
0–3	6.2
3–5	12.6
5–10	35.7
10–15	25.9
15–20	19.6
20–25	0.1

The Scott density of this powder was determined by the procedure of ASTM designation B212-48 (reapproved 1970) "Apparent Density of Metal Powders," and found to be 78.8 g/in^3.

(b) The tantalum hydride powder was heated under a high-purity, flowing argon atmosphere to 800°C and held for 2 hours at temperature to remove the hydrogen. Then the furnace was evacuated to about 10^{-3} torr, and the degassed tantalum hydride powder was further heated to 1320°C and held for one-half hour at temperature to preagglomerate the powder. Then this degassed and preagglomerated powder was cooled under vacuum for 2 hours, and then under helium during cooling to ambient temperature.

(c) The degassed and preagglomerated powder was milled and screened using a 200 mesh U.S. standard screen. The oversize material was further milled so that all powder was −200 mesh. The −200 mesh material was comprised of 55.6% by weight of −325 mesh powder by a screen test, had a Fisher subsieve size (FSSS) of 6.7 μ, and a Scott density of 72.0 g/in^3.

(d) The −200 mesh powder was heated in vacuum (about 10^{-3} torr) to 1450°C and held for 1 hour at temperature, then cooled under vacuum for 2 hours and finally under helium to ambient temperature to produce reagglomerated powder.

(e) The reagglomerated powder was milled and screened using a 35 mesh U.S. standard screen, with any oversize material remilled and rescreened until all powder was −35 mesh.

(f) The −35 mesh powder was blended, sampled, and tested. The Fisher subsieve size was 9.0 μ, Scott density was 62.1 g/in^3, and a screen test showed 54.6% by weight of −325 mesh material. The final tantalum powder was found to have a Hall flow time of 32 seconds when measured in accordance with "Standard Method of Tests of Flow Rate of Metal Powders," ASTM B218-48 (reapproved 1965), except that the test unit was modified to vibrate the Hall flow cup at a frequency of

3,600 cycles per minute and an amplitude of 0.024 inch. Chemical analysis of the final powder showed an oxygen content of 1,479 ppm (0.1479 weight percent), and a nitrogen content of 19 ppm (0.0019 weight percent).

The powder was pressed into individually weighed 2.010±0.020 gram compacts in a 0.261 diameter die to a green density of 7.0±0.05 g/cm^3 for determination of green strength. The compacts were each individually laid sideways under the anvil of a Chatillon Model LTCM Universal Tensile, Compression and Spring Tester provided with a flat anvil and base, and were crushed at a compression rate setting of 2.0. The pressure in pounds required to crush the compact was recorded as the green strength.

Four compacts were tested for green strength, and the data were averaged. In this test, if any compact had a value that is an outlier as determined in accordance with ASTM E178-61T, an additional anode was pressed and tested. (An outlying observation, or outlier, was one that appeared to deviate markedly from other members of the set in which it occurred.) The green strength of the compacts of the final powder averaged 19.5 pounds.

For the electrical tests, the powder was pressed into individually weighed 2-gram anodes with an embedded tantalum lead wire in a 0.261 inch diameter die to a green density of 7.2±0.05 g/cm^3. The anodes were sintered either for 30 minutes at 1800°C (optical) or for 30 minutes at 2000°C (optical) in a cold-wall, vacuum sintering furnace (10^{-5} torr absolute pressure), and then were tested for shrinkage in diameter during sintering and sintered density.

The sintered compacts were anodized to produce a dielectric surface film and tested for electrical properties including DCL, specific capacitance (CV/g), dissipation factor, and breakdown voltage (determined only on anodes sintered at 2000°C).

The electrical testing procedure involved anodizing the sintered anodes in 0.01% phosphoric acid in water at 90±2°C. Anodizing was carried out at a current density of 35 mA/g until 200 V was reached. The anodes sintered at 1800°C were held at 200 V for 2 hours. The anodes sintered at 2000°C were anodized using a current density of 35 mA/g to 200 V, and 12 mA/g from 200 to 270 V, and were held at 270 V for 1 hour.

The anodes, after anodizing, rinsing and drying, were first tested for DCL. A phosphoric acid solution was employed. The test conditions were as follows:

Anode Formation Voltage	Test Electrolyte Concentration	Test Voltage
200	10.0% H_3PO_4	140
270	0.01% H_3PO_4	240

The anodes were immersed in the test solution to the top of the anode and the proper voltage was applied for 2 minutes, after which time the DCL was measured.

After DCL measurements were completed, the anodes formed to 200 V were placed in a tray containing 10% phosphoric acid and permitted to soak 30 to 45 minutes, and the anodes formed to 270 V were washed for three to five minutes in running distilled water, dried 45 minutes at 105°±50°C in air, and then soaked in 10% phosphoric acid for 30 to 45 minutes. Thereafter, the capacitance was measured on the anodes immersed in 10% phosphoric acid by a type 1611B General Radio Capacitance Test Bridge with an ac signal of 0.5 V and a dc bias of 3 V. The dissipation factor also was determined from this bridge test.

Six similar anodes prepared from the same powder and sintered for 30 minutes at 2000°C, but not anodized, were measured for breakdown voltage (BDV). The BDV test was carried out by electroforming the anodes in an agitated 0.1% H_3PO_4 solution at 90°±2°C, with the forming voltage increased at the rate of 3 to 4 V/min until dielectric breakdown occurred.

The point of breakdown was established when the forming current of the anode increased by 100 mA over the current flowing at 100 V, or when scintillation occurred. A mean breakdown voltage was determined after elimination of outliers. (An outlying observation, or outlier, was considered one that appeared to deviate markedly from other members of the set from which it occurred.) Only one outlier per test lot was considered acceptable. The procedure of ASTM E178-61T, "Tentative Recommended Practice for Dealing with Outlying Observations," was followed.

The test results on the anodes of this example were:

	Anodes Sintered . . . .for 30 min at	
	1800°C	2000°C
Shrinkage in diameter, %	3.9	7.8
Sintered density, g/cm^3	8.07	9.01
Direct current leakage, μA/μFV x 10^5	24.3	56.6
Capacitance, μFV/g	3,917	3,061
Dissipation factor, %	18.2	14.4
Breakdown voltage, V	–	239

The data of this example demonstrate that this process produces an improved tantalum powder with a low Scott density, excellent flow properties, high green strength, and low oxygen content; when sintered, it exhibits low shrinkage and low density; and produces anodes having high capacitance and low dissipation factor when sintered up to 30 minutes at 2000°C. The direct current leakage and breakdown voltage are similar to those for anodes made of other high-purity, state-of-the-art tantalum powders.

Molten Double Salt Bath

C.F. Rerat; U.S. Patent 4,149,876; April 17, 1979; assigned to Fansteel Inc. describes a process of producing tantalum and columbium powders which comprises maintaining a liquid bath of a double salt of the formula:

$$R_2MX_7$$

wherein R is an alkali metal, M is a metal of the group consisting of tantalum or columbium and X is a halogen atom of the group consisting of fluorine, chlorine, or bromine, such as K_2TaF_7, and a diluent comprising an alkali metal halide selected from the group consisting of NaCl, KCl and mixtures thereof at a temperature above the liquidus temperature of the salt mixture but not exceeding 760°C in a closed reaction vessel.

A molten alkali metal is added to the bath at a rate of at least 0.2 kg/kg of the double salt per hour during a nucleation period until a desired reaction temperature between 760° and 1000°C is reached. The temperature rise during the nucleation period is at a rate of at least 10°C per minute. Thereafter additional molten alkali metal is added to the bath at a rate of at least 0.1 kg/kg of the double salt per hour during a growth period while providing forced cooling to the bath at a rate of at least 10 kcal/min/kg of the double salt. The liquid bath is continuously agitated, producing crystals of metal M and these crystals are collected in a lower portion of the reaction vessel. The rate of alkali metal addition and the rate of forced cooling are balanced to maintain the reaction temperature in the growth period within 50°C of the desired reaction temperature but in excess of 760°C.

Figure 7.1 shows a cross section of a reactor vessel **50** for conducting reduction of K_2TaF_7 using rapid feed rates for the sodium reducing metal and forced external cooling. The vessel has thermal insulation **52** and can be heated by electrical resistance heating elements **54**. Liner **56**, constructed of a material such as stainless steel, surrounds a retort **58**, which is constructed of a heat resistant alloy such as Inconel. Lid **59** of the retort has a flange sealed by gasket **62** to exterior port flange **100** of liner **56**.

The molten salt bath **63** is comprised of a metal fluoride (or chloride) salt, such as K_2TaF_7, and a selected amount of diluent salt such as NaCl or KCl, or a mixture of these. The liquid level **63L** of the salt bath and tantalum crystals **63C** that build up during the reaction also are illustrated.

The salt bath is initially molten before the sodium addition. In the specific reactor used, the amount of K_2TaF_7 in the molten bath ranged from about 500 to 1000 kg and the diluent salt from about 200 to 450 kg, while the weight ratio of diluent salt to K_2TaF_7 ranged from about 0.2 to 1.0, and the mol ratios from about 0.8 to 4.6. The specific amounts of diluent salt and K_2TaF_7 used, and their weight ratio depended on the specific particle size of tantalum powder to be produced. The higher amounts of diluent salt were used when the finest particle size tantalum powders were produced.

Figure 7.1: Apparatus for Producing Tantalum or Columbium Powders

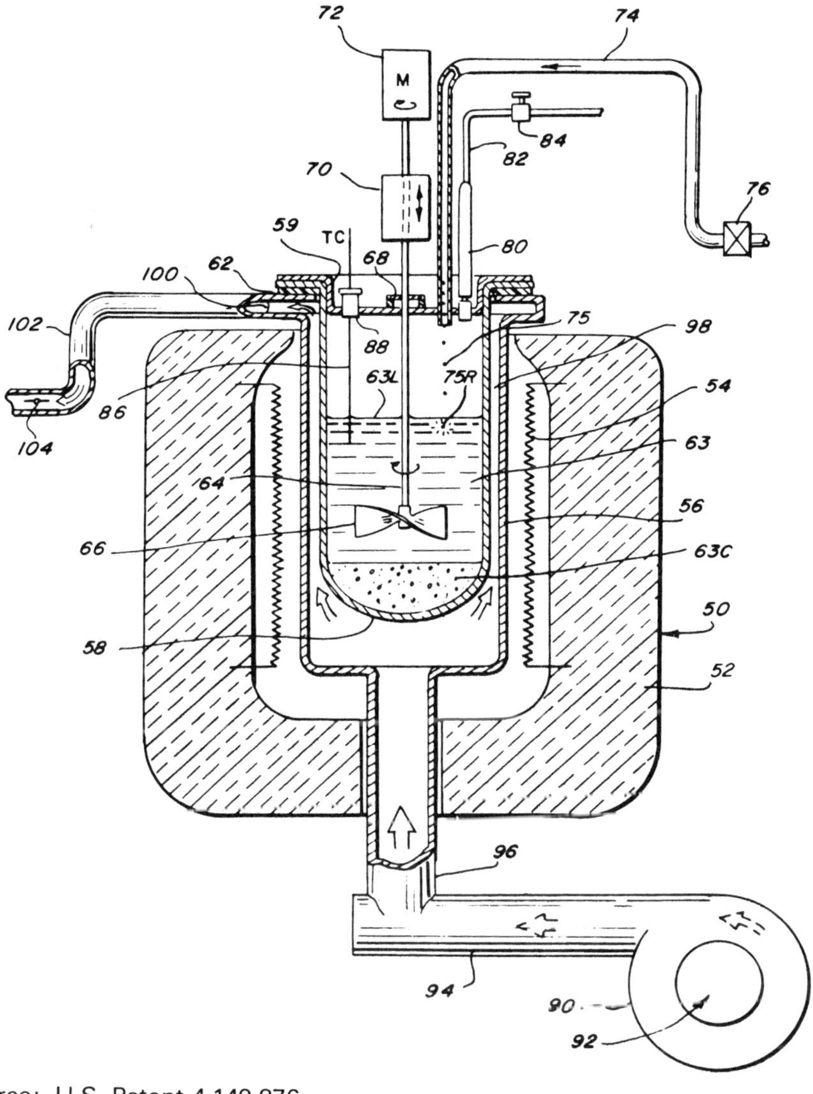

Source: U.S. Patent 4,149,876

When adequate volume was available in the retort, the entire amounts of K_2TaF_7 and diluent salt were added as solid salts to the reactor and were melted together. In some cases where higher amounts of diluent salt were desired, all of the K_2TaF_7 and a portion of the diluent salt were melted first, then the balance of the diluent salt was added as one or more supplementary additions.

Agitation of the bath is maintained during the reduction reaction by agitator **64**. The agitator has blades **66** with an outside diameter typically about one-half that of the inside diameter of the retort. The blades generally have a pitch in the range of about 30 to 60 degrees, typically about 45 degrees with respect to the agitator shaft.

The agitator extends vertically upward through seal **68** positioned on the retort lid to engage to vertical drive means **70** and rotary drive means **72.**

Agitation is provided continuously from the time the salt bath is molten until the completion of the reduction reaction. In preferred embodiments, the agitator speed is typically varied in the range from about 70 to 140 rpm during the run, or from about 0.05 to 0.50 rpm/kg of the molten salt mixture when the mixture is in the range of 500 to 1,000 kg. The slower speeds are used initially to obtain a homogeneous molten salt bath before beginning the sodium feed. Fastest agitator speeds are used during the nucleation period when the highest sodium feed rate is used, then slower speeds are employed during subsequent lower sodium feed rates, and also after completion of the sodium feed.

Although changing agitator speeds during the run is a preferred procedure, satisfactory results can be achieved in this process using a fixed agitator speed during the run.

Another preferred procedure is to intentionally and progressively raise the position of the agitator during the period when the sodium is added. This is done using the vertical drive means to raise the agitator continuously or in steps, typically about 2" per step, upward about 4" to 12" above its initial position. This is done to avoid plowing of the tantalum metal powder particles that settle and progressively build up on the retort bottom, and to compensate for the added bath volume resulting from the sodium additions.

Although progressive raising of the agitator is preferred practice, satisfactory results can be achieved in this process by proper initial vertical placement of the agitator in an essentially mean position.

Specific combinations of agitator design, rotational speed and change, and intentional vertical raising of the agitator are used in an effort to achieve a continuously homogeneous bath composition during the runs made under different process conditions within the scope of this process to achieve the desired final tantalum particle sizes.

Molten sodium at a temperature of about 100°-160°C is introduced through line **74** extending through the retort lid. Drops or a stream **75** of sodium react in the molten bath when it contacts K_2TaF_7 as illustrated at **75R**. The flow rate of sodium is controlled by valve **76** in the line from a heated sodium tank (not shown). Sodium condenser **80** is connected to the retort lid, and has a line **82** leading to a safety pop-off valve **84**.

Thermocouple **86** is immersed in the molten bath and extends through seal **88** to a temperature recorder and controller (not shown).

A means for forced external cooling of the retort is provided by blower **90**. Ambient air or other coolant such as liquid or gaseous CO_2 are introduced at inlet **92** in the blower. The coolant flow proceeds through duct **94** connected to extension **96** leading to the interior of liner **56**, then upward in the space **98** between the retort **58** and the liner to provide cooling to the retort. The coolant flow then proceeds through a collection manifold located at the top of the liner, then is vented through duct **102** containing a damper **104**, or it may be recirculated by providing a means for heat removal (not shown).

In the exemplary embodiments, forced air cooling on the external of the retort was generally used. With a 36" i.d. retort, air flow rates used to cool the exterior of the retort ranged from about 500 to 3,000 cfm. The highest rates were used in the process parameters employed to produce the finest tantalum powder particle sizes. Cooling of the retort can be further enhanced by providing fins or other suitable configurations on the exterior of the retort.

Forced air cooling and liquid carbon dioxide cooling are described in examples of the process. However, such cooling methods are not limiting, and this process also contemplates other means for forced cooling, such as coolant coils or a coolant jacket, sprays, and other methods.

When this process is used for producing columbium by sodium reduction of K_2CbF_7, the preferred ratio of diluent salt, such as NaCl, to K_2CbF_7 is about the same on a mol basis as for the ratio of NaCl to K_2TaF_7, namely about 0.8 to 4.6 as described above. The preferred sodium feed rates for reducing K_2CbF_7 are about 30% greater than described in the examples for sodium reduction of K_2TaF_7.

The abovedescribed means to obtain fine particle size tantalum metal powder with exceptional properties for the manufacture of anodes employed in the production of electrolytic capacitors can be equally well applied to the manufacture of somewhat coarser particle size powders to achieve the same added benefits of high productivity and high yield.

Anodes for Solid Electrolyte Capacitors

J.-C. Moulin, B. Escher and D. Prince; U.S. Patent 3,984,208; October 5, 1976; assigned to Societe Lignes Telegraphiques et Telephoniques, France have developed sintered anodes for solid electrolytic capacitors produced from a mixture of at least two powders the first being of pure film forming metal, i.e., tantalum and the second of a film forming metal, i.e., tantalum with addition of at least one of the elements: nitrogen, tungsten, molybdenum, vanadium and hafnium. The relative concentration of the added element is between 100 and 20,000 ppm by weight.

The mixing operation is carried out under the following conditions: the pure tantalum powder and the tantalum powder containing one or more of the aforesaid elements are disposed in a mixer; when necessary, the required quantity of binder is added thereto, in solution in a volatile solvent which is inert to the metals under consideration, the quantity of such solvent being so limited that the mixing can be carried out without fear of sedimentation in the mixer, but being sufficient to ensure that a fluid paste is formed. The mixing is continued until the evaporation of the solvent brings the paste thus formed to the appropriate viscosity for the pregranulation. The steps of manufacture are known to the person skilled in the art.

Experiments have shown that favorable results are also obtained by introducing at least two additives into the same powder. There may also be employed a number of powders which are different in respect to the nature of the additive which they contain.

One example is capacitors of which the anode is produced from a mixture of two powders, one of which is pure tantalum, while the other is tantalum containing molybdenum. This particular example, which is given as a nonlimiting illustration, has been chosen because, of various experiments carried out, it has given particularly useful results in regard to two characteristics of the capacitors, namely, the stability of the value of the capacitance as a function of temperature at low temperature, and the dissipation factor.

Valve Powders with High Surface Area

T.X. Mahy; U.S. Patent 3,829,310; August 13, 1974; assigned to Norton Co. has developed a process to produce tantalum and other valve metal powders in the form of high surface area, high capacitance powders of adequate purity, flowability, and pelletability, in high yield, which have improved pelleting properties, as measured by the ability to be pressed into a binderless compact having, in the case of tantalum, a green density of 5.0 g/cc with a crush strength of at least 100 pounds for a 1.8 gram compact and, in the case of other valve metals, a similarly low percent of theoretical density as green density.

The powder so produced has a Scott bulk density of less than 2.14 g/cc and distinctly advantageously less than 1.53 g/cc consistent with an oxygen content of no greater than 0.2 weight percent and consistent with one or more of the preceding objects.

Solid particles of potassium fluotantalate are thoroughly dried and then coated with molten form sodium reducing agent, the coating being accomplished under temperature conditions controlled to remain below a temperature which will initiate the exothermic reduction reaction between these materials. After completion of the coating, the charge of premixed double salt and reducing agent is then heated through the temperature at which the exothermic reduction reaction is initiated. The initiation of the exothermic reaction causes a rapid temperature rise.

Over 90% of the tantalum values of the salt charge are formed as elemental tantalum in particle form in the course of the exothermically driven temperature rise. The charge temperature is then stabilized in the range of 700°-1100°C, held for a time at this elevated temperature and then slowly cooled. During the temperature rise, tantalum metal particles nucleate throughout the mass of the premixed charge to form a coherent skeletal structure. This skeleton resists settling as the temperature is raised above the melting temperature of the salt mass, and a substantially homogeneous distribution of tantalum values throughout the reduction charge is formed and maintained. Tantalum metal particles so formed are highly structured and can be separated as a high yield of high surface area-high capacitance tantalum powder.

CHROMIUM AND TIN

Chromium-Chromium Carbide Coatings

J.F. Pelton; U.S. Patents 3,846,084; November 5, 1974 and 3,901,689; August 26, 1975; both assigned to Union Carbide Corporation describe a composite powder for use in producing articles or coatings having unique wear and frictional characteristics. It consists essentially of a chromium matrix with at least one chromium carbide taken from the class of carbides consisting of $Cr_{23}C_6$; Cr_7C_3; and Cr_3C_2 and each particle containing from about 0.2 weight percent to about 5.4 weight percent carbon.

Basically, the material is prepared by chemical reaction of an intimate mixture of a source of Cr and a source of C; temperatures of 1000°-1400°C are suitable for solid state reactions. Times of from about 1-50 hours are suitable. Temperatures in excess of 1500°C are required for production of the powder by melting referred to hereinafter. The principal reaction involved is:

$$xCr + yC \rightarrow Cr_xC_y \quad (1)$$

The principal product is $Cr_{23}C_6$, with minor amount of Cr_7C_3 and Cr_3C_2.

When oxygen is present in the Cr (as Cr_2O_3) or Cr_2O_3 is used as the Cr source, reaction (1) is preceded or accompained by:

$$Cr_2O_3 + 3C \rightarrow 2Cr + 3CO \quad (2)$$

The Cr formed in reaction (2) may react with C present in excess of the amount required to bring reaction (2) to completion to form Cr carbide by reaction (1). The process will be illustrated with reference to Figure 7.2.

The methods of this process produce a composite powder containing the desired amount of chromium carbide and chromium in which substantially each particle contains at least some chromium and chromium carbide. Examples of the possible distributions of the carbide phases in the powder particles are shown in Figures 7.2c through 7.2 e. For producing plasma or detonation-gun

Figure 7.2: Chromium-Chromium Carbide Coatings

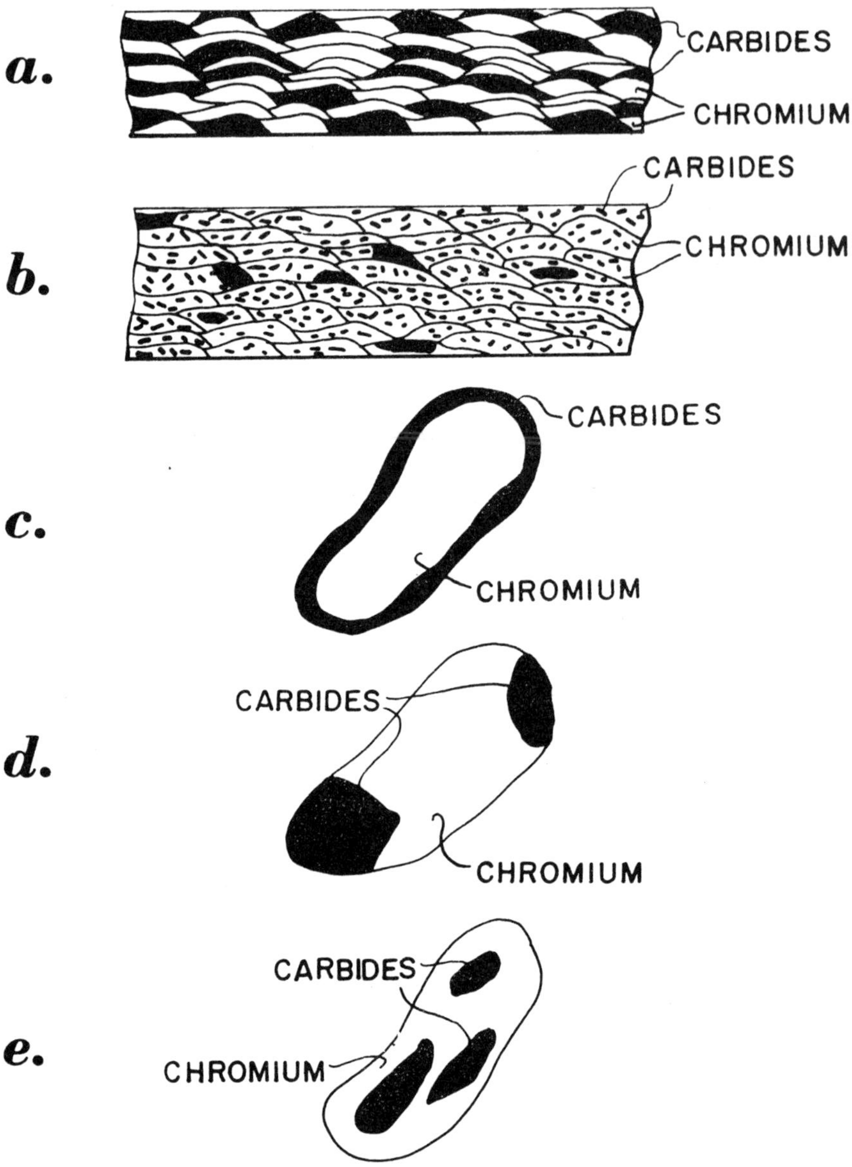

(a) Structure obtained by depositing mechanical mixture of chromium and chromium carbides.

(b) Structure obtained by depositing the powder of this process.

(c), (d) and (e) Possible distribution of the carbide phases in the powder particles.

Source: U.S. Patent 3,901,689

coatings, the exact composition of the carbide phases in the powder or the distribution of the carbide phases as shown in the figures are not important, only the total carbon content, since during deposition the particles become essentially completely molten. As the individual splats solidify during deposition, the carbides reprecipitate from the melt forming $Cr_{23}C_6$, Cr_7C_3, or Cr_3C_2, or a combination of these depending on the total amount of C present and the rate of solidification. The preferred composition results in a predominantly $Cr_{23}C_6$ dispersion.

The carbide distribution within the powder particle is a function of the method of production. When a mixture of solid carbon and chromium is heated in vacuum, the predominant form is that shown in Figure 7.2d because the carbon tends to react with the chromium surface closest to it. The finer and more uniform the distribution of carbon in the starting mixture, the more uniform the distribution of carbides around the surface of the chromium will be. The ultimate extension of this trend is achieved when a gaseous source of carbon is used either by directly supplying a hydrocarbon gas or by heating the solid carbon plus chromium in a hydrogen atmosphere (which results in a hydrocarbon gas). The carbide distribution which results is like that in Figure 7.2c. A distribution of carbon particles throughout the powder particle, Figure 7.2e, may result when a solid ingot of the proper total composition is reduced to powder.

The coating of this process is characterized by the presence in substantially every splat of both Cr and Cr carbide. As pictorially illustrated in Figure 7.2b, the relative amounts of Cr and Cr carbides will vary between splats as a necessary result of the use of powder with a range of partial sizes and adventitious difference in the degree to which each Cr particle is carburized and in the conditions to which the various particles are subjected in passing through the coating device. Nevertheless, the coating of this process is distinguished from that produced from a powder which is a simple mixture of Cr and Cr carbide, which is pictorially respresented in Figure 7.2a, in that the splats in the latter type of coating are each individually either all Cr or all Cr carbide.

Figure 7.2b is to be understood as being merely illustrative of one feature of the distribution of the carbides in the coating. Upon extraction by chemical methods of carbides from the process and examination of these carbides by optical and electron microscopy, it has been found that at least some, probably most, of the carbides are much finer than suggested by Figure 7.2b. The majority of the carbide particles were found to be of submicron size and most were predominantly in the shape of a lace-like network, suggesting that the coatings contained fine-grained interlocking, continuous networks of both carbide and Cr, the separation between the interstices of these networks being so small that they are not resolvable in optical microscopy.

The coatings produced with the powder of this process have a number of advantages in addition to general processing advantages which are associated with metal spray deposition.

1. Coatings are superior to those formed by the plasma deposition of commercial electrolytic chromium powder in that increased wear resistance and resistance to spalling are found, though there is minimal increase in hardness as measured by diamond pyramid indentations.
2. Coatings are superior to coatings in which nitrogen rather than carbon is the strengthening additive, in that carbide-strengthened material is much less brittle and much less prone to spalling.
3. In the laboratory lubricated rubbing wear test described in Example 1, the coatings of this process with a carbon content in the preferred range of 0.8-1.7 weight percent C, performed as well as or better than commercial electrolytic chrome plate.
4. Coatings of this process performed far superior to electrolytic chrome plate coating on internal trochoid surfaces in rotary combustion engines as in Example 4.

The following examples illustrate the process.

Example 1: 8,879 grams of Union Carbide Mining and Metals Division electrolytic chromium, screened through a 230-mesh sieve, was mixed with 200 grams of Fisher Scientific Company Norit A decolorizing carbon, similarly screened, and blended for 2 hours in a cone blender. A portion of this mixture was used to fill eight pans, each about 0.6 cm deep, so that each pan contained between 210 and 230 grams of the mixture. The pans were vertically stacked in a vacuum furnace so that there was about 0.4 cm clearance between pans.

The furnace was evacuated slowly to about 500-micron pressure and then more rapidly to about 0.5 micron, using an oil-diffusion pump. Power was then applied to tantalum strip heaters surrounding the stack of pans and the pans heated over a period of about 80 minutes to a temperature of 1080°C as indicated by a thermocouple in contact with the powder in the uppermost pan; system pressure was maintained below 50 microns during this period by adjusting the rate of heating. The powder was maintained at 1080°C for 4 hours, during which time the pressure gradually dropped to about 0.3 micron.

The furnace was then allowed to cool to room temperature with pumping continued. When the pans were removed from the furnace, the material was in the form of sintered cakes of a significantly more metallic appearance than the original powder mix. These cakes were crushed in a mechanical pulverizer until about 95% of the material was reduced to powder that would pass a 325 mesh screen. The balance of the original mixture of chromium and carbon powders was processed identically in four additional furnace runs.

The −325 mesh powders from the 5 furnace runs were individually analyzed for combined carbon, free carbon, and oxygen. All showed less than 0.1% free carbon, between 300 and 420 ppm oxygen, and 1.05-1.08% combined carbon. The distribution of carbides on the chromium was similar to that in Figure 7.2d.

The products of the five runs were blended together and used to produce coatings by deposition through a plasma torch. Coatings so produced, when separated from the substrates on which they were plated, analyzed 1.03-1.06% C. The wear resistance of these plasma-deposited coatings was measured using a Dow-Corning LFW-1 Friction and Wear Test Machine according to ASTM Standard Method D2714-68.

Coatings deposited 12 mils thick on the wear surface of milled steel wear blocks were ground to a final thickness of 5 mils and tested against carburized AISI 4620 steel rings (surface hardness 58-63 Rockwell "C") at 450 lb specific load for 5,400 ring revolutions at about 180 rpm; MIL-5606A hydraulic fluid was used as lubricant. Wear scar volumes, calculated from the projected scar areas and the known ring diameter, ranged between 24 and 49 $\times$ 10^{-6} cm^3.

Example 2: A powder analyzing 1.13% C prepared by the method described in Example 1 was plated onto test blocks using a detonation gun. Microstructural differences between these coatings and those formed by plasma deposition were observed consistent with the difference in method of coating formation. For wear-test conditions identical with those employed for the plasma-deposited materials, scar volumes of 15-19 x 10^{-6} cm^3 were measured on the detonation-gun coatings.

Example 3: Four hundred pounds of Cr_2O_3 was blended with 94.8 pounds of lampblack in a twin-shell blender and then more thoroughly blended in a vibratory ball mill. This product was then mixed with 9.5 pounds of cornstarch binder and enough water to make a mix suitable for forming briquettes in a standard briquetting press. It was then pressed into briquettes of about 2" maximum dimension mix and dried to remove excess water. The briquette mix charged to a large vacuum furnace in an 19" deep bed covered with graphite plates, was heated to 1000°C without letting the pressure exceed 5,000 μ, held one hour at 1000°C after the pressure had dropped below 2,000 μ, then heated to 1400°C and held at that temperature for 50 hours, at the end of which time the pressure had dropped to less than 150 μ.

A portion of this product was pulverized to −325 mesh size and found to contain 1.14% C and 460 ppm oxygen. The carbide dispersion in the powder was similar to Figure 7.2d. Wear samples formed from this material by plasma deposition and tested exhibited wear scars of 21-24 x 10^{-6} cm^3 volume.

Example 4: Plasma deposited coatings produced in a manner similar to Example 1 were applied to the interior trochoid surfaces of rotary combustion engines fitted with graphite-aluminum composite rotor apex seals. The engines were run in laboratory test standards and in test vehicles. The trochoids were made of several different types of materials and of two different sizes, examples of which are shown in the table below. Over 3,113 hours of test stand operation have accumulated on the small engine size and 331 hours of test stand and 7,000 hours of vehicle operation on the large engine size. In comparison with hard electroplated chromium the coatings of this process showed the following advantages:

(a) Essentially no wear of the coated surface has been observed and no roughening or "wash boarding," as occurs with electroplated chromium, has developed.

(b) The wear of the mating seal surface is approximately one-half that caused by hard electroplated chromium, which is greater than 0.005"/100 hr.

(c) Performance of the coating is less sensitive to surface finish than hard electroplated chromium. There was no appreciable difference in wear of either the coated surface or the seal surface between as-ground coating surfaces of 16 to 32 microinches rms and honed surfaces of 6 microinches rms. In comparison, a hard electroplated chromium surface must be finished to better than 6 microinches rms to perform satisfactorily.

(d) Finishing of the plasma-deposited coating is far simpler and may be cheaper since it can be used as-ground while a hard electroplated chromium coating must be ground, then etched to enhance the microcracked texture of the surface, and then honed to improve the surface finish. Because thickness control is better with plasma deposition than with electroplating, the amount of material that must be removed in finishing is also less.

(e) The performance of engines with the coatings of this process is far less sensitive to fluctuation in coolant temperatures than those coated with hard electroplated chromium.

(f) The performance of engines with the coatings of this process is far less sensitive to fluctuation in oil lubrication than that of engines coated with hard electroplated chromium. The latter require continuous addition of oil to the combustion chamber, but an engine provided with the coating of this process continued to perform satisfactorily when the oil addition was stopped.

(g) The cost of engines and the vehicles can be reduced using the coatings of this process because they can be applied directly to aluminum trochoid housings while the use of hard electroplated chromium requires a steel liner. This reduces not only the cost of the housing, but also the weight of the engine and therefore the cost of vehicle frame, suspension, etc. In addition, engine cooling is more efficient. With lower total vehicle weight fuel efficiency is increased.

Trochoid Type	Trochoid Size* (inches)	Coating Thickness (inch)	Test Time (hr)	Average Seal Wear Rate (in/100 hr)
Aluminum with				
Steel liner	9.5 x 7.1 x 2.75	0.015-0.017	800	0.0026
No liner	9.5 x 7.1 x 2.75	0.019-0.020	444	0.0026
Steel liner	11.5 x 8.6 x 2.75	0.018-0.015	220	0.0026
No liner	11.5 x 8.6 x 2.75	0.016-0.018	76	0.0026
Cast iron	9.5 x 7.1 x 2.75	0.010-0.0115	200	0.006

*Major axis x minor axis x width.

In related work *J.F. Pelton; U.S. Patent 3,881,910; May 6, 1975; assigned to Union Carbide Corporation* describes coatings produced from a mechanical mixture of chromium and chromium carbide powders. These are illustrated in Figure 7.2a. The abovedescribed process avoids certain limitations arising from mechanical mixtures.

Supercooled Stable Tin Particles

D.H. Rasmussen and C.R. Loper, Jr.; U.S. Patents 4,042,374; August 16, 1977 and 4,101,310; July 18, 1978; both assigned to Wisconsin Alumni Research Foundation describe a method for producing spherical metal droplets of micron size capable of supercooling from temperatures above the melting point temperature of the metal to temperatures far below their melting point, and which are characterized by long term stability in their metastable supercooled state. Such micron sized spherical particles find beneficial use as a heat transfer medium, as bearing elements in lubrication and as filler in the manufacture of high density plastics.

The micron sized spherical droplets of metal, e.g., tin and tin alloys are formed by (1) emulsification, in an inert atmosphere, of the metal in a stable carrier liquid while at a temperature above the melting point temperature of the metal and preferably at a temperature within the range of 20°-25°C above the melting point of the metal or alloy to a maximum temperature corresponding to the boiling point temperature of the carrier liquid or the decomposition temperature of the chemical ingredient reacted to oxidize the surfaces of the molten metal particles and more preferably 20°-50°C above the melting point, and (2) reaction, while in the emulsified state, with a chemical which provides the surfaces of the molten metal particles with a reaction product, preferably noncrystalline, which functions as a surfactant to maintain emulsion integrity of the emulsified spherical metal particles in their finely divided separated state.

As the carrier liquid, use can be made of a fluid characterized by (1) low volatility at elevated temperature at which the metal is maintained in the molten state for emulsification; (2) tolerance of an oxidizing atmosphere at temperatures at which the metal is maintained in a molten state for emulsification, such as at a temperature within the range of 250°-350°C for use with such metals as tin, bismuth, lead and alloys thereof, or higher temperatures for use with metals or alloys having a higher melting point temperature; (3) support for the chemical reaction between the molten metal particles and the oxidizing or other agent to provide for emulsion stability and maintenance of the separated relation between the finely divided metal particles in the emulsion.

Representative of fluids which can be used as the carrier fluid are such materials as polyhydroxy compounds, polyethers and silicone oils, and preferably polyphenyl ethers having 5 to 6 benzene rings, as represented by a polyphenyl ether marketed by Monsanto Chemical Company (OS-124), having a boiling point above about 500°C and which is stable to air oxidation at a temperature over 400°C.

As the surface reactant use is made of an oxidizing agent alone or in combination with an acid which catalyzes the oxidation reaction, which operates to raise the molten metal droplet to a charged state, or provides a suitable anion as a coating on the surface of the droplet to facilitate maintenance of the particles in the emulsified state in the carrier fluid.

The metal emulsions prepared in accordance with this process are characterized by long term stability in the metastable supercooled state. The metal droplets can be heated to temperatures above their melting point and then supercooled to 100°-130°C below their melting point before the droplets nucleate and crystallize.

Example 1: 2 g of pure tin, along with 0.1 g of terephthalic acid, and 5 cc of silicone oil were introduced into a pyrex tube. Argon was continuously circulated into the tube to purge the tube before heating and to maintain inert atmospheric conditions during reaction and while the tin was in the molten state. A furnace was adjusted for heating the ingredients in the tube to a temperature of about 25°C above the melting point temperature or 232°C for tin. When the desired temperature conditions were reached, and the tin was reduced to the molten state (about 10 minutes), the motor was activated to rotate the stirrer at high speed, such as at 28,000 rpm.

Under these conditions, the molten metal was sheared into a fine droplet dispersion and then 2 cc of a 10% by weight solution of di-tert-butyl peroxide in silicone oil was introduced with continued emulsification for a short period of time of about ½ to 1 minute. Upon removal from the furnace, a fine powder of tin settled to the bottom of the tube.

In a series of experiments, Example 1 was repeated except that instead of silicone oil, corresponding amounts of other fluids including polyphenyl ether having 5 to 6 benzene rings (OS-124) and silicone fluid (Dow-Corning 704) was used.

Example 2 to 5: The following examples illustrate this process with tin alloyed with lead and with bismuth. Lead was used as an alloying element with tin since, at the temperature of nucleation observed for pure tin, there is little, if any, solid solubility for lead in tin. Bismuth was used as an alloying element with tin because bismuth is soluble in the solid state to about 10% by weight at these same temperatures. These examples were performed to inquire about the effect of such alloying elements on supercoolings, of the formed emulsions, as a function of solute content.

The procedure of Example 1 was repeated but with an alloy of 80 parts by weight tin to 20 parts by weight of lead instead of pure tin as the metal component. In Example 2, copper isophthalate was used as the oxidant. In Example 3, silver isophthalate was used as the oxidant. In Example 4, benzoyl peroxide was used as the oxidant in the presence of isophthalic acid. In Example 5, di-tert-butyl peroxide was used as the oxidant in the presence of isophthalic acid.

The emulsions that were formed were of similar size, averaging 8.73 μm in diameter with a standard deviation of 2.20 μm. The supercoolings were uniform, depending somewhat on the surface coatings of the metal components.

High Density, Low-Melting-Point Metal-Containing Lubricant

G.L. Hurst; U.S. Patent 4,076,637; February 28, 1978; assigned to Tyler Corporation has developed a method for producing high density metal-containing lubricants and for comminuting metal into fine spherical particles.

According to the process, a solid metal is heated until it reaches a liquid state and is then thoroughly admixed into a carrier fluid which is thermally stable at the temperature of the molten metal. The metal disperses into fine globules which are held in suspension without any substantial tendency to coalesce back into larger particles. Furthermore, upon sufficient cooling, the globules harden and remain dispersed in the carrier fluid. The finely divided metal particles are homogeneously dispersed throughout the carrier fluid in the form of spherules of micron and submicron size. When lubricants are used as carrier fluids, the presence of the spherical metal particles causes an increase in the density of the lubricant without any appreciable loss of the lubricating qualities of the lubricant itself.

In one embodiment, the discrete spherical metallic particles suspended in the carrier fluid are separated from the carrier fluid by common washing, dissolving or evaporation techniques resulting in a metal powder consisting of the abovedescribed metal spherules.

As long as a sufficiently thermally stable carrier fluid is utilized, it is believed that any normally solid metal can be used in the process. When preparing metal-containing greases with common types of lubricating materials as carrier fluids, metals having relatively low melting points are preferred. The requirement of the metal component is that it be capable of changing phases from solid to liquid at temperatures which will not cause the lubricant or carrier fluid to decompose. Accordingly, low melting point metals, such as tin, bismuth, cadmium, lead, indium and gallium as well as alloys of these metals, for example, are preferred. The table below lists the composition and melting points of some of the preferred alloys which can be employed according to the process.

LOW MELTING POINT ALLOYS

Melting Point °C	Name	Composition, wt %				
46.5	Quinternary	Sn	Bi	Pb	In	Cd
	Eutectic	10.65	40.63	22.11	18.1	8.2
47	Quinternary	Bi	Pb	Sn	Cd	In
	Eutectic	44.7	22.6	8.3	5.3	19.1
58.2	Quaternary	Bi	Pb	Sn	In	
	Eutectic	49.5	17.6	11.6	21.3	
60.5		In	Bi	Sn		
		51.0	32.5	16.5		
70	Wood's Metal	Bi	Pb	Sn	Cd	
		50.0	25.0	12.5	12.5	
70	Lipowitz's Metal	Bi	Pb	Sn	Cd	
		50.0	26.7	13.3	10.0	
70	Binary Eutectic	In	Bi			
		67.0	33.0			
91.5	Ternary Eutectic	Bi	Pb	Cd		
		51.6	40.2	8.2		
95	Ternary Eutectic	Bi	Pb	Sn		
		52.5	32.0	15.5		
97	Newton's Metal	Bi	Sn	Pb		
		50.0	18.8	31.2		
98	D'Arcet's Metal	Bi	Sn	Pb		
		50.0	25.0	25.0		
100	Onion's or Lichten-	Bi	Sn	Pb		
	berg's Metal	50.0	20.0	30.0		
102.5	Ternary Eutectic	Bi	Sn	Cd		
		54.0	26.0	20.0		
109	Rose's Metal	Bi	Pb	Sn		
		50.0	28.0	22.0		
117	Binary Eutectic	In	Sn			
		52.0	48.0			
120	Binary Eutectec	In	Cd			
		75.0	25.0			
123	Malotte's Metal	Bi	Sn	Pb		
		46.1	34.2	19.7		
124	Binary Eutectic	Bi	Pb			
		55.5	44.5			
130	Ternary Eutectic	Bi	Sn	Zn		
		56.0	40.0	4.0		
140	Binary Eutectic	Bi	Sn			
		58.0	42.0			
140	Binary Eutectic	Bi	Cd			
		60.0	40.0			
183	Eutectic solder	Sn	Pb			
		63.0	37.0			
185	Binary Eutectic	Tl	Bi			
		52.0	48.0			
192	Soft solder	Sn	Pb			
		70.0	30.0			
198	Binary Eutectic	Sn	Zn			
		91.0	9.0			

(continued)

Melting Point °C	Name	Composition, wt %			
199	Tin foil	Sn	Zn		
		92.0	8.0		
199	White Metal	Sn	Sb		
		92.0	8.0		
221	Binary Eutectic	Sn	Ag		
		96.5	3.5		
226	Matrix	Bi	Pb	Sn	Sb
		48.0	28.5	14.5	9.0
227	Binary Eutectic	Sn	Cu		
		99.25	0.75		
240	Antimonial Tin solder	Sn	Sb		
		95.0	5.0		
245	Tin-silver solder	Sn	Ag		
		95.0	5.0		

Example 1: Nine parts of Wood's metal (50% bismuth, 25% lead, 12.5% tin, and 12.5% cadmium by weight) and one part of a silicone based grease (Dow Corning # 111) were deposited in a suitable container. The container was positioned on a hot plate and heated until the contents reached 80° to 100°C. The Wood's metal, which has a melting point of 70°C, became molten. The metal and grease were then thoroughly admixed using a blender type mechanical mixer. The liquid metal vanished into and was seemingly enveloped by the silicone grease. The resulting suspension contained discrete spherules of Wood's metal and did not display any tendency to precipitate. The average size of the discrete metal spherules was less than about 2 μ. The metal containing grease was then cooled and the resultant density was determined to be about 5 g/ml.

Example 2: The procedure of Example 1 was repeated substituting a hydrocarbon based grease (Micro Lube) for silicone grease. The petroleum based grease completely enveloped the liquid metal and the suspension appeared to be substantially stable. This experiment demonstrated that the unexpected acceptance of a molten metal once admixed with a lubricant carrier fluid, does not depend upon synthetic lubricant bases but occurs readily when petroleum based carrier fluids are employed. The resulting metal-containing grease had a density of approximately 5 g/ml.

Example 3: Equal weights of bismuth and Dow Corning # 111 silicone grease were combined together in the same container and heated until the bismuth became molten. The mixture was then admixed thoroughly until a homogeneous dispersion resulted. The consistency of the dispersion was almost the same as the consistency of the grease component alone. The resulting density of the metal-containing grease was about 1.8 g/ml and tiny spherules of tin having an average diameter less than 10 μ were present.

Example 4: Nine parts of Wood's metal (50% bismuth, 25% lead, 12.5% tin, and 12.5% cadmium by weight) and one part of a lubricant additive (STP) were placed in a container and heated until the metal became molten. Upon mixing, the oil additive completely enveloped the liquid metal and resulted in an apparently stable suspension. The resulting metal-containing grease had a density of approximately 5 g/ml.

TUNGSTEN

Tungsten and Tungsten Carbide Using Dry Hydrogen

W.H. Bleecker; U.S. Patent 3,850,614; November 26, 1974; assigned to Carmet Company describes the production of tungsten and tungsten carbide in powder form by chemically reducing to tungsten oxide a tungsten compound such as ammonium paratungstate containing less than 100 ppm total of the elements sodium, potassium, and calcium. Thereafter the tungsten oxide is chemically reduced to tungsten powder with dry hydrogen having a dew point not higher than −30°F. The reduction is accomplished by passing the oxide through a reduction zone while flowing dry hydrogen through the zone in a direction concurrent with the oxide. The tungsten powder may be carburized to produce tungsten carbide powder.

Example: Eight hundred fifteen pounds of ammonium paratungstate powder containing less than 100 ppm of the elements sodium, potassium, and calcium was calcined in the usual manner in an atmosphere of dissociated ammonia at 1000-1150°F. The resulting blue tungsten oxide had an oxygen deficiency of 0.53%, and an average particle diameter as measured by the Fisher Sub Sieve Sizer of 11-14 μ. The yield was 699.4 pounds of this blue oxide.

The tungsten oxide was loaded into molybdenum reduction boats and fed to a gas-fired reduction furnace. Atmosphere flow was maintained at 78 l/min of dry hydrogen, having a nominal dew point of −60°F. The reduction temperature rose from 1300° to 1650°F as the reduction boats advanced in the furnace. The flow of hydrogen was concurrent with the movement of the boats through the furnace. After reduction, the powder was screened through 200 mesh to remove coarse agglomerates and scale. This powder had an average particle size of 1.05 μ, and the yield was 558 pounds of pure tungsten powder.

The metal powder was blended in the usual manner with 36.5 pounds of pure carbon black and carburized at 2750°F for one-half hour in a graphite crucible. Typical analysis of the powder after carburizing was 6.11% total carbon, with a trace of free carbon being present. The particle size had increased to 1.25 μ.

The tungsten carbide powder obtained as above was blended and milled with 10% by weight of fine cobalt powder and 0.3% vanadium carbide for 70 hours in a ball mill with a hydrocarbon solvent. The milling media used in this operation were also made of a 10% cobalt-tungsten carbide alloy. A lubricant, e.g., paraffin was added to the powder blend to facilitate pressing.

The dried and lubricated powder was pressed into standard specimens of size 0.930" x 0.310" x 0.250", presintered at 1200°F, and final-sintered in vacuum for one-half hour at 2600°F. The resulting parts were ground, polished, and etched in the usual manner to reveal the microstructure of the alloy. The grain size, determined according to ASTM Procedure B390-64 was essentially 0.25-0.50 μ, with a maximum individual particle size of about 1.0 μ.

Tungsten metal powder made in accordance with the process differs considerably from the conventional fine powder in the 1.1-1.3 μ range. First, the submicron fraction is unexpectedly higher and, secondly, the larger particles are loosely bound agglomerates, with a lesser tendency to retain their size identity in the ball milling operation after carburizing. Any tendency toward particle growth in sintering via the previously mentioned mechanism can be further precluded by inhibitors such as tantalum carbide, titanium carbide, chromium carbide, and vanadium carbide in amounts of from 0.1-0.5%. The latter is preferred, used in amounts of 0.2-0.3%.

Tungsten powder production by this method is not pyrophoric and cemented carbides made therefrom contain only few particles as large as 1 μ.

Reclaiming Cemented Tungsten Carbide

Cemented carbide tools are made by consolidating extremely hard and extremely fine metal carbide particles together with a suitable binder or cement.

Typical of such tools are those made of tungsten carbide cemented with cobalt. Although the tungsten carbide tools are most widely used, hard carbides of other metals such as titanium, vanadium, chromium or molybdenum may also be used. All of the metals that produce hard carbides are relatively expensive metals.

When tools become worn or when tools are prepared with flaws that make them inadequate, it is desirable to reclaim the expensive metal carbide.

A.G. Hartline, III, J.A. Campbell and T.T. Magel; U.S. Patent 3,953,194; April 27, 1976; assigned to Allegheny Ludlum Industries, Inc. describe a rapid, complete and inexpensive method for reclaiming cemented carbide.

The first stage in the process is catastrophic oxidation of the cemented carbides from which metal carbides are to be reclaimed. It is effected at a temperature of at least 1100°F in the presence of oxygen. The process may be improved or hastened if a high partial pressure of oxygen is involved. The temperature must be at least 1100°F but it may be as high as desired, limited only by the characteristics of the material being oxidized. The oxidation may also be hastened by breaking the cemented carbide material into small pieces although milling to a fine powder is not required.

Oxidation of the cemented carbide material also involves oxidation of the cement. For example, a material made of tungsten carbide cemented with cobalt, when subjected to the oxidation step, will result in a tungsten oxide-cobalt oxide mixture as well as some intermetallic compounds of oxygen, cobalt and tungsten. Oxidation is effected until substantially all of the carbon has been removed. Ordinarily, oxidation of a metal carbide can be effected at any temperature wherein the metal itself can be oxidized completely. Tungsten carbide cemented with cobalt can be oxidized readily at 1400°F to remove substantially all of the carbon, but if it is oxidized at temperatures substantially higher than 1600°F, tungsten oxide begins to vaporize from the cemented metal carbide as it forms.

The second stage of the process is subdivision of the metal oxide, which is readily accomplished in that the oxidized cement does not hold the oxidized carbide particles together firmly. In fact, the cement oxide also may be readily subdivided. Subdivision can be accomplished by simple crushing or milling by any of the known techniques, and with very minimal amount of crushing or milling the oxides are reduced to a fine powder. Ordinarily, if the cement oxide is to be separated from the metal oxide, the separation will be effected on the subdivided powders.

The third stage is reduction of the metal oxide or the combined metal oxide-cement oxide powder to an oxygen content of at most 0.5 weight percent. Reduction is effected by contacting the metal oxide powder or the mixed metal oxide-cement oxide powders with a reducing gas at conditions to effect substantially complete reduction of all of the oxides. Reducing gases may include hydrogen, hydrocarbons such as methane, or carbon monoxide; hydrogen is preferred. When reduction is effected with carbon monoxide or hydrocarbon gas, depending upon the conditions chosen, the metal may be partially or completely carburized during the reduction process.

The conditions maintained during the reduction process will depend upon which metal is being treated. By way of example, tungsten oxide may be reduced at temperatures as low as 1290°F when contacted with hydrogen. The cement oxide is reduced at a substantially lower temperature and will always be reduced at a temperature at which the metal oxide is reduced. For example, cobalt oxide may be reduced with hydrogen at temperatures as low as 575°F. Some metal oxides, most notably titanium oxide, cannot be reduced and carburized at low temperature.

For example, the combined reduction and carburization of titanium oxide by treating it with a mixture of hydrogen and methane requires temperatures as high as 3000°F to be completed. At such high temperatures, any of the common cement metals melt and produce a useless product. Thus, reduction and carburization of titanium oxide cannot be effected while it is mixed with the cement oxide powder.

The fourth stage of the process involves carburization of the metal powder, which is effected by contacting the reduced metal with carbon that is available to react with it at conditions wherein the reaction between metal and carbon can be effected. Available carbon may be carbon per se or carbon compounds that will decompose to provide carbon available to react with the metal. By way of example, lamp black or powdered graphite may be intimately mixed with the metal powder or metal oxide powder, or the powder may be contacted with hydrocarbon gas, hydrocarbon-hydrogen gas mixtures or carbon monoxide.

Carbon dioxide is not a source of available carbon. Carburization of tungsten or of tungsten oxide can be effected by contacting with carbon monoxide or with methane or with a hydrocarbon gas-hydrogen mixture at temperatures of at least 1600°F. Carburization can also be effected by heating the metal powder mixed with powdered graphite or carbon black at temperatures of at least 2200°F in the absence of air. Carburization of tungsten-cobalt metal powders or their mixed oxides should be effected at temperatures between 1600° and 2365°F because temperatures higher than 2365°F cause undesirable grain growth.

The result of the process is a stable metal carbide compound alone or mixed with a suitable cement. The product may be used per se as a suitable material to be consolidated into cemented metal carbide tools by methods known to the art, or it may be mixed with virgin material to produce cemented metal carbide products meeting all the specifications that products from the virgin material alone can meet. Except for an oxidizing furnace, no extra equipment is necessary to effect the process other than that required for the production of cemented metal carbide tools from virgin material. The cost of metal carbide reclaimed according to this process has been less than half of the cost of the virgin material.

Whiskers from Catalyst and Oxide

C. Tomita; U.S. Patent 3,798,023; March 19, 1974 produces tungsten whiskers ranging in diameter from about 0.5 μ to about 3 μ and in length from about 20 μ to about 300 μ by interdispersing tungsten oxide and a catalyst, placing the resultant interdispersion in a hydrogen atmosphere and raising the temperature to between about 950° and 1100°C. The catalyst particles should range in size from about 0.02 μ to about 5 μ. The catalyst content should range from about 1 x 10^{-3} to about 2 atomic percent of the total weight of the interdispersion. Suitable catalysts are nickel, palladium, platinum and their halides.

The following examples illustrate the process

Example 1:

Raw Materials	Grams
WO_3	500
$NiCl_2$ (catalyst)	0.5

The interdispersion was carried out by mechanically mixing the granules of the two materials. The rate of flow of hydrogen gas was 300 l/h and the temperature was held at 1000°C. About 350 g of whiskers varying in diameter from 0.5 to 1 μ and in length from 20 to 200 μ were produced.

Example 2:

Raw Materials	Grams
WO_3	500
$PdCl_2$ (catalyst)	0.5

The mixture of WO_3 and $PdCl_2$ was interdispersed as described above and the compounds were reduced at a temperature of 1000°C by hydrogen gas flowing at a rate of 300 l/hr. About 350 g of tungsten whiskers varying in diameter from 1 to 1.5 μ and in length from 20 to 300 μ were obtained.

Example 3:

Raw Materials	Grams
WO_3	500
Pt powder (catalyst)	0.5

The interdispersion of WO_3 and Pt was produced as above. The reaction was carried out at 1000°C, and tungsten whiskers varying diameter from 1 to 3 μ and in length from 20 to 100 μ were obtained. The quantity of whiskers obtained was about 200 g.

No attempt was made to use catalysts having a narrow range of sizes. In all cases, the catalyst particles ranged in size from about 0.02 μ to about 5 μ. Although up to 5 atomic percent of catalyst can profitably be used in the manufacture of tungsten whiskers, it is preferable not to go above about 2 atomic percent. The reason is that at the higher concentrations of catalyst the whiskers produced tend to be shorter and to be angular.

Although only chlorides have been given as examples of catalysts, any halide of the specified metal reducible by hydrogen in the specified temperature range could be used.

Acid Washing of Doped Oxides

I. Koseki and M. Seimiya; U.S. Patent 4,101,309; July 18, 1978; assigned to Tokyo Shibaura Electric Co., Ltd., Japan have developed a method for preparing tungsten materials in which tungsten powder obtained by reducing tungsten oxides having doping agents added thereto is washed first with hydrochloric acid having a concentration of not lower than 4% by weight and then with hydrofluoric acid having a concentration of not lower than 3% by weight, and thereafter dried. The washing steps may be carried out by using a mixture of such hydrochloric acid and hydrofluoric acid as specified above. Articles formed by treating tungsten materials obtained by this process are subjected only to a low high temperature deformation.

Next, several samples of tungsten powder shown in Table 1 (Samples 10, 11,

14, 15 and 21) were pressed, sintered, swagged, and drawn by conventional techniques to obtain tungsten wires having a diameter of 0.39 mm. In order to examine their high temperature deformation, five specimens were taken from each of the wires thus obtained, and compared according to JIS H4460-1970 entitled "Thermal Deformation Test." The results are given in Table 2.

Table 1

Sample No.	Washing Fluid	Concentration (%)	Residual Doping Agents (ppm) Al	Si	K	Impurities (ppm) Fe	Ni	SEM Rating	Remarks
1	HCl	1	60	2300	120	40	25	C	Control
2	"	2	50	2100	117	35	23	C	"
3	"	3	45	2000	100	25	20	C	"
4	"	4	37	1800	93	19	11	C	"
5	"	5	39	1790	90	20	13	C	"
6	"	10	38	1850	91	18	12	C	"
7	HF	1	36	1380	99	42	29	C	"
8	"	2	30	900	95	39	28	C	"
9	"	3	19	300	85	26	27	B	"
10	"	4	18	295	82	25	27	B	"
11	"	5	22	285	85	24	26	A	"
12	"	6	20	290	87	25	25	A	"
13	"	10	20	298	83	23	26	A	"
14	HCl HF	3 4	18	290	84	24	20	B	"
15	HCl HF	4 3	20	292	85	19	10	A	This invention
16	HCl HF	4 5	19	289	86	19	11	A	"
17	HCl HF	3 5	20	290	88	22	19	B	Control
18	HCl HF	3 10	18	291	87	21	21	B	"
19	HCl HF	4 1	30	1030	90	19	10	C	"
20	HCl HF	7 1	26	870	93	20	9	B	"
21	HCl HF	7 3	17	293	85	18	9	A	This invention
22	HCl HF	7 6	16	190	78	17	9	A	"

(Note:
In this table, HCl and HF stand for hydrochloric acid and hydrofluoric acid, respectively.)

Table 2

Sample No.	Specimen No.	Measured Value (mm)
10	1	11.5
	2	13.0
	3	10.0
	4	12.5
	5	10.0

(continued)

Sample No.	Specimen No.	Measured Value (mm)
11	1	10.0
	2	9.5
	3	9.5
	4	10.5
	5	9.0
14	1	9.5
	2	11.0
	3	10.5
	4	9.0
	5	11.0
15	1	9.5
	2	8.5
	3	9.0
	4	8.0
	5	8.5
21	1	8.5
	2	9.0
	3	8.0
	4	9.0
	5	8.5

As is evident from Table 1, the samples washed with HCl alone (Samples 1 through 6) contained a large amount of residual doping agent and a considerably large amount of silicon (Si) in particular, resulting in unsatisfactory washing. But, as long as its concentration was 4% or higher, HCl seemed to be effective in removing impurities. Upon observation with a scanning electron microscope (SEM), the samples washed with HCl alone, regardless of its concentration, were undesirably rated as C.

The samples washed with HF alone (Samples 7 through 13) seemed to be satisfactorily freed of unnecessary doping agents when HF had a concentration of 3% or higher. But, regardless of its concentration, impurities were not sufficiently removed by washing with HF alone. Upon observation with SEM, the samples washed with HF having a concentration of 5% or higher were rated as A in spite of the presence of impurities, presenting such a desirable condition that the residual doping agents were distributed uniformly.

For Samples 15, 16, 21 and 22 which were prepared in accordance with this process, good results were obtained upon both chemical analysis and observation with SEM. Though some other samples (Samples 14 and 17 through 20) were also washed successively with HCl and HF or with a mixture thereof, the washing effect was insufficient because the concentration of either HCl or HF was below the specified lower limit. Some of the unsatisfactorily washed samples were properly selected and again washed under the same conditions, except that the washing time was doubled. As a result, the washing effect showed a slight improvement, but failed to present a striking contrast.

As can be seen from Table 2, the specimens made of Samples 15 and 21 (prepared in accordance with this process) underwent only a very low degree of high temperature deformation and showed little variation in the amount of deformation among the specimens. On the other hand, the specimens made of Samples 10, 11 and 14 (controls) underwent a high degree of high temperature deformation and showed great variation in the amount of deformation.

Although a tungsten material suitable for use as filaments for common electric tubes and bulbs was exclusively discussed in the above example, it has been confirmed that similar results can also be obtained with respect to tungsten materials doped with K_2O and SiO_2 and thus suited for use as heaters for vacuum deposition as well as tungsten materials doped with, in addition to Al_2O_3, K_2O, and SiO_2, a proper combination of Co, Fe, and Sn for the purpose of improving their workability.

Tungsten Carbide and Cobalt

W. Scheithauer, Jr. and J.E. Ritsko; U.S. Patent 4,070,184; January 24, 1978; assigned to GTE Sylvania Incorporated describe a process comprising forming an aqueous slurry of water and solids consisting essentially of a refractory metal carbide and a suitable matrix metal or metal alloy in the desired ratio, the water and solids being in the weight ratio of from about 1:2 to about 1:4.

The method continues with attritor milling the slurry for from about 1 to about 10 hours, removing the slurry from the milling and forming a solid concentration of from about 70 to about 90% by weight, adding from about 1 to about 3% by weight, based upon the solids, of a water-soluble relatively long chain polyglycol to the slurry and spray drying the resulting slurry at a temperature sufficient to remove the water to form an improved powder consisting essentially of the refractory metal carbide, the binder and the polyglycol. The powder contains spherical particles having a relatively narrow size distribution and is capable of being pressed into shapes having an improved green strength and upon sintering the relative amount of grain growth is reduced.

Example: The following charge is added to an attritor mill that contains WC 13 Co balls:

WC powder, 5,460 parts
Co powder, 540 parts
H_2O, 2,000 parts

The mill is adjusted so that the agitator shaft turns at 200 rpm. Milling time can vary from 1 to 10 hours. For this particular grade which contains 9% cobalt, and a medium particle size WC, 1 hour is sufficient time. Milling times have to be increased as cobalt content is decreased and more importantly when finer WC powders are used.

After the appropriate milling time is reached, the slurry is discharged from the mill. This generally requires the addition of some H_2O to thin the slurry and rinse the mill. During discharging the slurry is passed through a 400 mesh screen. This allows for the removal of contamination that may have been introduced and any chips from the milling balls.

Water is decanted from the screened slurry to obtain the desired solids concentration for spray drying. Generally, this ranges from 70-90%, and for this example of WC-9% Co a solids concentration of 80% is used.

Next the slurry is transferred to the spray dryer feed tank. It is heated, to about 50°C, and agitated while the Carbowax 6000 addition is made. This addition is generally 1-3%. For this grade it is preferably 2%. At this point the spray drying process begins. A suitable spray drier is a Proctor-Schwartz spray tower with two-fluid top-nozzle atomization. Some of the important drying parameters are air pressure of 20 psi, an inlet drying temperature of 200°-230°C and an outlet temperature of 100°-130°C.

After drying the product is spherical and free flowing and ready for subsequent use. Some properties which distinguish it from conventional powders are listed below.

	Spray Dried with Carbowax 6000	Conventional Powders with Paraffin Wax
Hall flow rate, sec/50 g	20.00	27.00
Bulk density, g/cc	3.80	4.10
Green strength after compacting at 20 ksi, psi	1,350.00	520.00

Electrode Using Tungsten Carbide and Silver

K. Mund; U.S. Patent 3,943,005; March 9, 1976; assigned to Siemens AG, Germany has developed a method for the manufacture of an electrode material containing silver and tungsten carbide for electrochemical cells, particularly fuel cells and storage cells. Tungsten carbide in powder form is placed in a solution of a tungstate; then silver tungstate is precipitated by adding thereto a silver salt and the silver tungstate in the tungsten carbide-silver tungstate mixture is reduced to silver and tungsten, and subsequently the tungsten is carburized to form tungsten carbide.

In one embodiment, a suspension of tungsten carbide particles having a mean grain size of less than about 1 μ is suspended in an aqueous solution of an alkali metal tungstate, preferably sodium metatungstate ($Na_2O \cdot 4\,WO_3$). The amount of tungsten, considered as elemental tungsten, in the tungsten carbide particles is preferably equal to the tungsten in the tungstate solution. To such solution and while it is being vigorously agitated, an aqueous solution of silver nitrate is slowly added, preferably dropwise, thereby precipitating silver metatungstate which envelopes the existing tungsten carbide particles and joins them together to form particles of large size, in the range of 10-80 μ.

After the agitation is stopped, the supernatant liquid is decanted, and the precipitate dried. Then the silver metatungstate in such precipitate is reduced to metallic silver and tungsten in a flowing stream of hydrogen at a temperature in the range of about 400° to about 600°C, preferably about 540°C. Thereafter such metallic tungsten is carburized by treatment with flowing carbon monoxide at a temperature in the range of about 800° to 1000°C, preferably about 840°C, thus producing the silver-containing tungsten carbide particles having an average size in the range of 10 to about 80 μ.

SPRAY POWDER

Spray Drying in Solvent-Binder System

J.M. Laferty, Jr., J.E. Ritsko and D.J. Port; U.S. Patents 4,028,095; June 7, 1977; and 3,973,948; August 10, 1976 both assigned to GTE Sylvania Inc. produce free flowing powders, by spray drying a slurry of finely divided particles of a metal having a melting point of about 500°C in a solvent-binder system to produce agglomerates, wherein the binder is a soluble compound of the metal. These agglomerates possess sufficient green strength to be screened and exhibit higher apparent densities than comparable powders agglomerated with conventional organic binders. When these powders are heated in a reducing atmosphere above the decomposition temperature of the binder, the binder converts to base metal and harmless by-products, such as nitrogen and water, thus avoiding contamination of the product, equipment and work area usually associated with conventional organic binders.

Example: Four slurries are prepared by first dissolving appropriate quantities of MoO_3 in 28% NH_4OH solutions to form ammonium molybdate solutions. These solutions are then diluted with water to obtain about 2.5 gallons each. The four solutions contain equivalent amounts of 5.9, 11.2, 11.2 and 20% MoO_3, respectively on a weight percent solids basis. To the third solution is added 0.45% of a polyethylene glycol binder (Carbowax 6000), and 0.175% stearic acid, on a weight percent solids basis. Forty pounds of molybdenum powder having a particle size of less than 10 μm are then added to each solution and the solutions mixed to form a slurry.

The slurries are all spray dried under identical conditions, i.e., the solutions under continuous agitation are fed into one inlet of a two fluid nozzle at the top of a spray drying chamber, while air is fed into the other inlet at a pressure of about 37 psi. The drying air enters the chamber at a temperature of about 230°C and exits at about 130°C. The unagglomerated particles, so-called cyclone fines, are held for recycling, while the chamber products are subjected to a standard sieve analysis. The products are then presintered at 1000°C for 4 hours in H_2 to convert the binder to Mo and evaluated by sieve analysis, apparent density and Hall flow measurements in order to investigate strengthening and densification.

The products are then sintered at 1060°C for 4.5 hours in H_2 and evaluated as at presintering. Results are shown in the table below. Also shown in the table are comparative results for the same molybdenum powder processed with the organic binders used in Lot. No. 3, labeled Prior Art.

Lot No.	1	2	3	4	Prior Art
MoO_3 (weight percent solids)	5.9	11.2	11.2 with binder	20	0
SPRAY DRIED (GREEN) PROPERTIES					
Sieve Analysis	Percent Retained				
+100 mesh	4%	8	6	28	9
+200	30	36	24	42	31
+325	32	30	27	17	28
−325	34	26	42	13	32
Apparent Density g/cc	2.20	2.26	2.30	2.26	1.95
PRESINTERED PROPERTIES					
Sieve Analysis	Percent Retained				
+60	1	1	1	5	9
+200	31	39	29	62	32
+325	43	40	44	23	25
−325	25	20	26	10	34
Apparent Density g/cc	2.26	2.26	2.20	2.32	—
Hall Flow, Sec.	37	35	39	33	—
SINTERED PROPERTIES					
Sieve Analysis	Percent Retained				
+60	—	—	—	—	8
+200	34	43	33	53	33
+325	36	36	37	31	27
−325	30	21	30	16	32
Apparent Density g/cc	2.26	2.26	2.20	2.38	1.97
Hall Flow, Sec.	33	35	37	32	45

The results indicate that the apparent density of the process product is substantially higher than that of the prior art product, both after spray drying and after sintering. Thus, the volume of material to be handled is reduced and processing time is decreased. The flow properties, as indicated by Hall flow measurements, of the sintered product are also improved over that of the prior art product subjected to identical presintering and sintering conditions. It will be seen that Lot No. 3, which includes a portion of organic binder, also exhibits improved density and flow properties.

R.F. Cheney, F.J. Mower and C.L. Moscatello; U.S. Patents 3,909,241; September 30, 1975 and 3,974,245; August 10, 1976; both assigned to GTE Sylvania Inc. describe another process for producing free flowing powders including the steps of entraining these powders in a carrier gas, feeding them through a high temperature reactor at a substantially uniform flow rate so that interparticle contact and coalescence is substantially avoided and at a feed rate such that at least the outer surfaces of a substantial number of particles are melted during their time of exposure to the high temperature zone of the reactor. After passing through the reactor, the particles are then cooled at a rate sufficient to solidify at least the outer surfaces of the particles prior to their contact with a solid surface or with each other.

Because they were melted while entrained in a carrier gas, the solidified particles are substantially spherical, have smooth surfaces and thus have excellent flowability. In addition, the solidified particles have the same general size range as the starting material, but, depending on the porosity of the starting material, may have a smaller mean particle size, due to densification during melting. This densification is advantageous in that it leads to increased efficiency in coating operations.

The free flowing powders of this process are primarily useful in coating applications, such as flame spray applications, but are also useful in other applications where flowability, apparent density or fine mean particle size are important considerations.

In accordance with a preferred embodiment, materials in finely divided particulate form (less than about 40 μm) are agglomerated such as by spray drying in a slurry with a binder, and classified to obtain a desired particle size range. At this stage, the agglomerates are porous, irregular in shape and have a rough surface. They are then processed as above resulting in conversion to smooth, substantially spherical particles, to make powders having apparent densities of 40% of theoretical density or more of the material. If the agglomerates consist of more than one type of particle, of more than one metal or ceramic or a combination thereof, these materials will react or alloy together during melting, to produce prealloyed powders or homogeneous composite particles.

Beneficial chemical reactions may also be carried out during melting. For example, by introducing hydrogen into the hot zone or by mixing carbon into the starting powder material, oxides may be reduced to low levels. Addition of carbon or boron may be used to form carbides or borides.

Example 1: Molybdenum powder is agglomerated by spray drying an aqueous slurry of 70 weight percent solids molybdenum, 2 weight percent Carbowax 6000 and 0.25 weight percent polyvinyl alcohol. The slurry is fed through one inlet of a two fluid nozzle into a commercially available spray dryer at a rate of 4 gallons per hour (114 pounds of slurry per hour) while heated air is fed into the other inlet. Inlet air temperature is 400°C and outlet air temperature is 165°C.

The spray dried powder is fired for approximately 7 hours at 1000°C to remove the organic binders and to strengthen the agglomerate particles. The fired powder is then separated into size fractions by screening. The size ranges obtained are −100+200, −170+200, −200+325 mesh, and −270+325 mesh, Standard U.S. Sieve.

Each size fraction is fed separately through a commercially available plasma torch into a water cooled collection tank. A mixture of 126 ft^3/hr of argon and 70 ft^3/hr of hydrogen is fed to the plasma torch. The torch power is about 28 kVA. Nitrogen gas is fed to a powder feeder at a rate of 7 ft^3/hr to entrain the powder and then is fed through the torch. The nitrogen provides a nonreactive atmosphere as well.

The product collected is then examined. Product size and yield information is shown in the table below.

	Feed Size			
	-100+200	-170+200	-200+325	-270+325
Feed weight (g)	616	267	980	572
Run time (min)	90	40	75	61
Torch power (kVA)	28.2	27.4	24.5-26.3	28.8

	Weight							
	Grams	%	Grams	%	Grams	%	Grams	%
Product								
+200	80	16.8	2	0.8				
-200+325	333	70.1	170	69.7				
-325+400	27	5.7	40	16.4				
-400	35	7.4	32	13.1				
Total	475	100	244	100				
+270					25	2.5	3	0.6
-270+325					442	43.4	39	7.9
-325+400					267	26.2	113	23.0
-400					285	28.0	337	68.5
Total					1019	100	492	100

The effect that melting has on densifying the particles is shown by the decrease in particle diameter. The −200+200 mesh feed drops to 83% below 200 mesh. The −270+325 mesh feed decreases to 91.5% below 325 mesh.

Measurements on apparent density show an increase from 1.8 g/cc for the spray dried feed to 5.4 g/cc for the product. Flow by a Hall flowmeter according to ASTM specification B213-48 in which the time for 50 g to flow through a standard orifice is measured. Flow for the spray dried feed is 41 seconds and for the product is 11-12 seconds.

Example 2: A Mo-15 weight percent W powder is prepared by spray drying molybdenum powder as a slurry with water and binder and fired as in Example 1. The spray dried and fired product is classified and fed to the plasma gun. Argon as the carrier gas is fed at the rate of 0.8 ft^3/ft hr and the plasma gas is at 28 ft^3/hr. Gun current is 550 amps and gun voltage is 28 V. The product is a Mo-15W alloy powder with an apparent density of 6.22 g/cc and a flow time of 9 seconds for the −325 mesh fraction.

In yet another related process, *R.F. Cheney, D.J. Port and J.R. Spencer; U.S. Patent 3,881,911; May 6, 1975; assigned to GTE Sylvania Inc.* have found that binder-related deficiencies of contamination, low apparent density and low agglomerate strength can be overcome by heating the agglomerates above the vaporization temperature of the binder to remove the binder and thereafter heating the powders to a still higher temperature causing the subparticles of the agglomerates to sinter or fuse together by diffusion bonding. This second temperature must not exceed ½ of the melting point of the material since that would result in the agglomerates themselves sintering or fusing together to form an unusable cake or sponge.

The sintering of the subparticles restores or improves the strength which the agglomerates had prior to removal of the binder, and also causes a densification of the agglomerates, thus improving the apparent density. By controlling this sintering within the permissible range, the apparent density can be intentionally adjusted over a considerable range.

For successful strengthening of the agglomerates, it is also necessary to control the relative sizes of the subparticles of the agglomerates. In general, these subparticles should be no larger than 1/5th the agglomerate diameter and generally 1/10th or less. This is important because if the subparticles are too large, heating under the conditions necessary to avoid appreciable sintering of the agglomerates will result in agglomerate strengths insufficient to permit handling after the sintering operation. Accordingly, the subparticles should be as small as possible, e.g., 1/100th the agglomerate diameter or less.

For particles within the above size range, sintering results in a product which is free flowing binderless powder having sufficient strength to withstand considerable handling, and being useful in flame spraying applications.

Where size classification of the product is desired, it may be carried out at any stage of the processing found to be convenient.

Oxygen-Rich Plasma Spray Powders

W.D. Lafferty, R.F. Cheney and R.H. Pierce; U.S. Patent 4,146,388; March 27, 1979; assigned to GTE Sylvania Inc. have found that the hardness and consequently the wear resistance of molybdenum flame spray coatings may be substantially increased by incorporating at least about 0.5 weight percent of oxygen into the molybdenum plasma spray powder, either as dissolved oxygen or oxide, or as a second phase containing oxygen or oxide or as a surface oxide or in some combination of these forms, prior to plasma spraying The plasma sprayed coatings so formed exhibit improved hardnesses over those of "pure" molybdenum plasma sprayed coatings typically containing from 1 to 2 weight percent oxygen.

Such oxygen-rich molybdenum powders are formed by passing molydenum particles through a plasma, such as that formed by commercially available plasma spray guns, while the particles are in contact substantially with free oxygen or an oxide of molybdenum or a precursor of an oxide molybdenum. As used herein, the term "precursor" means a compound which on heating above a critical temperature will convert to an oxide of molybdenum, for example, the various ammonium molybdates which decompose essentially instantaneously at plasma temperatures.

Such ammonium molybdates include ammonium dimolybdate, ammonium paramolybdate, ammonium tetramolybdate, ammonium polymolybdate and normal ammonium molybdate. Other suitable precursors include by way of example, molybdenyl sulfates, molybdenyl chlorides and, in the presence of oxygen, molybdenum disulfide and molybdenum pentachloride.

According to one embodiment, the oxygen content of the powder may be controlled by bleeding controlled amounts of a free oxygen containing gas such as air into the plasma such as by aspiration through a variably-sized orifice in the plasma equipment housing.

According to another embodiment, the oxygen content of the powder may be further controlled by bleeding controlled amounts of a reducing gas such as hydrogen into the plasma, or subsequent to passage through the plasma, by heating the powder in a reducing atmosphere such as hydrogen to reduce molybdenum oxides to molybdenum, or by washing the powder with an oxygen leaching agent such as ammonium hydroxide.

According to still another preferred embodiment, ammonium molybdate is selected to be the precursor, which also serves as a binder for molybdenum particles in particle agglomerates, conveniently produced for example, by spray drying molybdenum particles in an aqueous ammonium molybdate solution as taught in U.S. Patent 4,028,095

In accordance with still another preferred embodiment, such ammonium molybdate-containing agglomerates may be mixed with molybdenum powder particles in order to further control the amount of oxygen incorporated into the molybdenum plasma spray powder.

Coatings produced from such plasma spray powders combine the hardness of wire-sprayed coatings and the inexpensive processing and flexibility of plasma spray coatings. By controlling the amount of oxygen in the molybdenum powder, the hardness of the resulting coatings can be controlled.

Increase in Bulk Density by Nickel Addition

In related work, *D.J. Port and R.F. Cheney; U.S. Patent 3,907,546; September 23, 1975; assigned to GTE Sylvania Inc.* have found the bulk density of a flame spray powder of agglomerates of finely divided molybdenum or molybdenum alloy particles is increased by adding from 0.1 to 0.5 weight percent of finely divided nickel particles to the agglomerates and heating the agglomerates at a temperature of from 1050° to 1300°C for from 1 to 7 hours. This results in increased flame spray coating rates due to the increased weight of metal sprayed per unit of time.

In accordance with a preferred embodiment, materials in finely divided particulate form (less than about 40 μm) are agglomerated such as by spray drying in a slurry with a binder, and classified to obtain a desired particle size range. At this stage the agglomerates are porous. For example, for agglomerates having sizes below 170 mesh bulk densities typically range from about 2.2 to 2.7 g/cc. The agglomerates are then processed as above, resulting in substantial increases in bulk density. For example, for the above particle sizes bulk densities as high as 3.6 g/cc can be achieved for molybdenum particles while substantially retaining agglomerate integrity and thus free flowing characteristics.

Cobalt-Molybdenum Mixed Metal Agglomerates

J.A. Powers, M.B. MacInnis and D.J. Port; U.S. Patent 4,011,073; March 8, 1977; assigned to GTE Sylvania Inc. produce a flame spray powder of cobalt and molybdenum mixed metal agglomerates by spray drying a slurry of molybdenum metal particles and cobalt oxide particles in an ammonium molybdate solution to produce agglomerates. Thereafter the agglomerates are heated in a reducing atmosphere to convert the oxide particles to cobalt metal and the binder to molybdenum metal.

Example: This example used cobalt oxide and molybdenum metal powder with ammonium molybdate as the binder. The feed materials were:

Material	Pounds
Cobalt oxide	100
Molybdenum metal powder	15.4
Molybdenum trioxide	15
Ammonium hydroxide (28%)	8
Water (4 gallons)	33.4

The molybdenum trioxide was dissolved in the ammonia and diluted with the water to form an aqueous solution of ammonium molybdate. The molybdenum metal powder and cobalt oxide were added to make the slurry. The initial cycle was made up in two equal batches and spray dried. The chamber product (agglomerates) from the two batches were classified, by screening, to remove agglomerates with the particle size range between 200 and 325 mesh. The outsized agglomerates and the cyclone product (fines) from the two batches (52 pounds) were slurried in 8 liters of water and again spray dried and classified.

A total of four cycles of spray drying, classifying and reslurrying of out-sized material was made. The agglomerates with a particle size range between 200 and 325 mesh were fired at 900°C for 3½ hours under hydrogen to give the final product.

The following table gives the particle size distribution of the chamber product after the first spray drying and after the fourth.

Particle Size Range of Agglomerates/Cycle	First	Fourth
+170 mesh, %	38	60
-170+200 mesh, %	13	16
-200+270 mesh, %	15	10
-270+325 mesh, %	14	7
-325 mesh, %	20	12

The fact that after four cycles, only 12% of the agglomerates were <325 mesh, is evidence that the binding action of the ammonium molybdate is preserved through recycling when a nonreactive form of cobalt is used.

Molybdenum and Alumina Plasma Flame Spray Powders

D.J. Port, W.D. Lafferty and R.F. Cheney; U.S. Patent 3,960,545; June 1, 1976; assigned to GTE Sylvania Inc. have found that plasma flame spray coatings exhibiting improved hardness and wear resistance can be produced from flame spray powders comprising spray dried agglomerates of finely divided molybdenum and alumina powder subparticles, wherein the diameters of the agglomerates are at least two times the maximum diameter of the subparticles.

The agglomerates of powder particles of molybdenum or alloys of molybdenum contain at least 50% by weight of molybdenum together with alumina particles in the amount of from 1 to 50% by weight of the flame spray powder, and preferably from 5 to 20% by weight of the flame spray powder. Agglomeration of the finely divided starting powders may be by any of several techniques known in the art although in a preferred embodiment such agglomeration is achieved by spray drying a slurry of the starting powders with a suitable binder.

Agglomeration conditions should be chosen in order to achieve agglomerate diameters at least two times, and preferably four times the maximum diameters of the particle diameters. By way of example, molybdenum starting powder having particle sizes from about 1 to 10 μm and preferably 2 to 3 μm, and alumina particles having sizes less than 10 μm and preferably less than 1 μm are agglomerated to produce agglomerates 90% within the range of -200 (74 μm) to $+325$ (44 μm) mesh.

A suitable binder material for spray drying to produce such agglomerates is a thermally decomposable compound of molybdenum, such as ammonium molybdate. This may be conveniently formed in the case in which an aqueous slurry is utilized for spray drying by the addition of molybdenum trioxide and ammonium hydroxide to the slurry. These constituents combine in aqueous media to form ammonium molybdate which has been found to be a suitable spray drying binder. Following spray drying the resulting agglomerates may be classified according to size by conventional techniques such as screening to produce size fractions suitable for the intended flame spray powder application.

A typical size fraction which has been found suitable for producing coatings having superior hardness and wear resistance is 90% between −200 and +325 mesh. Following such size classification, the powder is advantageously subjected to a sintering step at conditions of time and temperature sufficient to substantially increase the bulk density of the agglomerates, and therefore of the flame spray powder, but insufficient to result in substantial sintering together of agglomerates, which would result in an unusable mass, cake, or sponge of material. Such conditions are not a necessary part of this description, but are well within the skill of the art to effect.

Example: Two sample powders having the compositions of 6 weight percent alumina remainder molybdenum and 19 weight percent alumina remainder molybdenum were prepared by the following procedure. The molybdenum powder had a size range from about 1 to 10 μm and the alumina had a size nominally less than 1 μm. These powders were suspended in water-base slurries together with molybdenum trioxide and ammonium hydroxide in the proportions listed in Table 1.

Table 1

Mo-6% Al_2O_3	**Mo-19% Al_2O_3**
2.5 lb Al_2O_3	10 lb Al_2O_3
7.5 lb MoO_3	7.5 lb MoO_3
42.5 lb Mo	35 lb Mo
2 gal hot water	2 gal hot water
½ gal NH_4OH	½ gal NH_4OH

The slurries were each spray dried under identical conditions, that is, about 600°F inlet temperature, about 340°F outlet temperature, and about 41 psi atomization air pressure. Following spray drying the dried powders were finish screened to −200+325 mesh and that size fraction was sintered at 1050°C for about 3½ hours. The bulk properties of the powders which were used for plasma spray coating appear in Table 2.

Table 2

	Mo-6% Al_2O_3	**Mo-19% Al_2O_3**
Sieve Analysis, %		
+170	0 %	0 %
+200	0 %	0 %
+270	41 %	30 %
+325	36 %	44 %
-325	23 %	26 %
Bulk density	2.26 g/cc	2.10 g/cc
Hall flow	33 sec/50 g	38 sec/50 g

Powders were then plasma flame sprayed onto a mild steel plate for metallographic examination. Conditions were as follows. The plasma gun was operated at 600 amps and 32 volts with argon and hydrogen plasma gas flow rates at 32.5 and 18.3 l/min respectively, and argon feed gas flow rate at 2.6 l/min. A powder feed rate of 473 cc/hr and a gun-to-substrate distance of about 100 mm were utilized. Metallographic examination revealed dense coating with Al_2O_3 particles substantially uniformly dispersed in a molybdenum matrix.

Wear testing samples were then prepared as follows. The surfaces of 9 standard size test blocks made of Spartalloy 2-60 cast iron (containing about 3.7 weight percent carbon, 2.5 weight percent silicon, 0.6 weight percent manganese, and 0.6 weight percent chromium) were prepared by grit blasting with 36 mesh grit garnet. Using the above spray parameters for preparation of the metallographic samples, the two powders were sprayed onto three blocks each to a thickness of about 0.25 mm and machine ground to a thickness of about 0.15 mm. Using identical procedures, three blocks were sprayed with a pure molybdenum powder.

The flame sprayed molybdenum surfaces were then ground by hand with 600 grit abrasive powder with final grinding being in the contemplated direction of the wear test. All of these samples were then cleaned per ASTM specification D-2714. The samples were then wear tested on a standard friction and wear test machine against rings also made from Spartalloy 2-60 cast iron. The tests were run for 30,000 cycles at a rate of 200 cycles per minute at an applied load of 136.4 kg. Stoddard solvent, a low viscosity hydrocarbon solvent, was used as a coolant during testing, such that test temperatures were about 40°C initially, climbing to about 70°C by the termination of testing.

Changes in frictional force, width of the wear scar in the coating, and weight loss of both the test block and the ring were recorded. This data, showing the superiority of the coatings of this process as compared to the pure molybdenum coatings are presented in Table 3.

Table 3

Coating	Wear Scar Width (mm)	Volume Loss (mm^3)
Mo	2.3	0.35
Mo-6% Al_2O_3	1.8	0.12
Mo-19% Al_2O_3	1.6	0.14

Diamond pyramid hardness at 300 g (DPH_{300g}) of the coatings were also measured. The molybdenum-6 weight percent alumina coating is 157 DPH_{300g} units harder than the pure molybdenum coating while the molybdenum-19 weight percent alumina coating is 42 DPH_{300g} units harder than the 6 weight percent alumina coating and 199 units harder than the pure molybdenum coating.

Tungsten Carbide-Cobalt Agglomerated Powder

R.F. Cheney, W.D. Lafferty and G.J. Long; U.S. Patent 4,025,334; May 24, 1977; assigned to GTE Sylvania Inc. produce tungsten carbide-cobalt agglomerated flame spray powder by spray drying a slurry of particles in an aqueous cobalt nitrate solution. The agglomerates are classified according to size and out-of-size agglomerates are recycled. The classified agglomerates are heated in flowing hydrogen, to reduce the nitrate to cobalt metal, and then sintered to strengthen the agglomerates for subsequent flame spraying.

Out-of-size agglomerates can be deagglomerated by reslurrying them in water (resulting in dissolution of the cobalt nitrate), and repeating the spray drying cycle again, thereby avoiding the need for a separate binder removal step.

Elemental carbon and/or tungsten powder may be added to the slurry prior to spray drying, or to the spray dried agglomerates prior to reduction and/or sintering, in order to adjust or compensate for shifts in the stoichiometry of the WC during processing. Cobalt may be introduced as an insoluble chemically reducible compound of cobalt, such as cobalt oxide, which can then be reduced subsequently when the soluble cobalt nitrate binder is reduced.

Example 1: 44 pounds of tungsten carbide, 6.77 pounds of cobalt oxide and 8.2 liters of water were milled for one-half hour in a commercially available attritor mill. For this purpose, tungsten carbide balls were used as the milling medium. 3.34 pounds of cobalt nitrate were then dissolved in this milled slurry and the slurry was spray dried in a Proctor-Schwartz dryer using a two-fluid nozzle. The inlet air temperature was 600°C and the outlet air temperature was 360°C. The resulting spray dried agglomerates were screened into two size fractions, the first being within the range of −325 mesh to +20 μm and the second within the range of −170 to +325 mesh. The oversize powder was reslurried for another spray drying cycle.

The agglomerates within the desired size fractions were then heated in flowing hydrogen at 725°C for 3½ hours to reduce the cobalt oxide and cobalt nitrate to metallic cobalt. The reduced agglomerates were then further heated in hydrogen at 1230°C for one-fourth hour to strengthen and densify the agglomerates by sintering. The powder was then passed through a 100 mesh screen to break up cakes of agglomerates which had formed. Hall flow was measured of 50 g samples for both fractions and was 32 seconds for the −325 mesh fraction and 15 seconds for the −170+325 mesh fraction. Such Hall flow values represent good to excellent flowability for such flame spray powders.

Example 2: The procedure of Example 1 was repeated except that the carbon level was lowered from 5.3 to 4 weight percent by adding tungsten powder to the starting material. Thus, the starting materials were 159 pounds of WC,

35.5 pounds of cobalt oxide, 61 pounds of tungsten and 10.5 pounds of cobalt nitrate. Again the resulting powder exhibited good flowability as evidenced by Hall flow values on 50 g samples of 14 seconds and 30 seconds for the −325 and −170+325 mesh fractions, respectively.

OTHER METALS AND MIXED SYSTEMS

Niobium-Zirconium-Titanium Capacitor Powder

G.J. Villani; U.S. Patent 3,849,124; November 19, 1974; assigned to Norton Company has developed a powder for making electric capacitors or the like comprising a ternary niobium-zirconium-titanium alloy. The alloy is selected as to composition in the range of 20 to 40% niobium, 30 to 60% zirconium, and 20 to 40% titanium. The preferred proportions are 50% zirconium, 25% niobium, and 25% titanium. It is treated to produce and retain the beta (body-centered-cubic) phase. The resultant product affords high capacitor stability and low leakage approaching the characteristics of the more expensive tantalum at a capacitance cost comparable to or better than that of niobium.

Example 1: Several alloys of nominal compositions listed in Tables 1-3 below were made by arc melting 50-100 gram alloy buttons in argon atmosphere using a nonconsumable tungsten electrode. The buttons were formed from niobium chips and crystal bar zirconium and titanium with raw material specification as in Example 2. Each alloy button was melted four times and then heated to about 80% of the melting point and held at that temperature for 2 hours. The buttons were then sliced with a diamond wheel saw. One slice of each button was chemically analyzed and the remainder were cold rolled from three-sixteenths inch to a thickness of about 0.015 inch.

The rolled sheets were cut into panels, degreased, chemically polished, rinsed and annealed above 1000°C at about 80% of their respective melting points for 30 minutes in a vacuum of about 10^{-5} torr. The panels were anodized at 1 mA/cm^2 of surface in 0.01% H_3PO_4 (at 25°C in one test and at 92°C in a second test) to 200 volts and held for 30 minutes. The anodized panels were rinsed and dried and electrically tested in a wet cell (1% H_3PO_4 electrolyte) as formed and again after a later heat treatment of 300°C for 30 minutes in air.

Figure 7.3 is a ternary phase diagram of the system showing data points used. The results are shown in Table 1 for 25°C anodization and in Table 2 for 92°C anodization where DCL is dc leakage in μA/cm^2 of anode surface tested, DF is dissipation factor (percent), C is capacitance in microfarads (μF) or millimicrofarads (mμF)/cm^2 of tested anode surface and ΔC is capacitance change (percent) due to heat treatment:

$$\frac{\text{C after - C before}}{\text{C before}} \times 100$$

Figure 7.3: Ternary Composition Diagram (Atomic Percent)

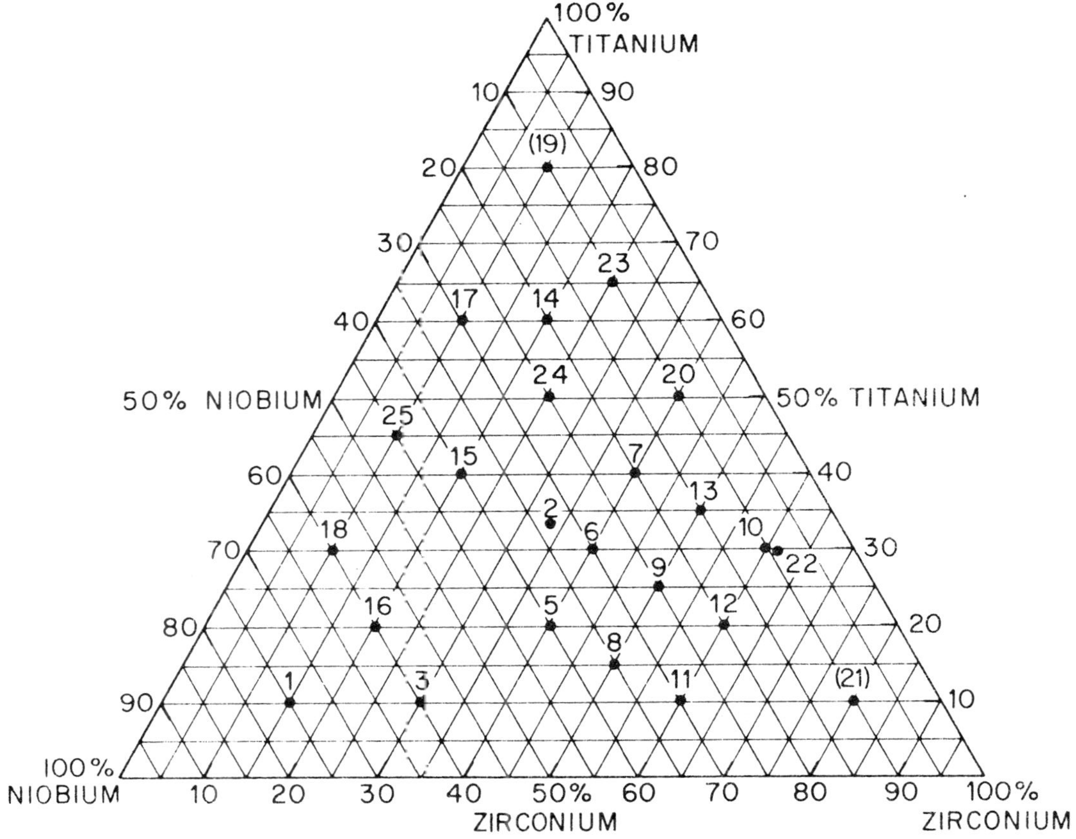

Source: U.S. Patent 3,849,124

Table 1

Data Point	Nominal Composition Nb-Zr-Ti (atom %)	DCL ($\mu A/cm^2$)	DF (%)	C ($m\mu F/cm^2$)	ΔC (%)
24	25-25-50	<0.02	1.6	56	70
9	25-50-25	<0.02	1.7	52	30
23	10-25-65	0.02	1.8	53	36
7	20-40-40	0.02	2.0	52	81
12	20-60-20	0.02	2.1	59	9.3
11	30-60-10	0.04	1.7	44	3.9
14	20-20-60	0.03	2.0	55	88
15	40-20-40	0.05	1.9	56	86
6	30-40-30	0.05	1.7	55	74
2	33.3-33.3-33.3	0.11	2.0	55	6.7
17	30-10-60	0.08	4.5	61	86
25	45-10-45	0.12	3.6	101	152
8	35-50-15	0.30	1.4	35	1.2
18	60-10-30	0.33	2.4	64	128
16	60-20-20	0.53	3.1	59	78
10	10-60-30	0.58	2.3	50	29
13	15-50-35	0.63	2.6	48	61
1	75-15-10	1.6	7.6	–	–
5	40-40-20	3.3	4.6	–	–
3	60-30-10	4.5	6.8	–	–
22	10-60-30	18	5.8	–	–
20	10-40-50	46	7.2	–	–
Niobium	–	0.06	2.5	78	72
Tantalum	–	<0.02	2.1	68	3.0

Table 2

Data Point	Nominal Composition Nb-Zr-Ti (atom %)	DCL ($\mu A/cm^2$)	DF (%)	C ($m\mu F/cm^2$)	ΔC (%)
9	25-50-25	0.02	1.2	54	0.3
12	20-60-20	0.02	1.5	54	39
13	15-50-35	0.02	1.7	49	38
2	33-33-33	0.02	1.5	41	74
6	30-40-30	0.02	2.1	45	0.7
22	10-60-30	0.02	2.2	54	88
10	10-60-30	0.06	2.5	56	31
24	25-25-50	0.14	2.2	44	30
5	40-40-20	0.09	1.5	34	0.5
16	60-20-20	0.04	1.6	48	94

(continued)

Table 2: (continued)

Data Point	Nominal Composition Nb-Zr-Ti (atom %)	DCL ($\mu A/cm^2$)	DF (%)	C ($m\mu F/cm^2$)	ΔC (%)
15	40-20-40	0.09	1.5	36	110
1	75-15-10	0.06	3.4	56	105
25	45-10-45	0.13	3.1	81	181
18	60-10-30	0.33	1.7	56	172
17	30-10-60	2.1	2.2	47	91
11	30-60-10	7.4	4.4	–	–
7	20-40-40	9.5	7.5	–	–
3	60-30-10	22	5.0	–	–
14	20-20-60	37	11.2	–	–
23	10-25-65	high	5.7	–	–
8	35-50-15	high	10	–	–
20	10-40-50	high	13	–	–
Tantalum	–	0.02	1.6	54	2.2
Niobium	–	high	will not anodize		–
19	10-10-80	–	–	–	–
21	10-80-10	–	–	–	–

Microstructure of the heat treated buttons was examined. In general the structures were equiaxed with subgrain boundaries and dislocation etch pits reminiscent of a strained body-centered-cubic phase quenched from a high temperature. There was no evidence of second phase transformation except in data points 19 and 21; these had fine precipitated structures within the equiaxed structure.

The data points in Tables 1-2 are grouped from lowest leakage to highest leakage. Table 3 shows a further series of experimental results indicating reproducibility of the Table 2 data.

Table 3

Data Point	Nominal Composition Nb-Zr-Ti (atom %)	DCL ($\mu A/cm^2$)	DF (%)	C ($m\mu F/cm^2$)	ΔC (%)
2-1	33.3-33.3-33.3	<0.02	1.5	41	74
2-2	33.3-33.3-33.3	0.02	1.6	46	96
2-3	33.3-33.3-33.3	0.02	1.6	45	120
2-4	33.3-33.3-33.3	0.03	1.8	48	44
2-5	33.3-33.3-33.3	0.02	1.8	43	50
5-1	40-40-20	0.09	1.5	34	0.5
5-2	40-40-20	0.17	1.3	36	0.0
5-3	40-40-20	0.04	1.4	36	1.0

(continued)

Table 3: (continued)

Data Point	Nominal Composition Nb-Zr-Ti (atom %)	DCL ($\mu A/cm^2$)	DF (%)	C ($m\mu F/cm^2$)	ΔC (%)
6-1	30-40-30	0.02	2.1	45	0.7
6-2	30-40-30	0.02	2.1	50	29
6-3	30-40-30	0.02	2.0	49	1.3
7-1	20-40-40	9.5	—	—	—
7-2	20-40-40	12	—	—	—
7-3	20-40-40	5.8	—	—	—
9-1	25-50-25	<0.02	1.2	54	0.3
9-2	25-50-25	0.02	1.9	52	7.8
9-3	25-50-25	<0.02	1.4	54	8.2
10-1	10-60-30	0.06	2.5	56	31
10-2	10-60-30	0.06	2.1	48	194
10-3	10-60-30	0.02	2.0	54	152
22-1	10-60-30	0.02	2.2	54	88
22-2	—	0.03	2.3	50	90
12-1	20-60-20	<0.02	1.5	54	39
12-2	20-60-20	<0.02	1.5	56	40
12-3	20-60-20	0.02	1.6	56	41
15-1	40-20-40	0.09	1.5	36	110
15-2	40-20-40	0.08	1.6	46	14
15-3	40-20-40	0.04	1.6	46	44
16-1	60-20-20	0.04	1.6	48	94
16-2	60-20-20	0.02	1.8	42	107
16-3	60-20-20	0.16	2.2	39	118
24-1	25-25-50	0.14	2.2	44	30
24-2	25-25-50	0.29	2.3	44	198
24-3	25-25-50	5.7	—	—	—

Example 2: Several alloy buttons were formed by melting together niobium, zirconium and titanium with the following purities: Nb (machining chips), 99.94%; Zr (crystal bar), 99.91%; and Ti (crystal bar), 99.94%.

The nominal compositions of the alloys are given in Table 4. The buttons were hydrided, ground to powder and dehydrided. The powders in varying size cuts were sintered above 1000°C into porous compacts of approximately 1 g each, cooled and anodized in 0.01% H_3PO_4 electrolyte at 92°C temperature with a current density of 63 mA per anode to a formation voltage for 2 hours. As a control, niobium anodes were similarly formed at 25°C at current density of 50 mA per anode.

The anodized compacts were impregnated with manganese nitrate and pyrolyzed to form manganese dioxide electrolyte. The pyrolysis temperature was 275°C for 8 minutes. Three impregnation dips followed by pyrolysis were used. Dip time was 3 minutes. After MnO_2 impregnation, the anode (anodic oxide film) was reformed at 2 V/min in the 0.01% H_3PO_4 at 92°C to break-down. The reformation break-down is given in Table 4 as V_R.

Similar capacitor samples were reformed to 35 V. A counter-electrode was added and the devices were tested at 20 V for capacitance and leakage. The results are given in Table 4.

Table 4

Data Point	Nominal Composition Nb-Zr-Ti (atom %)	Powder Size*/ Sinter Temp. (°C)	V_R (volts)	C (μF)	L/C (μA/μF)
2-1	33-33-33	F/1100	40	23.3	0.39
2-2	33-33-33	F/1200	40	16.5	0.07
2-3	33-33-33	F/1300	40	12.4	0.51
2-4	33-33-33	C/1300	40	5.3	0.26
5	40-40-40	F/1300	65	11.7	0.22
6-1	30-40-30	F/1100	47	12.0	0.17
6-2	30-40-30	C/1200	47	5.4	2.3
6-3	30-40-30	C/1300	52	4.8	0.35
6-4	30-40-30	C/1100	—	18.0	0.08
9-1	25-50-25	F/1200	40	8.6	0.14
9-2	25-50-25	F/1200	63	8.9	0.06
9-3	25-50-25	C/1300	38	4.4	0.11
9-4	25-50-25	C/1300	65	4.4	0.05
12-1	20-60-20	F/1100	80	7.8	0.06
12-2	20-60-20	C/1200	62	4.5	0.13
12-3	20-60-20	C/1300	62	4.5	0.13
13	15-50-25	F/1100	63	14.2	0.09
15-1	40-20-40	F/1100	55	11.2	2.0
15-2	40-20-40	C/1200	58	9.9	1.6
15-3	40-20-40	C/1300	42	4.7	2.5
16-1	60-20-20	F/1300	56	15.4	—
16-2	60-20-20	C/1400	58	5.8	1.1
22	10-60-30	F/1100	65	13.1	0.31
24	25-25-50	F/1200	67	11.5	0.65
Control	100-0-0	A/2050	85**	15.3	0.10

*F = -325 mesh, plus 5 microns; C = -100+325 mesh; A = -140 mesh, plus 5 microns (FAPD).
**25°C test for 100-0-0, 85°C for others.

Solid capacitor samples of compositions noted in Table 5 of −325 mesh powder size were put through life tests of 200 to 1,000 hours and resultant life test data charted in Table 5. In the life test, the leakage currents, capacitance

Table 5

Initial and final median or average values at 25° C. and 92° C.

Data point	Nominal composition At. percent Nb-Zr-Ti	Leakage (microamps)				Capacitance (μF)				Dissipation factor, percent			
		$I_{25°}$	$I_{85°}$	$F_{85°}$	$F_{25°}$	$I_{25°}$	$I_{85°}$	$F_{85°}$	$F_{25°}$	$I_{25°}$	$I_{85°}$	$F_{85°}$	$F_{25°}$
2	(33.3-33.3-33.3)	[a] 6.3	[a] 37	[a] 38	[a] 3.0	16.5	21.6	18.9	16.1	17.0	19.3	14.5	16.9
5	(40-40-20)	2.6	13	13	1.9	11.7	13.1	12.5	11.3	7.2	4.8	7.5	2.8
6	(30-40-30)	1.7	10	6.5	0.7	15.0	17.5	15.0	12.9	12.2	10.7	10.2	5.8
9	(25-50-25)	0.9	3.0	3.5	0.5	8.8	10.1	9.9	8.3	7.3	5.8	7.2	3.3
12	(20-60-20)	0.5	3.5	2.0		7.8	8.7	8.2		4.0	7.3	6.0	
13	(15-50-25)	1.2	7.2	9.2	1.5	14.2	15.7	14.4	11.2	10.3		9.2	3.2
15	(40-20-40)	[b] 19.0	[b] 55	[b] 49		10.5	15.0	11.4		21.0	21.0	15.5	
16	(60-20-20)	[c] 0.4	[c] 2.2	[c] 3.8		15.4	19.3	17.8		22.0	18.0	13.3	
22	(10-60-30)	4.1	24	75	9.0	13.1	15.3	10.9	9.3	8.0	5.8	8.4	2.6
24	(25-25-50)	7.5	23	37	7.9	11.5	13.4	12.4	8.1	17.0	8.1	12.1	3.9
Niobium (i.e. 100-0-0)		1.6	2.4	4.4		15.3	17.5	15.0		15.0	16.0	25.0	

[a] 1 of 9 samples tested at data point 2 was a failure (i.e. Leakage in rising to the milliampere range).
[b] 1 failure out of 4 samples tested at data point 15.
[c] 2 failures out of 3 samples tested at data point 16.

and dissipation factor are measured at 25°C. Then the temperature is raised to 85°C and the capacitors retested. Temperature is held at 85°C for an extended period, the capacitors retested and then the temperature is dropped to 25°C for retest. Initial (I) and final (F) values of these parameters are shown in Table 5. In some instances the "final" value at high temperature is a median or average value.

Nitrogen Reduction of Vanadium Pentoxide

M. Vojkovic; U.S. Patent 3,857,695; December 31, 1974; assigned to International Minerals & Chemical Corp. describes a method for producing nitrided alloys containing about 50 to 82% by weight vanadium, together with minor amounts of iron, carbon and nitrogen, and at most about 1% by weight oxygen. It comprises establishing an intimate admixture of a complex comprising vanadium pentoxide and an iron hydroxide or salt, together with finely divided carbon in an amount in the range of about 90 to 120% of the carbon stoichiometric to react with the complex to reduce the reducible metal values in the complex with the production of carbon monoxide.

The admixture is maintained in an enclosed reaction zone at a temperature in the range of about 1250° to 1350°C for a period of 20 to 120 minutes in the presence of a reducing atmosphere consisting of ammonia or mixtures of hydrogen and nitrogen having mol ratios in the range of about 2:1 to about 4:1, while continuously removing oxygen-containing gases from the reaction zone. The vanadium pentoxide and iron salt are reduced and a nitrided alloy is formed. The alloy is cooled to a temperature below that at which spontaneous ignition of the alloy occurs on exposure to atmospheric air and the nitrided alloy is recovered.

The nitrided alloys produced in accordance with this process will contain an amount of vanadium which is essentially dependent upon the vanadium content of the initial admixture reduced. Products having a vanadium content in the range of about 50 to 82% are suitable and preferred. Slightly higher vanadium contents are achievable and lower vanadium contents are of course possible, but not preferred.

Recovery of Cadmium and Yttrium from Industrial Waste

T. Sakagami, S. Kakumoto, K. Okawa and K. Miyashita; U.S. Patent 3,915,691; October 28, 1975; assigned to Matsushita Electric Industrial Co., Ltd., Japan describe a method of treating an industrial waste liquid containing a metallic powder selected from the group consisting of powders of metals, metallic compounds and mixtures thereof which comprises:

1. Mixing an aqueous waste liquid containing the metallic powders with a water-soluble high molecular weight compound selected from the group consisting of poly-

vinyl alcohol, methyl cellulose, carboxy methyl cellulose, sodium alginate and sodium cellulose xanthogenate to form a liquid mixture.

2. Transferring the liquid mixture onto a movable belt to spread it over the surface of the belt.
3. Subjecting the liquid mixture to temperatures between 100° and 300°C to evaporate the water in the liquid and to concentrate the mixture.
4. Cooling the concentrated mixture to form a solid film of the high molecular weight film-forming organic compound, which film contains the metallic powders embedded in the solidified organic high molecular weight film.

Reclaiming the metals from the concentrated compound can be done by burning the organic compound to leave the metals as residue, for instance, oxide or sulfide of metals.

Treatment of Cyanide Complex with Dry Hydrogen

R.A. Paris, P.A. Amblard and A.C. Rousset; U.S. Patent 4,111,686; September 5, 1978 provide a method of producing metals of the groups III-A, IV-A, VI-A and III-B of the Periodic Table.

The process also provides a method of making yttrium, the lanthanides (elements of atomic numbers 58 to 71 inclusive), titanium, zirconium, hafnium, chromium, gallium, indium or thorium and alloys thereof in high purity.

The process is based in principle on the reduction by hydrogen of the metal cyanide complex of the metal or the metals in question.

If the cyanide complex is monometallic (i.e., containing only a single metallic element) its reduction by hydrogen yields one metal. If on the contrary the cyanide complex is polymetallic (i.e., containing several different metallic elements) its reduction yields an alloy.

In the case of monometallic cyanides, they can be obtained by reacting hydrocyanic acid HCN with a metal hydrate according to the general reaction formula:

(1) metallic hydrate(hydroxide or oxide) + hydrocyanic acid → cyanide complex + water

The named metals are produced by reducing the cyanide complex with hydrogen according to the general reaction formula:

(2) cyanide complex + hydrogen → metal + hydrocyanic acid

The hydrocyanic acid thus obtained, with minor losses, in reaction (2) is used again for reaction (1), making a cycle which indirectly reduces the metal hydrate (hydroxide) by hydrogen, which often is impossible to do directly. In the production of alloys, polymetallic cyanide complexes are reduced by hydrogen according to the general reaction formula:

(3) polymetallic cyanide + hydrogen → alloy + hydrocyanic acid

In this case again the hydrocyanic acid which is recovered in reaction (3) can be used to obtain polymetallic cyanides.

The reduction by hydrogen of a polymetallic cyanide complex according to reaction (3) yields an alloy in the form of a fine powder each granule of which is an alloy particle and not just a juxtaposition of granules of each of the component metals. Thus the process yields an alloy formed by a homogeneous phase (solid metallic solution or definite combination) or a mixture of phases conforming to that predicted by the formula. In other words the uniform alloy obtained has the same structure as that which is obtained by fusion and cooling of the metals, with the condition, of course, that the proper thermal treatments have been carried out so that the phases appear under identical allotropic forms.

The powders so produced can be used in powder metallurgy. The production of sintered-metal bearings and like parts is facilitated by the high-purity powders according to this process.

The advantages of this process are numerous. First of all the temperatures necessary are relatively low. They range between 700° and 1300°C according to the metal, which is substantially lower than the temperatures necessary for the prior art methods. It is also much easier to avoid reaction with the receptacles at these temperatures, and purer products are obtained.

In the case of refractory metals and alloys the cyanide reduction process yields a powder whose granule size is dependent on the final temperature and also on how long this temperature is maintained. This means that metallic materials can be produced which are particularly adapted to sintering.

The economical advantages inherent in this process are a result of using hydrogen as reducing agent, hydrogen being particularly inexpensive and of avoiding preparation of the constituent elements of the alloy individually, this operation being often very difficult and expensive when the metals are refractory or very reactive, as are the metals of the lanthanum series.

Reduction of Metal Carboxylates

R.K. Jordan; U.S. Patent 3,955,961; May 11, 1976 has developed a process for the production of metals of atomic numbers 23 through 30 wherein a metal carboxylate selected from at least one of oxalates, formates, acetates, carbamates and malonates containing less than 12% water is reduced in an atmosphere containing hydrogen or carbon monoxide, or mixtures thereof, at a temperature in the range of from about 150° to about 800°C.

By the hydrogenation of anhydrous carboxylates of metals of atomic number 23 through 30, ideally at 300° to 500°C, metals containing practically no oxides can be obtained. Further, as compared to the same process utilizing the corresponding dihydrate of the metal carboxylate, a savings of 30 to 40% in energy required for the process can be realized. Still further, as a result of a process for the direct production of anhydrous formates the metals may be made so economically as to enable their competitive use in alloying and in general use.

Exothermic Reduction Reaction of Valve Metals

H. Vartanian; U.S. Patent 3,992,192; November 16, 1976 has developed a process for producing metal powders of agglomerated form having a uniform specific surface area for various size fractions thereof in a size spectrum extending from 3 μ to 10 mesh by heating a charge of a salt source of the metal, precoated with reducing agent, to initiate an exothermic reduction reaction and continuing application of heat after termination of the exotherm heat output to modify powder size.

The reaction charge is agitated substantially throughout and further heating periods follow to maintain homogeneity. The charge may be diluted with inert salts to modify the necessary minimum hold temperature and the properties of resultant powders. The metal may comprise one or more of tantalum, columbium, zirconium, vanadium, hafnium, tungsten, titanium, thorium, chromium, yttrium, rare earths, germanium, manganese, beryllium, boron, iron, nickel and platinum group metals.

Referring to Figure 7.4, a preferred embodiment of the powder production process is shown in a flow chart of consecutive steps applied illustratively to a K_2TaF_7-NaCl-Na charge but having general application to other ingredients. The chart also includes optional additional steps useful for some grades of powder and a last step of sintering to produce porous blocks such as electrolytic capacitor anodes. The steps are:

A. Drying and Mixing and B. Coating: The metal-containing salt, e.g., K_2TaF_7, is charged into a reactor vessel. Optionally a diluent salt, e.g., NaCl, may also be added.

The reactor vessel is sealed and then heated to an elevated temperature above 100°C, all the while stirring the salt charge. Molten sodium is then rapidly added to the charge with inert gas atmosphere. The vessel heating is controlled to maintain a temperature that keeps the sodium molten, but is sufficiently low to avoid initiating the exothermic reaction between the sodium and K_2TaF_7. The charge is agitated, through use of an internal stirrer within the

Figure 7.4: Flow Chart of Powder Production Process

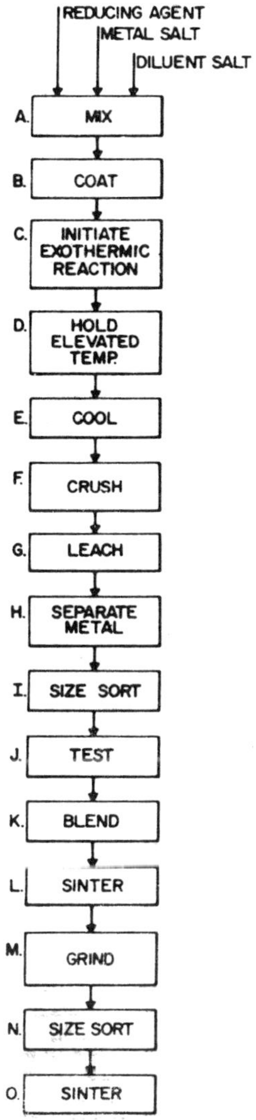

Source: U.S. Patent 3,992,192

vessel or by tumbling the vessel, to ensure homogeneous admixture of the charge during the addition of sodium. The molten sodium permeates through the charge and coats salt particles therein through this mulling operation. The stirring is such as to minimize the layer of excess sodium, if any, which would otherwise form on top of the charge.

C. Initiate Exothermic Reaction: External heating is then increased to raise charge temperature to a level which initiates an exothermic reaction, typically 200°-400°C for sodium and K_2TaF_7. The reaction proceeds very rapidly and raises the charge temperature. Agitation of the charge is continued throughout the reaction. Temperature rise due to the exotherm levels off at 600°-800°C in five to ten minutes from initiation of the reaction and the reduction reaction is essentially completed at this point.

D. Hold Elevated Temperature: External heating is however continuously applied thereafter to raise and hold an elevated temperature for a hold period of fifteen minutes to four hours. Agitation of the charge is continued throughout at least an initial portion of the hold period, preferably throughout the hold period.

The time and temperature of the hold period are adjusted to control particle size, leakage and flow properties of metal produced in the reaction. The percent of fine particles (−325 mesh) obtainable from the charge is reduced with increased time and temperature of hold. This reduction of fines may improve flow properties of the main mass. The hold period is typically up to three hours at 900°-1000°C, when producing tantalum powder for electrolytic use.

The charge agitation prevents formation of skeletal metal-salt or metal structures larger than about 0.5″ and limits them to smaller coarse powder agglomerates within the charge and continuously recirculates these powder agglomerates and smaller metal powder particles initially formed in the exothermic reaction together with K_2TaF_7 and NaCl diluent (if used) so that homogeneous distribution of all species is fostered. The recirculation action includes lifting of materials and does not involve substantial chopping or grinding.

The metal powders and powder agglomerates are suspended within the molten salt mass and recirculated therein by the agitation to prevent gradations of reactant concentration at different height levels within the melt.

E. Cooling: Stirring and external heating are terminated, and the charge is cooled. The internal stirrer, if any, is preferably withdrawn from the charge before cooling so that such stirrers will not be immobilized.

F.-H. Crushing, Leaching, Metal Separation: The reaction products are crushed to produce particles of the reaction products. The particles are leached through several cycles to remove salt products of reaction from the metal, leaving metal powders having a spectrum of size distribution therein.

I.-K. Size Sort, Test and Blend: After leaching the powder, extremely fine particles may be removed. Portions of the powder are processed through to a prototype of final usage to determine powder properties and lots may be blended to achieve the desired stock. This process allows elimination or reduction of much of the usual sorting, testing and blending in certain instances.

L.-N. Sinter, Grind and Screen, i.e., Thermal Agglomeration: The natural agglomeration of the metal powder and its flow properties may be enhanced by presintering, a known process per se which comprises lightly sintering the powder into a block and grinding the block. The resultant agglomerated powder may be screened to utilize desired fractions.

O. Sinter or Other Final Usage: The powders may be compressed into blocks and sintered or poured into a mold and sintered there to form a coherent porous structure, usable as filters, anodes or cathodes of electrolytic cell devices, catalysts or catalyst carriers, resistors and other devices dependent on controlled porosity and high internal surface area. In some usages, such as capacitor anodes, the sintered block is electrolytically anodized to form a dielectric oxide coating over the internal surfaces of its pore structure and then impregnated with electrolyte, contacted with a counter-electrode and packaged to provide a capacitor.

Reduction of Gaseous Halides with Seeding

E.K.A. Svanstrom; U.S. Patent 3,832,157; August 27, 1974; assigned to Rederiaktiebolaget Nordstjernan, Sweden describes a process for producing a powdered metal product selected from the group consisting of V, Nb, Ta, Cr, Mo, W and Re by the reduction of a gaseous halide of the metal in a reaction chamber. The process is accelerated in the reaction chamber by means of a seed-initiating reaction, which comprises injecting into the gaseous halide a small but effective amount of preheated hydrogen heated to at least the reaction temperature while feeding the gaseous halide into the reaction chamber.

It is not necessary to add crystallization seed substances or seed-forming ones as such, but it is also possible to form or initiate such seeds in the reaction mixture. One can, for instance, add hydrogen as a seed-initiating gas preheated to reaction temperature in small but effective amounts, the addition taking place continuously or by pulses to the reaction so that locally conditions for a more rapid reaction to grains of metal or metal compounds are achieved, which grains serve as crystallization seeds.

It is also possible to bring the reaction mixture to considerably increased temperature to initiate the formation of seeds, for instance, by electrical discharges or sparks so that a more rapid reaction to grains of metal or metal compound takes place locally, the grains also in this case serving as crystallization seeds, especially at the production of metal carbide powder. As an example of such a carbonaceous compound, carbon tetrachloride can be mentioned.

Example 1: A mixture of $TiCl_4$, CH_4, H_2 and WCl_6 in the molar ratio 1:1:5:0.009 was introduced in gas phase into a reaction chamber which was maintained at a temperature of 1300°C. From the titanium carbide formed at the reaction 92% was obtained in the form of a powder having a total carbon content of 20.03%, of which free carbon was 0.50%.

When the reaction was carried out under the same conditions with the same gas mixture which, however, did not contain a small but effective amount of WCl_6 as seed-forming substance, about 5% of the titanium carbide was obtained in the form of a powder.

Example 2: The test in Example 1 was repeated but, instead of WCl_6, a small but effective amount of $MoCl_5$ was used as seed-forming substance. 70% of the titanium carbide formed was obtained in the form of a powder. The powder contained totally 20.20% carbon, of which 0.55% was free carbon.

Example 3: A gas mixture of $TiCl_4$, CH_4, H_2 and CCl_4 in a molar ratio of 1:1:10:0.01 was reacted in a reactor at a temperature of 1300°C. Titanium carbide was formed, of which 57% was obtained in the form of a powder. In this case, the powder, however, was well-defined crystals different from the powders obtained in the previous examples, the crystals having quite another form in comparison with the grains of the powder obtained when WCl_6 and $MoCl_5$ were used as seed-forming substances.

Magnesium Granules and Alloys

I.A. Barannik, G.I. Belkin, V.I. Borodin, N.A. Voronova, O.N. Romanenko, A.P. Bogdanov, V.S. Belosludtsev, V.D. Marfin, N.Y. Shish, A.F. Shevchenko, I.Y. Emelyanov, J.N. Boiko, M.N. Kljuchnik, N.V. Khaber, V.G. Raskatov, A.F. Trukhin, A.B. Kondratenko, V.D. Yazev, V.S. Alontsev, V.A. Startsev and J.A. Sergeev; U.S. Patents 3,881,913; May 6, 1975 and 3,969,104; July 13, 1976 have developed a method of producing magnesium-containing mixtures, in the form of granules of a particular shape. They are made by the introduction of a salt to liquid metal at a temperature of 670°-720°C, heated to a temperature of 670°-730°C, composed of metal chlorides and fluorides not reducible by magnesium, taken separately or in combination, and having an initial crystallization temperature below that of the granulated metal, and a density at the temperature of 670°-730°C equal to 0.95-1.2 of the density of the metal.

The molten mixture is atomized by centrifugal force with subsequent cooling of the thus produced discrete particles by air flow. The mixture obtained in producing granules of magnesium or its alloys comprises in weight percent: 65-97 of granulated magnesium, 0.3-16 of magnesium and alkali metal chlorides, 0.1-10 of one of the fluorides, up to 3 of alkaline-earth metal chlorides, and up to 6 of magnesium oxide. The above mixture can also comprise in weight percent: 60-97 of granulated magnesium alloy, 0.3-16 of magnesium and alkali metal chlorides, 0.1-10 of one of the fluorides, 0.3-10 of alkaline-earth metal chlorides, and not over 5.0 of magnesium oxide.

The method is universal because it permits the production of a wide range of magnesium-containing granulated materials able to meet the requirements of diverse consumers, metallurgical, foundry and chemical industries. Thus, mixtures obtained during the production of granules of magnesium and its alloys are successfully used in desulfurization and inoculation of cast iron with a higher efficiency than with the use of magnesium powders produced by the milling method.

The mixtures produced according to this process are looser in nature, have a bulk weight of 0.9 to 1.2 g/cm^3, carry no explosion hazard, while the temperature of their self-ignition is 400°-500°C higher than that of magnesium powders produced by the milling method.

Example 1: A chamber-enclosed centrifugal granulator is employed to pulverize liquid electrolytic magnesium having a temperature of 720°C with a rotational speed of the perforated cup of 1,000 rpm. In the course of being poured into the granulator, the magnesium receives 0.5% by weight of the salt addition heated to a temperature of 730°C. The salt addition is of the following composition in percent by weight: 10.0 $MgCl_2$, 48.0 KCl, 39.08 NaCl, 2.0 $CaCl_2$, 0.02 MgO.

Liquid particles of the magnesium-salt mixture flung through the holes in the rotating cup by the action of the centrifugal force are cooled in their flying trajectory by the counter air flow drawn through the chamber by means of a fan.

After becoming solidified, the granules produced thereby consisted in 80% of elongated fused particles in the form of commas or strings whose diameter in the thickest portion was 0.4 to 2 mm and the length 1 to 15 mm.

The above product has been analyzed for the content of chlorion (Cl) (an indicator of the residual content of chlorides in granules of magnesium) and magnesium oxide. The results of the analysis revealed that this product included percent by weight: 0.21 Cl, 2.8 MgO and 0.1 F.

Example 2: The process conditions are similar to those described in Example 1, but the amount of the salt addition is increased to 2.1% by weight.

Granulated magnesium produced thereby consisted of 60% of elongated granules shaped as described in Example 1, while the remaining 40% were spherical or ellipsoidal granules 0.4 to 2.5 mm in diameter coated with a thin salt film. Granulated magnesium produced thereby had a content of 1.1% by weight: 1.1 Cl and 3.2 MgO.

Example 3: The process conditions are similar to those described in Example 1, but the amount of the salt addition is increased to 2.6% by weight.

Granulated magnesium produced thereby consisted of 55% of spherical or ellipsoidal granules, while the remaining 45% were elongated granules shaped as described in Example 1. Granulated magnesium produced thereby had a content in percent by weight of 1.5 Cl and 2.5 MgO.

Example 4: The process conditions are similar to those described in Example 1, but the amount of the salt addition is increased to 20% by weight.

Granulated magnesium produced thereby included 1.5% of elongated granules, while the remaining granules were of regular spherical or ellipsoidal shape. The granules are coated with a thin salt film. Apart from metal granules the product contained about 6% of spherical salt addition particles. The results of the chemical analysis of the product obtained in percent by weight are 9.5 Cl, traces of MgO.

Example 5: The process conditions are similar to those described in Example 4, but the salt addition contains 0.2% by weight F.

Granulated magnesium produced thereby included 95% of ball-shaped granules, 0.4 to 2.5 mm in diameter. The results of the chemical composition analysis in percent by weight are 8.3 Cl, 0.03 F, MgO not found.

Example 6: Several versions of granulated magnesium and its alloys produced by similiar methods as described in the examples were subjected to a comparative test of properties. Actual compositions of the samples are given in Table 1.

Table 1

	Sample 1	Sample 2	Sample 3	Sample 4	Sample 5
	.... Content in Weight Percent.				
Composition					
Granulated Mg	96.0	80.0	–	89.1	92.5
Granulated Mg alloy	–	–	96.0	–	–
Chlorine ions	0.21	9.5	0.24	5.2	4.0
Fluorine ions	0.1	–	0.1	0.1	0.15
Magnesium oxide	2.8	–	3.0	–	0.1
Shape of particles	elongated	spherical ellipsoidal	elongated	spherical ellipsoidal	spherical 85% elongated 15%
Ignition temp, °C	650	1000	620	920	900
Bulk weight, g/cm^3	0.83	1.0	0.91	1.2	0.98

The materials were tested for corrosion resistance in a humid atmosphere for which purpose they were kept above water in an desiccator for 12 hours. After the 12-hour period of keeping the samples above water they were dried and their moisture content was determined by the change of their weight before and after the treatment in the desiccator. Samples No. 2 and 4 revealed a weight increase of 12-15%, samples No. 1 and 3, 3%; sample No. 5, 3.5%.

Samples No. 2, 4 and 5 were tested in the desulfurization of cast iron. The process of desulfurization was carried out in a 60-ton capacity ladle by injecting magnesium or magnesium alloy granules into the cast iron by compressed air with the help of a tuyere.

The amount of the product used in all tests was equal to 0.35 kg/ton of cast iron. The initial sulfur content in cast iron, 0.045% by weight. The results of desulfurization are given in Table 2.

Table 2

Results of Desulfurization	Sample No. 2	Sample No. 4	Sample No. 5
Final sulfur content in cast iron in wt %	0.03	0.03	0.026
Character of desulfurization	*	*	**

*vigorous boiling with liberation of a large amount of smoke
**process reaction less vigorous

Electric Arc to Prepare Titanium, Zirconium and Nickel

R.R. Akers; U.S. Patent 3,975,184; August 17, 1976; assigned to Westinghouse Electric Corporation has developed a process for producing high quality powders wherein an electric arc which is struck between an electrode and the surface of a pool of molten material rotates under the influence of a magnetic field to thereby free liquid particles from the surface of the pool. The particles produced are quenched in an adjacent inert atmosphere or solidified in a vacuum and become powder. This powder is then collected conveniently.

It would be advantageous if material which comes in contact with the molten particles were not reactive, so that relatively pure, high quality powders could be produced. It would also be advantageous if a nonconsumable electrode, especially of the arc moving type, could be utilized as part of the apparatus.

In the process the electrode is disposed in a chamber for producing powder. The chamber may comprise suitable metallic walls which may be water cooled in some embodiments. A crucible or similar means is provided in proximity to the electrode and the crucible is provided with molten material or solid feed stock which may range from low or medium quality machine shop chips or scrap to high quality material. The main electrode may supply the heat for melting or an auxiliary device may be provided which melts these materials to place them in a molten state thus forming a relatively shallow pool of molten material.

The shallow pool of molten material is electrically connected in circuit relationship to one terminal of a suitable power supply. The previously described electrode is connected in circuit relationship to a power supply terminal of different voltage value. Consequently, an electric arc may be struck between the nonconsumable electrode and the surface of the shallow pool of molten material. In some embodiments, a magnetic field generating means may be provided proximate the tip of the nonconsumable electrode and/or proximate to the crucible and/or proximate a portion of the furnace wall.

The magnetic field producing means, if an electromagnet, may be provided with electrical current of sufficient magnitude to produce magnetic field lines of flux in the gap or space between the electrode and the surface of the molten pool. The magnetic lines of flux may be made to have a strong component generally parallel to the bottom surface of the electrode and to the top surface of the pool. The electric arc may interact with the generally perpendicular magnetic lines of flux to cause the arc to move across the surface of the pool in a generally rotational manner.

The plowing effect of the arc may cause particles to splash away from the surface of the molten pool. In addition, other phenomena such as arc spin, centrifugal force, arc pressure and JXB force (Lenz's law) may also significantly contribute to the overcoming of the surface tension of the pool of molten material to free droplets of molten material which may then be accelerated to or otherwise provided to a region outside of the area between the electrode and the pool. The particles may be quenched by a suitable atmosphere thus providing solid granules of high quality powder.

Figure 7.5 shows an embodiment of the process utilizing a water cooled, nonconsumable electrode **E1**. It should however be understood that a graphite electrode may also be utilized under certain circumstances. There is provided a chamber or furnace member **10** having a nonreactive inner wall **12**, which may be for example a stainless steel inner wall. By nonreactive it is meant that the materials will not react with the commonly known powder products produced in the furnace. There is provided in the inner region of the chamber or vessel **10** a crucible **14**.

In a preferred embodiment the crucible **14** is water cooled, electrically and thermally conductive and nonreactive at the interface with the molten material produced during powder production. Adjacent to the wall of the cooled crucible a skull or liner of hardened or semihardened molten material will usually form a barrier or interface which will prevent contamination. A water cooled copper crucible supplied with an electromagnetic field producing means is utilized in the preferred embodiment. Disposed above the crucible **14** is the nonconsumable electrode **E1**.

The **E1** may be supported in or by a member **30** which may be outside of the furnace or chamber **10**. In some embodiments of this process, the electrode **E1** may be a rotating electrode and the apparatus **30** may be used generally to cause the electrode **E1** to rotate, spin or move generally parallel to the surface **SF** of the pool **P**. In a preferred embodiment the length **L** to diameter **D** ratio of the crucible **14** is small to encourage maintenance of a shallow molten pool **P**.

In other embodiments of the process, the apparatus **30** is mechanically utilized to move the electrode **E1** into and out of the chamber in a vertical manner for striking the arc, establishing the arc length, allowing for material to be fed for installation, maintenance and the like. The electrode **E1** may preferably be of the Durarc type. There may be provided a suitable sealing means **28** in a region where the electrode **E1** passes through the top cover **26** of the chamber **10**. There may be also provided an auxiliary heating unit **29**. A suitable power supply **31** is provided to be connected in electrical circuit relationship with the crucible **14** and the electrode **E1** so that the previously described arc **A** may be struck therebetween.

Figure 7.5: Electric Arc to Prepare Titanium, Zirconium and Nickel

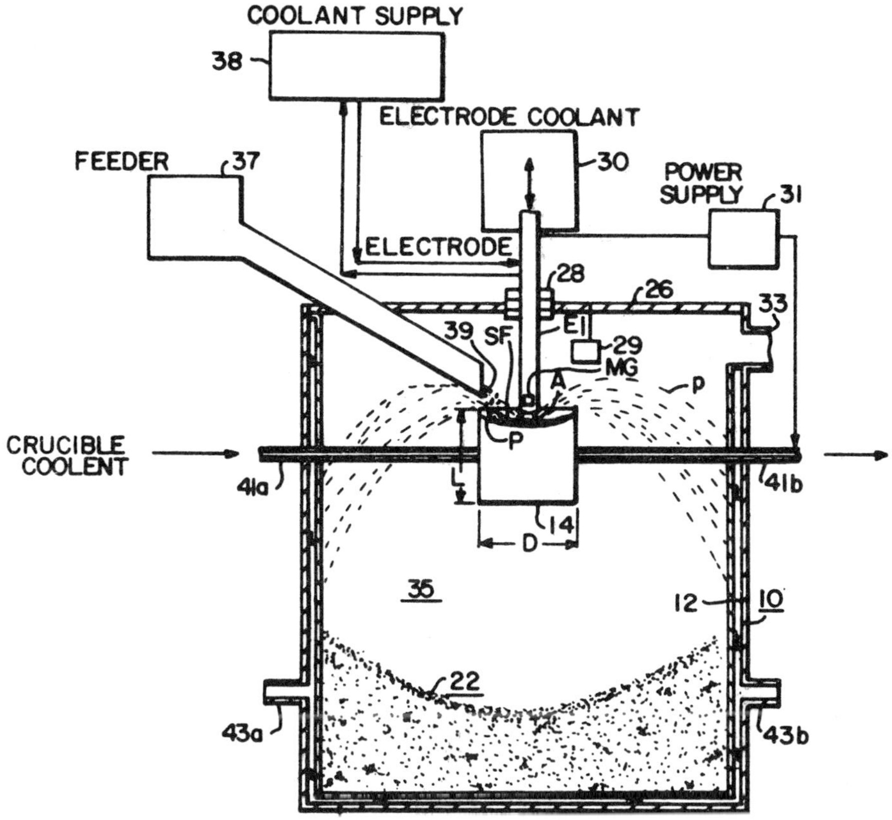

Source: U.S. Patent 3,975,184

There is within the crucible **14** a pool **P** of liquid molten material such as metal. It is on the surface **SF** of this pool **P** that the arc **A** plays. There may be provided an input port **33** to the chamber **10** where a vacuum fixture may be attached for evacuating the chamber or maintaining a subatmospheric pressure or through which gas **35** may be introduced into the chamber **10**. The chamber **10** may be of the sealed variety so that the gas **35** may not escape therefrom. The gas **35** may be pressurized in some embodiments.

There may also be provided a feeder **37** with means for providing feedstock material **39** to the vicinity of the pool **P**. A coolant supply **38** may be present which supplies electrode coolant such as water to the electrode **E1** with a suitable return. The crucible **14** may be externally cooled with a sufficient and suitable coolant such as water by way of input and output ducts **41a** and **41b**. In a similar manner, the chamber **10** may have hollow or ducted walls therein such that cooling paths **43a** and **43b** may be provided thereto for cooling the walls of the chamber **10**.

The particles **p** are shown being expelled or accelerated away from the surface **SF** of the melt **P**. In the region between the wall **12** and the surface **SF** of the pool **P**, the particles may be quenched by the gas **35** so that the particles strike the wall **12** as solid, semisolid or liquid particles, whereupon the particles may fall downwardly due to the force of gravity, cool, and collect, for example, at region **22**. It is desirable to produce the particles **p** as efficiently as possible. An auxiliary heater **29** may be provided to heat the melt **P** to remove collar from around the edge of the melt **P** so that the particles **p** may have a generally unobstructed passage to the wall **12**.

In addition, the material supplied to the melt by the feeder **37** may, in some embodiments of the process, be preheated and supplied to the crucible **14** as melted or molten material. In one embodiment of the process, it is envisioned that machine shop chips, broken turbine blades or the like may be provided as scrap pieces by way of the feeder **37** or otherwise to the molten pool **P** where they may be melted by the heat of the pool **P** or otherwise. This allows for the quick and inexpensive recovery of scrap for the production of high quality powder.

It has been found that the distance from the point where the particles are expelled from the surface **SF** of the pool **P** to the wall **12** should be sufficiently large to allow the particles to be quenched or cooled and solidified before they strike the wall. In some embodiments however, it is desirous to utilize the apparatus to produce splat or splatter material that is flat material with large relative surface area. This may be accomplished by allowing liquid particles to strike the wall.

It has also been found that it is advantageous to place the electrode **E1** as close to the melt as possible because the magnetic flux density around the bottom of the electrode **E1** is more intense closer to the tip surface of the electrode **E1**. This means that the rotation of arc spots in a space local to the surface **SF** of the melt **P** will be increased due to the relatively high magnetic flux density. The faster the arc **A** and the arc spots are accelerated, the more force will be applied to splash, throw or otherwise accelerate particles **p** away from the surface **SF** of the pool **P**.

Arc spots are defined as those volumes of melt near the surface **SF** of the pool **P** where the arc root attaches at any instant of time. It has also been found that it is advantageous to have the pool **P** as shallow as possible. This maintains the volume of molten material low, consequently the energy of the arc which is applied to expel particles **p** from the surface of the pool **P** is utilized almost exclusively for that purpose and not for moving the entire volume and mass of the melt **P**. It has been found that it is not necessary to move the entire volume of the pool **P** to achieve powder production. To the contrary, if the melt **P** is of a large volume, the pool **P** may oscillate causing large waves and perturbances to exist within the pool **P** thus leading to less efficient production of powder.

It has also been found to be advantageous to use decreased or lowered arc current while concurrently using an increased component of magnetic field current. In this embodiment the magnetic field producing means is in the electrode **E1**. The component of arc current is decreased to reduce the heating of the pool **P** which in turn causes a shallower pool **P** to be formed. Most of the energy of the arc **A** therefore is utilized in freeing particles **p** from the surface **SF** of the pool **P**. Since the electrical current flowing in the magnetic field generating apparatus or electromagnet **MG** is related to the amount of electromagnetic flux produced, more flux will be produced if the magnetic fluid is strengthened.

The increase of magnetic field current has a compensating effect for the decrease of the arc current. Consequently, the force associated with the rotating arc **A**, namely the JXB force is maintained or enlarged or raised even though arc current is lowered. The pool **P** may be kept shallow by supplying coolant material to the crucible **14** through the ducts **41a** and **41b**. It has been found that the pressure of the gas **35** within the chamber **10** affects the formation of powder particles. The effect may be twofold.

In one instance, the pressure of the gas causes the arc **A** to be relatively narrow and dense causing high speed very forceful local pool surface agitation. In a second instance, the pressure is exerted against the molten particles **p** as they move through the quenching gas **35** consequently causing the particles to be solidified more quickly. It has also been found that it is advantageous to keep the particles as impurity-free as possible.

It is advantageous in many instances to avoid having reactive agents or elements in the chamber or furnace. It is advantageous to provide a nonconsumable electrode **E1** constructed of generally copper alloy material as well as a copper crucible **14** and a stainless steel side wall **12** for the furnace **10**. In another embodiment of this process the electrode **E1** may also be rotatable.

The furnace **10** is adapted for the production of powders of titanium, zirconium, high temperature nickel-based superalloys, high temperature ferrous alloys and other materials. In the preferred embodiment, the side walls **12** of the furnace may be stainless steel to provide for a nonreactive surface. Also, it has been found advantageous in some instances to provide a gas **35** which is chemically reactive with some materials which may be found on scrap material for the purpose of removing these contaminating materials from the scrap material.

As an example, a hydrogen atmosphere may be utilized under certain circumstances to reduce the oxygen content of stainless steels. It may also be advantageous to provide a gas **35** to chemically react with the molten particle **p**. It is desirable in some embodiments of this process to provide a plasma jet for the auxiliary heater **29**.

Welding Powder for Build-Up Welding

H. Beyer and U. Buran; U.S. Patent 3,890,137; June 17, 1975; assigned to Goetzewerke-Friedrich Goetze AG, Germany have found that a welding powder comprising a first component selected from the group consisting of molybdenum powder, tungsten powder, and a mixture of molybdenum powder and tungsten powder and a second component of at least one element selected from the group consisting of oxygen, nitrogen, hydrogen and carbon can be used as the material for build-up welding and that this material produces excellent wear-resistant surfaces. The oxygen, nitrogen, hydrogen or carbon is present in the metal of the welding powder partly in an unbound, dissolved state.

The amount of the first component is between about 50 to 100% by weight, preferably 80 to 95% by weight. When a mixture or an alloy of tungsten and molybdenum powders is used, the weight percent of each metal in the mixture may vary from 0 to 100, but it is preferred to use mixtures containing 50 to 100 weight percent molybdenum, with the remainder tungsten.

The average particle size of the tungsten and molybdenum powders that are used in the welding powder of this process can be between about 40 to 200 μm, preferably about 60 to 75 μm. The final welding powder with its various constituents also preferably has this average particle size. There is intimate contact with no porosity between these constituents of the coating with another and with the surface of the base material.

When the oxygen is present in the welding powder, the oxygen content of the welding powder is between about 0.001 and about 12% by weight, preferably 3.9 to 4.5% by weight, based on the weight of the welding powder. Generally, at least 1 weight percent of this oxygen is unbound, and the welding powder contains from about 0.00001 to about 0.12 weight percent, preferably 0.039 to 0.045 weight percent, unbound oxygen.

The oxygen is introduced into the welding powder by spraying the melted powder in an air stream and the desired amount of unbound, dissolved free oxygen is obtained in the welding powder by the partial pressure of the oxygen. The presence of free oxygen makes the welded layer harder.

When nitrogen is present in the welding powder, the nitrogen content of the welding powder is between about 0.001 and about 0.5% by weight, preferably 0.003 to 0.006% by weight, based on the weight of the welding powder. Generally, at least 1 weight percent of this nitrogen is unbound, and the welding powder contains from about 0.00001 to about 0.005 weight percent, preferably 0.00003 to 0.00006 weight percent, unbound nitrogen.

The nitrogen is introduced into the welding powder by spraying the melted powder in a nitrogen atmosphere and the desired amount of unbound, dissolved free nitrogen is obtained in the welding powder by the pressure of the nitrogen. The presence of free nitrogen makes the welded layer harder.

When hydrogen is present in the welding powder, the hydrogen content of the welding powder is between about 0.001 and about 0.5% by weight, preferably 0.003 to 0.01% by weight, based on the weight of the welding powder. Generally, at least 1 weight percent of this hydrogen in unbound, and the welding powder contains from about 0.00001 to about 0.005 weight percent, preferably 0.00003 to 0.0001 weight percent, unbound hydrogen.

The hydrogen is introduced into the welding powder by spraying the melted powder in a hydrogen atmosphere and the desired amount of unbound, dissolved free hydrogen is obtained in the welding powder by the pressure of the hydrogen. The presence of free hydrogen makes the welded layer harder.

When carbon is present in the welding powder, the carbon content of the welding powder is between about 0.01 and about 2.5% by weight, preferably 0.1 to 1.5% by weight, based on the weight of the welding powder. Generally, at least 30 weight percent of this carbon is unbound, and the welding powder contains from about 0.003 to about 0.8 weight percent, preferably 0.03 to 0.5 weight percent, unbound carbon. The carbon can be introduced into the welding powder as amorphous or graphitic carbon.

Carbon makes the layer harder and gives to it better lubrication properties. Either only one of the elements of oxygen, nitrogen, hydrogen or carbon can be provided in the welding powder, or any combination of two or more of these elements can be provided, including all four elements. Preferably, when providing only one element, that element is oxygen; and when providing more than one element, the elements are oxygen and carbon.

The welding powder can optionally contain copper or copper alloys, in an amount of about 0.01 to about 20% by weight, preferably 5 to 10% by weight. The copper or copper alloy can be mixed or alloyed into the welding powder.

Porous Sintered Filter for Powder Recovery

R.L. Moss; U.S. Patent 4,080,431; March 21, 1978; assigned to PPG Industries, Inc. has found that finely divided refractory hard metal can be recovered substantially free of adsorbed volatile inpurity, e.g., unreacted metal halide, in one step without the problem of classification by particle size and without the need for prolonged calcining. According to this process finely divided borides, carbides, and nitrides of metals of Groups III-VI of the Periodic Table are formed in a reactor at high temperatures, i.e, boride, carbide, and nitride forming temperatures, by reaction in the vapor phase of a vaporous metal halide with a vaporous source of boron, carbon, or nitrogen.

The finely divided product is withdrawn from the reactor entrained in a gaseous reactor effluent stream and is separated from the effluent stream by passing the stream through a porous metal or refractory filter in a filtration zone, the temperature of the product and the effluent stream being maintained above about 200°C until they leave the filtration zone. The product is transferred from the filtration zone to a heated receiver purged with inert gas, e.g., argon. The filter removes substantially all of the finely divided product from the effluent stream, thereby avoiding classification of the product by particle size and the need for a final bag filter.

Temperatures in the reactor may vary between about 300° and 3000°C, and the powdery product is separated from the effluent gas stream at temperatures between about 200° and 1500°C or higher, for example between about 200° and 600°C, more typically between about 250° and 370°C. Some cooling of product and effluent stream may occur between the reactor and the filter unit, but the temperature is maintained at least at 200°C at all times until after the product and the effluent gas are separated.

Separation of product from the effluent stream at high temperatures, e.g., at least about 200°C, rather than at lower temperatures lowers the amount of adsorbed impurity on the product. Purging of the receiver serves to limit the diffusion of gaseous impurity from the filter unit to the receiver, but if the receiver is much cooler than the filter unit, an excessive volume of purge gas may be required to limit diffusion adequately. Thus, it is preferable to heat the receiver during the period when it is open to the filter unit to at least about 200°C, preferably to at least the temperature in the filter unit or higher.

The ultimate particles of the products produced by the vapor phase processes are typically submicron in size and range from about 0.01 to about 1.0 μ. The finely divided product that is collected is typically in the form of agglomerates of ultimate particles, typically between about 0.1 and 10 μ or more in nominal sectional diameter.

Submicron titanium diboride (TiB_2) having a B.E.T. [Brunauer, Emmett, and Teller, *Journal of Americal Chemical Society,* 60, 309 (1938), N_2 adsorption method] surface area of 9.1 m^2/g, at least 50 weight percent of its particles in range of about 0.05 to 1.0 μ in nominal sectional diameter, about 95 number percent of its particles below about 1.0 μ and containing coproduced submicron titanium carbide was prepared by vapor phase reaction of $TiCl_4$, BCl_3, and 1,1,2-trichloroethane in accordance with the process described in *Offenlegungsschrift* No. 25 23 423. Powdery TiB_2 entrained in the reactor effluent gas stream was separated from the effluent stream at 277°-305°C and 4.0 lb/in^2 gauge in a 7-bayonet filter unit.

The gas flow rate was 300 actual $ft^3/hr/ft^2$ of filter area, and TiB_2 was accumulated at the rate of 0.78 $lb/hr/ft^2$ of filter area. Substantially no TiB_2 was detected in the effuent gas stream flowing from the filter unit. The filter material was Mott sintered nickel having a wall thickness of 1/16″ and a nominal spherical particle retention rating of 5.0 μ. The filter housing was vibrated two to three times per minute for about 1 second each time, and one bayonet was cleansed by a reverse flow of hydrogen for 0.5 second every 250 seconds in rotation.

The product dislodged from the filter bayonets dropped into an argon purged receiver which was maintained at 285° to 299°C during the collection period, about 5 hours, and then allowed to cool. An argon atmosphere was maintained in the receiver during cooling. Upon analysis, the product from the receiver was found to contain 0.08 weight percent chlorine.

General Processes and Applications

MIXED METAL SYSTEMS

Composite Metal Powder

K. Tamura and T. Takeda; U.S. Patent 3,963,811; June 15, 1976; assigned to National Research Institute for Metals, Japan have developed a process for producing a composite metal powder which comprises mixing a powder of a metal or alloy with a melt of a metal or alloy of a different class from that of the metal or alloy powder and atomizing the resulting molten mixture by jetting a high speed jet stream of water against the stream of the mixture to thereby obtain a composite metal powder in which the particle surface of the metal or alloy powder is coated with a layer of the originally molten metal or alloy. It is possible according to the process to use an exceedingly great variety of combinations of the nucleate metal or alloy powder and the coating metal or alloy. Preferred combinations are illustrated in the following table.

Nucleate Metal or Alloy	Coating Metal or Alloy	Composition (%)	
Copper*	Aluminum**	–	–
Chromium or ferrite-type ferrochrome	Iron	Cr	15-25
Ni-Cr-Fe***	Iron	Cr	12-25
		Ni	4-20
Ni-Ti-Al-Mo-Cr-Co	Ni	Ti	1-5
		Al	1-5
		Mo	1-5
		Cr	10-20
		Co	10-20

(continued)

Nucleate Metal or Alloy	Coating Metal or Alloy	Composition (%)	
Ni-Ti-Al-Mo-Cr	Ni-Co	Ti	1-5
		Al	1-5
		Mo	1-5
		Cr	10-20
		Co	10-20
Copper-tin	Copper	–	–
Copper	Tin	–	–

*2 to 10%, preferably 5%.
**98 to 90%, preferably 95%.
***Austenite-type ferrochrome.

The metal powder (or alloy powder) used as the nucleus of the composite metal powder is that having preferably a particle size not greater than that passing through a 200-mesh sieve, and more preferably not greater than that passing through a 350-mesh sieve.

The amount of metal or alloy powder fed to the melt varies depending upon the dimension of the powder to be added. However, this amount is preferably not more than about 50% of volume, and more preferably 20-30% by volume.

For atomizing the continuous stream of the mixture of the melt and the metal or alloy powder to obtain a composite metal powder, a high speed fluid is jetted against the continuous stream of the molten mixture. Water, air, nitrogen, argon and mixtures of air and steam are suitably used as the high speed fluid. However, the usable fluid is not limited to these fluids. Particularly preferred is water.

The high speed fluid is conveniently jetted at a speed of 30 to 200 m/sec and in an amount of 20 to 400 l/min.

Any of the known mixing procedures may be used in mixing the metal or alloy powder with the melt. However, it is convenient to carry out the mixing in the following manner. The melt is caused to flow as a continuous stream, and preferably as a thin layer. The metal or alloy powder is then dropped continuously into the melt from a slit disposed above the continuous stream of the melt.

Figures 8.1a through 8.c are views illustrating the procedures for mixing the melt and the powder in accordance with the process. Figure 8.1a illustrates one mode in which the metal or alloy powder is mixed in with the melt flowing in one direction. Melt **1** flows at a rate of 100-200 g/sec, e.g., in a given direction in a vessel **2** formed from a refractory such as alumina, etc., and after its level has been adjusted by means of a slit **3** provided in the vessel **2** at a point just prior to where the mixing is performed, it is admixed with a metal or alloy powder **5** that has been dropped uniformly from a metal or alloy powder

Figure 8.1: Apparatus for Mixing Melt and Powder

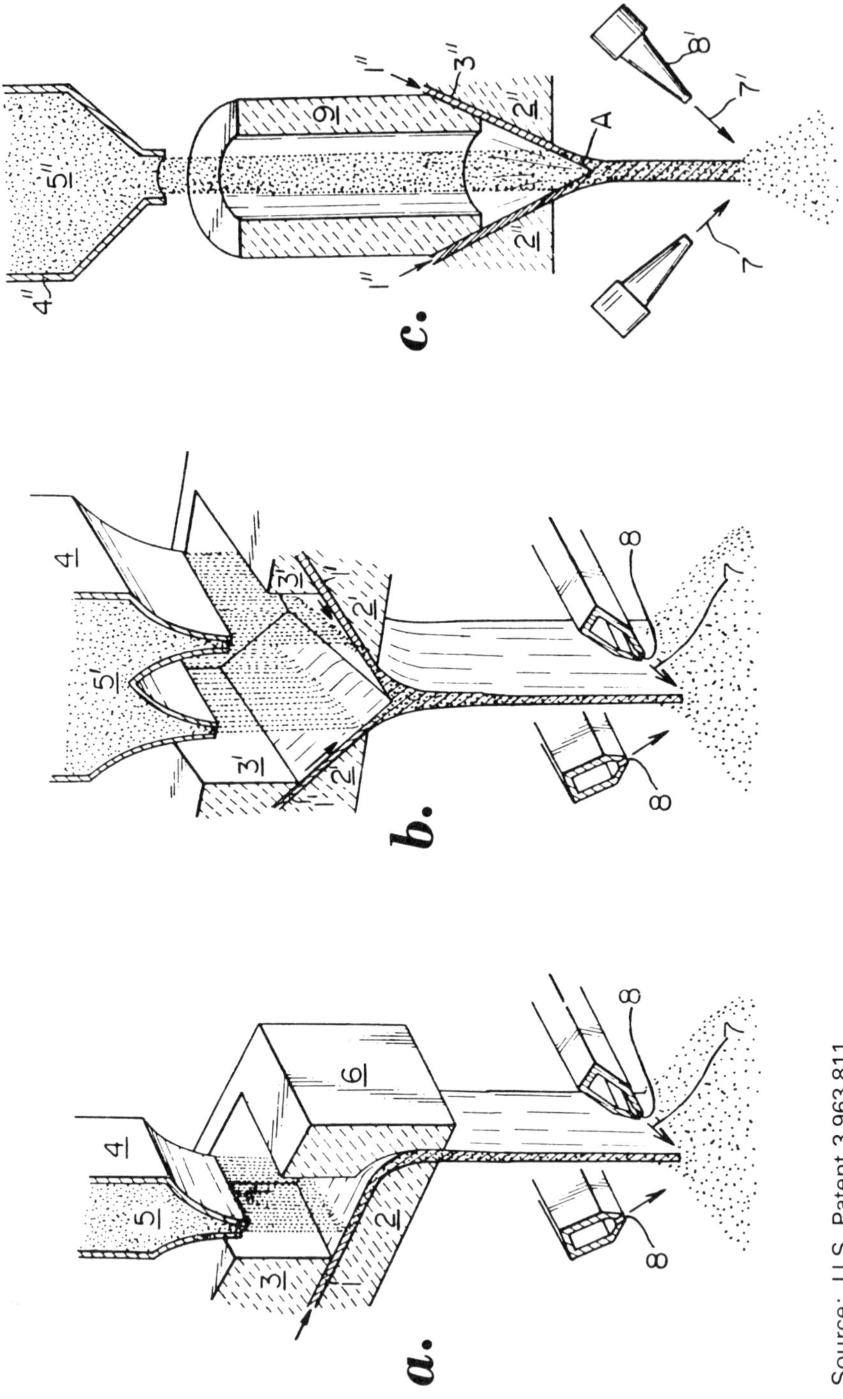

Source: U.S. Patent 3,963,811

feeding tank **4** disposed above the aforesaid vessel **2**, after which the resulting mixture is made to flow down in ribbon fashion by means of a baffle **6**.

This mixture stream is then atomized and cooled with a high speed jet stream **7** or **7′** (Figure 8.1c) from a plurality of nozzles **8** or slits **8′** (Figure 8.1c) to form a composite metal or alloy powder, i.e., a powder having as its nucleus the metal or alloy powder and as its coating the originally molten metal or alloy. The plurality of nozzles or slits that jet the high speed fluid are so disposed that they either incline in the same direction as that of the stream of molten mixture or intersect perpendicular thereto so that the jet stream jetted from the nozzles or slits intersects the stream of molten mixture at a given point.

Figure 8.1b illustrates a mode wherein the mixing is carried out using a plurality of slits for the melt and a plurality of powder feeding tanks. Melt **1** flows through the V-shaped vessel **2′** and, after its level has been adjusted by means of the respective slits **3′** provided at that point just before the part where the mixing is to be performed, is mixed with the powder **5′** that has been uniformly dropped onto the respective melt surfaces from the two feed ports of the powder feed tank **4′** disposed above the aforesaid vessel **2′**, following which the two mixture streams are brought together at the angulate bottom of the vessel and allowed to flow down in ribbon fashion. This mixture stream is atomized and cooled in the same manner as hereinbefore described to obtain the composite powder.

The dropping port of the powder feeding tank shown in Figures 8.1a and 8.1b can be of slit form having a rectangular or wavy shape or in the form of orifices for ensuring that a uniform mixing of the powder and melt stream is achieved.

Figure 8.1c illustrates a mode in which the vessel is of conic shape. Melt **1″** flows down over a vessel **2″** of refractor of material such as alumina formed in conic shape and passes through a slit **3″** formed by means of a spindle **9** to come together at point **A**. A powder feeding tank **4″** is disposed above the point **A**, and powder **5″** is allowed to drop uniformly from feeding tank **4″** to become admixed with the melt. The so obtained mixture stream is then atomized and cooled by means of a high speed fluid jetted from either a circular slit or annularly disposed nozzles **8′** to form the composite powder.

According to this method, the amount of melt fed can very readily be adjusted by a vertical movement of the spindle **9**. On the other hand, when the adjustment of the amount fed of the melt by a vertical movement of the spindle is not necessary, it is possible to provide a plurality of orifices at this part instead of the spindle. For avoiding the solidification of the melt at the slit portion, in all of the foregoing modes, preheating of this portion is performed at the time of starting the operation of the apparatus.

Cu-Ni-Co Dispersion-Strengthened Metals

A.V. Nadkarni and E. Klar; U.S. Patent 3,884,676; May 20, 1975; assigned to

SCM Corporation describe a process wherein a self-contained powder metal mixture of alloy and oxidant is enclosed within a sealed container and the mixture is dispersion strengthened by internal oxidation to form a hard, sintered cake therein; thereafter it is hot-worked and coalesced by extrusion from the same sealed container.

The powder metal mixture of powdered alloy and oxidant is intimately intermixed wherein the alloy comprises matrix metal and solute metal and the oxidant comprises heat-reducible metal oxide and hard, refractory metal oxide. Sufficient oxidant is combined with the alloy in approximately stoichiometric proportions for complete oxidation of the alloy solute metal. A residue of heat-reducible metal oxide produced during internal oxidation becomes dispersion strengthened during coalescence and extrusion whereby a dispersion-strengthened metal product is produced without removing the dispersion-strengthened metal mixture from the container.

The preferred powder alloy comprises a relatively noble matrix metal having a negative free energy of oxide formation at 25°C of up to 70 kcal/g atom oxygen, and a solute metal having a negative free energy of oxide formation exceeding that of the relatively noble matrix metal by at least about 60 kcal/g-atom of oxygen at 25°C. The preferred oxidant comprises an intimate mixture of heat-reducible metal oxide having a negative free energy of formation at 25°C of up to about 70 kcal/g-atom of oxygen, and finely divided hard, refractory metal oxide having a negative free energy of formation exceeding the negative free energy of formation of the heat-reducible metal oxide by at least about 60 kcal/g-atom of oxygen at 25°C.

The heat-reducible metal oxide is present in the oxidant in an amount sufficient for complete oxidation of the solute metal in the alloy. The hard, refractory oxide in the oxidant is present in substantially the same equivalent elemental proportion as the solute metal in the alloy, and both are of a particle size suitable for dispersion strengthening of the oxidant residue resulting from the internal oxidation.

After internal oxidation, the oxidant residue comprises particles of in situ residue of heat-reducible metal oxide and particles of hard, refractory metal oxide uniformly distributed therein and the residue of heat-reducible metal oxide is intimately dispersed within the alloy powder. The dispersion-strengthened metal mixture is eventually coalesced and consolidated by hot-working to form a solid metal workpiece whereby the residue of heat-reducible metal is dispersion strengthened by the hard, refractory metal oxide and forms an integral part of the dispersion-strengthened resulting workpiece.

To achieve the proper proportion of oxidant, about 0.1 to about 10 parts by weight of oxidant are employed per 100 parts of alloy to be internally oxidized. The exact proportions depend on the solute metal to be oxidized, the concentration of solute metal in the alloy, and the oxygen content of the oxidant. The heat-reducible metal oxide is present in substantially stoichiometric proportions for internally oxidizing all of the solute metal in the alloy. At least about 0.1 weight parts of oxidant are combined per 100 weight parts of powder alloy, desirably between about 0.1 to 20 weight parts of oxidant, and preferably about 0.1 to 10 weight parts of oxidant are combined with about 100 weight parts of powder alloy. Preferably, the proportion of the heat-reducible metal oxide and the hard, refractory metal oxide is predetermined so that the composition of the oxidant residue and internally oxidized alloy are substantially identical in the dispersion-strengthened metal after the internal oxidation step is completed.

The amounts of such oxidants to be added may be determined by the stoichiometric amount of oxygen required to oxidize the solute metal completely, but preferably excessive amounts of oxidant are not used so as to avoid leaving more than about 0.1% of reducible oxygen in the internally oxidized metal mixture.

Thus, the internally oxidized metal mixture would not lose greater than about 0.1% by weight after subjecting the same to a hydrogen reduction test. The hydrogen reduction test may be applied to a representative sample of internally oxidized metal mixture by removing the same from the sealed container after the internal oxidation step is completed. The sintered mass is then pulverized, screened, weighed, and then subjected to a hydrogen atmosphere at a temperature of 1600°F for one-half hour. The measured weight loss preferably is no greater than about 0.1% based on the total sample tested.

The metal moiety of the heat-reducible metal oxide in the oxidant preferably is the same metal as the matrix metal present in the alloy to be internally oxidized, although the heat-reducible metal oxide moiety can be different. For instance, alloy matrix metal/oxidant heat-reducible metal oxide combinations include:

ALLOY MATRIX METAL	OXIDANT HEAT-REDUCIBLE METAL OXIDE
copper	cobalt oxide, nickel oxide, copper oxide
nickel	cobalt oxide, nickel oxide, copper oxide
cobalt	cobalt oxide, nickel oxide, copper oxide

Similarly, the hard, refractory metal oxide in the oxidant preferably is the same as the solute metal oxide formed in the alloy during internal oxidation of the alloy, although the refractory metal oxide in the oxidant can be different from the solute metal oxide in the internally oxidized alloy. For example, solute metal oxide/oxidant hard, refractory metal oxide combinations include:

ALLOY SOLUTE METAL OXIDE	OXIDANT HARD, REFRACTORY METAL OXIDE
Al_2O_3	Al_2O_3, BeO, ZrO_2, ThO_2
BeO	Al_2O_3, BeO, ZrO_2, ThO_2
ZrO_2	Al_2O_3, BeO, ZrO_2, ThO_2
ThO_2	Al_2O_3, BeO, ZrO_2, ThO_2

In accordance with this process, the stoichiometrically proportioned alloy and oxidant are contained within a sealed container preparatory to the internal oxidation step. The container must be adaptable for both high temperature internal oxidation and subsequent high temperature extrusion.

The following examples illustrate the process. All parts are by weight and all temperatures are in degrees Fahrenheit, unless otherwise stated.

Example 1: *Preparation of the Alloy Powder*—Electrolytic tough-pitch grade copper rods were melted in an inert refractory crucible in an induction heating furnace under reducing conditions at a temperature of about 2300°F. Metallic aluminum shavings were introduced into the molten copper in the proportion of 0.33% by weight of the resulting molten metallic mass.

The molten solution of aluminum in copper was then super-heated to 2400°F, atomized through an atomizing aperture in a jet of nitrogen to yield an atomized copper-aluminum alloy powder which substantially all passed a 100-mesh U.S. sieve indicating that the average particle size was less than about 140 μ.

Preparation of the Oxidant—One hundred parts of commercially available cuprous oxide (Cu_2O) with an average particle size of about 1 to 2 μ were mixed with 4.1 parts of $Al(NO_3)_3 \cdot 9H_2O$ dissolved in water to form a slurry of cuprous oxide in aluminum nitrate solution. The solution of aluminum nitrate was slurried with cuprous oxide particles, and stirring was continued with mild heating at 200°F until the water had evaporated and the mixture had become almost dry. The mixture was then heated at a temperature of about 500°F for one-half hour to decompose the aluminum nitrate into aluminum oxide. The resulting agglomerate was then ground to form fine oxidant powder which passed a 325-mesh sieve. The resulting oxidant powder comprised 77.43% Cu_2O, 22.01% CuO and 0.56% Al_2O_3 by weight.

Preparation of Powder Metal Mixture—The alloy powder prepared earlier was thoroughly mixed with the oxidant powder prepared above in the proportion of 2.12 parts of oxidant to 100 parts of alloy powder to provide a powder mixture of alloy and oxidant. Mixing was achieved in a ball-mill.

Internal Oxidation—The alloy powder-oxidant mixture was then charged into an internal oxidation vessel. The oxidation vessel was a cylindrical copper container having an overall length of 24 inches, a diameter of 8 inches, and a wall thickness of one-eighth inch. Approximately, 234 pounds of the foregoing alloy-oxidant powder mixture were charged into the container, purged with argon gas, and thereafter sealed leaving a pin hole opening for pressure release.

The alloy oxidant powder mixture was then brought to a temperature of about 1750°F and maintained at this temperature for about 30 minutes to effect internal oxidation of the alloy powder.

At the end of the 30-minute internal oxidation period, substantially all of the aluminum in the alloy powder was oxidized to Al_2O_3 and substantially all of the cuprous oxide in the oxidant had been reduced to metallic copper. The particles of internally oxidized alloy comprises 99.37% by weight of copper plus negligible amounts of impurities and 0.63% by weight of Al_2O_3 and the oxidant residue comprised 99.37% copper particles and 0.63% Al_2O_3 particles. The overall internally oxidized metal powder composition comprised 98.21% internally oxidized alloy powder and 1.79% oxidant residue.

Thermal Coalescence and Extrusion—The container of internally oxidized metal mixture was then placed in a ram-type extrusion press and was extruded to form extrudate in the shape of cylindrical bar stock having a diameter of about 1.125 inches. This corresponds to an extrusion ratio of about 50:1 (i.e., the ratio of the cross-sectional area of the can to the cross-sectional area of the extrudate).

The bar stock was about 99.37% copper having dispersed throughout 0.63% (or about 1.5% by volume) of Al_2O_3 particles. The bar stock had a density of about 99.3% of the theoretical density, an electrical conductivity of 88% IACS, a tensile strength of about 72,000 psi, an elongation of 19% using ASTM Test E-8 (for a test specimen 0.16 inch in diameter and 0.65 inch gauge length) and a Rockwell hardness of about 75 units on the B scale. All property measurements were made at room temperature.

A sample of the bar stock was reduced by 50% in cross-sectional area by cold swaging whereby the tensile strength became 80,000 psi, the elongation 13%, Rockwell B hardness 84 units, and conductivity 86% IACS.

Example 2: The procedures of Example 1 were repeated except that the 8-inch diameter copper can was replaced by a 1.25-inch diameter copper can. The extrusion was carried out in an extrusion chamber of 1.38-inch diameter at an extrusion ratio of 30:1 yielding a 0.250-inch diameter rod. Such a rod had an electrical conductivity of 86.7% IACS, a tensile strength of about 73,000 psi, and an elongation of 19.8% in a gauge length of 0.650 inch.

Example 3: Dispersion-strengthened metal bar stock is formed as indicated in Example 1 by oxidizing 100 parts of a powdered alloy of 98.72% iron and 1.28% aluminum with 3.865 parts of pulverulent oxidant comprising 3.800 parts of iron oxide and 0.065 parts of aluminum oxide, but with the provision that a steel container is utilized for internal oxidation and extrusion. The resulting dispersion-strengthened bar stock has increased tensile strength and hardness at elevated temperatures or after annealing as compared relative to bar stock of a similar iron-aluminum alloy which had not been internally oxidized.

In related work *A.V. Nadkarni; U.S. Patent 3,893,844; July 8, 1975; assigned to SCM Corporation* recrystallizes alloy powder prior to internal oxidation to increase the grain size of the alloy to at least about ASTM (E-112) Grain Size No. 6

to reduce the grain boundary area in the alloy powder. Thus, the recrystallized alloy powder provides substantially improved structural properties in the dispersion-strengthened metal product.

Flaked Metal Powders

M. Megelas; U.S. Patent 4,172,720; October 30, 1979; assigned to United States Bronze Powders, Inc. has developed a method of making flaked metal powders having a narrow particle size distribution, a whiter color and a very high sparkle effect. A heterogeneous liquid system comprising an inert liquid and a lubricant and including a finely divided metal is subjected to attrition in an enclosure in which there are a plurality of attritive elements.

An agitator is moved through the elements to displace those in its path. In this method, the weight ratio of attritive elements to finely divided metal is between 70:1 and 90:1; the weight ratio of finely divided metal to lubricant is between 100:1 to 20:1; and the weight ratio of inert liquid to finely divided metal is between 0.5:1 to 2.5:1. Flaked Al, Cu, brass, stainless steel, nickel, cupronickel powders and the like are obtained by this method.

Example: A flaking apparatus as described in U.S. Patent 3,995,815 was used. The total volume of the container used was 2 gallons. The speed setting for the rotating arm throughout the present test series was kept at 185 rpm to standardize the test conditions. Other speed settings could also be used with slight modifications in the other ratios as may be appreciated by anyone skilled in the art.

The inert fluid used was Varsol, a petroleum distillate fraction having a specific gravity of approximately 0.779. The lubricant used was stearic acid to produce leafing pigments. The feed material used was either atomized or cut foil as described in the abovementioned patent. The attritive elements' size used was also standardized to reduce the number of parameters under consideration. The size was ⅛" or 3.175 mm steel balls.

The time was varied between 5 and 120 minutes. In all cases, it was kept at not more than 120 minutes, as other tests done with longer times produced products which were unsuitable for the present purpose of obtaining a high sparkle. The series of tests made according to the procedure is tabulated in the table below.

Test No.	Balls to Metal (weight)	Liquid to Metal (weight)	Ball to Liquid (volume)	Metal to Lubricant (weight)	Metal	Milling Time (minutes)	Cm^2/g water coverage (−400 mesh)	Colormaster MEECO (whiteness)
1	89	2.167	6.413	20	Al	15	2940	65.31
2	87	2.167	6.269	20	Al	15	2820	66.19
3	85	2.167	6.125	20	Al	15	4940	69.53
4	83	2.167	5.981	20	Al	15	4500	68.03
5	78	2.167	5.621	20	Al	15	4920	68.70

(continued)

Test No.	Balls to Metal (weight)	Liquid to Metal (weight)	Ball to Liquid (volume)	Metal to Lubri-cant (weight)	Metal	Milling Time (minutes)	Cm^2/g water coverage (−400 mesh)	Colormaster MEECO (whiteness)
6	75	2.167	5.404	20	Al	15	2325	63.72
7	70	2.167	5.044	20	Al	15	2560	65.03
8	87	2.0	6.792	20	Al	15	3100	69.59
9	85	2.0	6.636	20	Al	15	3075	66.75
10	75	2.0	5.856	20	Al	15	1980	63.03
11	85	2.0	6.636	20	Al	5	1575	60.63
12	85	2.0	6.636	20	Al	15	2625	69.41
13	85	2.0	6.636	20	Al	30	4650	73.07
14	85	2.0	6.636	20	Al	60	12000	68.94
15	85	2.0	6.636	20	Al	90	17400	69.55
16	85	2.0	6.636	20	Al	120	18300	62.72
17	85	2.0	6.636	20	Al	15	2430	64.28
18	85	2.0	6.636	20	Al	15	2370	61.82
19	85	2.0	6.636	20	Al	15	2835	64.18
20	85	2.0	6.636	20	Al	15	2445	65.71
21	89	2.0	6.947	20	Al	15	1770	63.41
22	83	2.0	6.480	20	Al	15	1860	64.91
23	78	2.0	6.09	20	Al	15	1965	60.11
24	70	2.0	5.47	20	Al	15	1995	62.85
25	85	1.0	13.27	20	Al	15	2790	69.21
26	78	1.0	12.18	20	Al	15	2670	61.77
27	85	0.25	53.08	20	Al	15	5010	67.83
28	78	0.25	48.70	20	Al	15	3840	69.86
29	85	2.0	6.64	20	Ni	15	4470	—
30	85	2.0	6.64	20	Brass	15	1350	—

Ball Diameter ⅛" = 3.0 mm
Temperature = 37°-50° C.

Test No. 1 was repeated by varying the metal to lubricant ratio from 20:1 to 40:1 to 60:1 to 80:1 to 100:1. No appreciable differences were observed in the resulting product. Test No. 2 was repeated by varying the attritive elements to inert liquid ratio from 3:1 by volume to 53:1 by volume or from 19.5:1 to 340:1 by weight. No appreciable differences were observed in the resulting product.

Briquette for Spot Hardening of Powdered Metal Parts

S.W. McGee and C.T. Shork; U.S. Patent 4,172,719; October 30, 1979; assigned to Amsted Industries Inc. have developed a process for locally hardening a powdered metal part for increased wear resistance, during the normal sintering operation, thereby eliminating the need for more costly induction hardening or other methods of localized hardening.

The process involves compacting a hardening material and an infiltrating material into a briquette having a shape corresponding to an area of the zone to be hardened of a powder metal part, placing the briquette on the part, and sintering the part and the briquette in an endothermic atmosphere furnace at a temperature of about 2075°F for about eight minutes at temperature. The powder metal part, as it comes from the furnace, is locally hardened to a depth of about 0.050" in the range of about 30-57 Rockwell "C."

The briquette is a powder metal part formed of a plurality of components each having a specific function. Essentially the briquette will include (a) a component which will promote hardenability in the ferrous metal part, comprising one or more constituents such as manganese aided by nickel, (b) a component to promote initial fluidity in the briquette, such as phosphorus, (c) a transport component, such as copper, to carry the hardenability component into the ferrous metal part, (d) a matrix forming component, such as graphite aided by manganese, to confine the hardening reaction to the specific area of the zone and (e) the components are compacted to a density to limit a depth of penetration of the infiltrating materials. Phosphorus also may aid this limiting function. It may also be advantageous to include a lubricating component such as zinc stearate which will facilitate the mixing of the components and compaction of the briquette.

A portion of the manganese is utilized in the form of powder of 99.5% purity. However, pure manganese tends to oxidize to form MnO and it is this portion of the manganese in the form of MnO which works with the graphite to form the matrix. Most of the manganese is furnished as an element of a prealloyed bronze powder which has been atomized and annealed. This portion of the manganese is that which, aided by nickel to minimize brittleness, contributes to the hardenability of the ferrous powder metal part. The bronze alloy also contributes the desired amount of copper, the transport component.

Nickel, which contributes to hardenability along with the manganese and phosphorus and which counteracts the tendency of the manganese to cause brittleness, is added as a carbonyl type powder with a subsieve mesh size of 3 μ maximum. A small amount of nickel is also furnished by the bronze alloy powder.

The graphite is a natural flake powder which is 97% pure and has been sifted to a subsieve mesh size of 3 μ maximum. The phosphorus is furnished as a phosphorus-iron alloy powder which has been shotted and milled and has a mesh size of about −150. Pure phosphorus, as is well known, is not entirely safe to handle, has a short shelf life and, having a high vapor pressure, evaporates quickly. It therefore is generally used in an alloy which tends to stabilize it. The phosphorus could be alloyed with iron, or with copper and/or nickel, but iron is preferred because it is the least expensive.

Example: Ferrous powder metal parts were spot hardened by means of a hardening briquette having the following composition:

Constituent	Percentage By Weight of Mix	Analysis-Percent	Mesh Size U.S. Sieve Series or as noted
Phosphorus-iron alloy powder, shotted & milled	3.0	P: 24.0 Fe: 75.5 Si: 1.0*	−150
Nickel powder, carbonyl type	3.0	Ni: 99.7 Fe: 0.15* Co: 0.15*	Subsieve size 5 microns*

(continued)

Constituent	Percentage By Weight of Mix	Analysis-Percent	Mesh Size U.S. Sieve Series or as noted
Natural flake graphite powder, sifted	1.0	C: 97.0 ** Si: 3.0 *	Subsieve size 3 microns*
Manganese powder, electrolytic	1.0	Mn: 99.5 Fe: 0.05 *	−150
Prealloyed bronze powder, atomized & annealed	92.0	Cu: 92.0 Fe: 4.5 Mn: 3.0 Ni: 0.5	−100

* maximum
** minimum

The above composition breaks down into the chemical analysis shown in the following table which also shows the percentages of the functional components together with an indication of the function performed by each component.

Constituent	Percentage by Weight of Mix	Percentage of Functional Component by Weight of Mix	Function of Component
Phosphorus	0.72	0.72	Promotes initial liquid-solid diffusion; fluidity
Graphite	0.97		Matrix to localize zone of hardening
Manganese converted to MnO	1.00	1.97	
Manganese from bronze	2.76		Promote hardenability
Nickel	3.45	6.21	
Copper	84.64	84.64	Transport medium
Iron	6.41	6.41	Not essential
Si-Co	trace	trace	Not essential

The bronze powder once again was supplied premixed with about 1% by weight of zinc stearate and all of the ingredients were mixed together for about fifteen minutes. The apparent density of the mix was about 2.6 g/cc with a Hall flow rate of 2 minutes, tap, typical. The mix was compacted at about 30 tsi into a briquette of the desired shape and of a sufficient thickness to provide a contact pressure of 25 g/in^2.

The sintering was carried out by placing the workpiece, with the hardening briquette thereon, in a sintering furnace having an endothermic atmosphere. The preferred hearth temperature is 2075°F with a normal range of control of ±15°F. It has been found that for specific applications, spot hardening temperatures may be as low as 1990°F and as high as 2350°F depending on the desired sintered properties. The parts showed an apparent hardness, in the area under the hardening briquette, of 50 Rockwell C to a depth of 0.050". The particle hardness of the hardened portion was about 60 Rockwell C converted from 100 g Knoop.

n-Heptane and n-Hexane for Solvent Removal

D.H. Graham and R.F. Preckel; U.S. Patent 4,172,866; October 30, 1979; assigned to Hercules Inc. describe a process for removal of residual processing solvent from propellant casting powder particles, following their manufacture. It comprises liquid drying the particles by contacting them with an organic liquid miscible with the processing solvent but exhibiting substantially no solvent action for the powder particle ingredients with substantially no penetration of the particles. Normally liquid hydrocarbons are exemplary liquid drying solvents, n-heptane and n-hexane being preferred.

In preferred practice the liquid drying solvent is maintained in substantially complete wetting relationship with the particle surfaces until emplacement of the particles for casting. This eliminates safety hazards involved in conventional heat drying of the green powder and subsequent handling of the resulting dry material.

Partial Solidification for Scrap Alloy Purification

M.C. Flemings, R. Mehrabian and D.R. Geiger; U.S. Patent 3,840,364; October 8, 1974; assigned to Massachusetts Institute of Technology have developed a method of producing purified powders and of refining metal alloys from alloy scrap and the like. The alloy is brought to a state which is partly liquid and partly solid and in mixture form, preferably by heating the alloy to create a melt and then cooling the melt to permit partial solidification thereof. The mixture is vigorously agitated to create a slurry, the temperature of the slurry being controlled to provide a predetermined percentage of liquid-solid therein.

Then the liquid is separated from the solid either by selectively draining the liquid or by producing a stream of the slurry and injecting into the stream a gas in the form of a high velocity jet, which atomizes and solidifies the liquid to form a particulate that is generally smaller than the solid portion of the mixture (i.e., the primary solid) and can be separated therefrom by screening, for example.

To prepare the slurry, the melt is partially solidified while it is vigorously agitated by one or more rotating propeller blades **1**, in Figure 8.2. As distinct from some prior-art processes, it is not simply the bulk liquid that is agitated, but the entire liquid-solid mixture which is designated **2** and in which the liquid is designated **3** and the primary solid **4** (the particles **4** shown are, of course, of exaggerated size for purposes of the present explanation). The vigorous agitation causes the liquid-solid mixture to behave as a fluid slurry to fractions solid as high as about 0.5. A combination of the vigorous agitation and relatively slow rate of heat removal from the solidifying slurry causes it to be essentially isothermal and to have uniform distribution of fraction solid throughout.

Figure 8.2: Apparatus for Scrap Alloy Purification

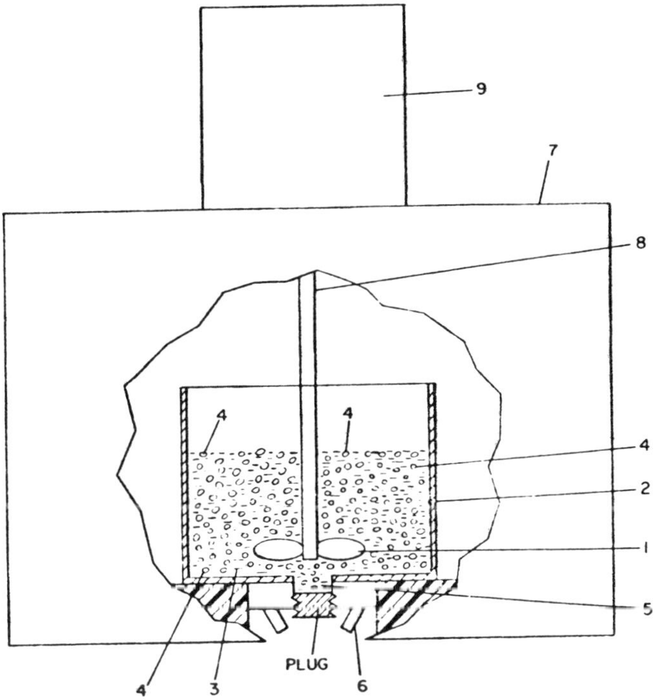

Source: U.S. Patent 3,840,364

The structure of the solid grains that form during this vigorous agitation is often very different from the usual dendritic structure that forms during usual solidification of castings and ingots. The solid particles in some cases approach small spheroids, and this is an important aspect of the processes described herein. Because the fine dendritic structure is absent, the segregated liquid is much easier to separate from the solid than in prior-art methods—there is, ideally, no liquid in small interdendritic "pockets." In other alloys, the primary solid forms as a "faceted" phase, without obvious dendrites, even when no agitation is employed. Agitation however, causes alloys even of this latter-mentioned type also to behave as a slurry suitable for the refining treatment discussed herein.

Figure 8.2 is a schematic illustration of the basic heating and mixing apparatus used in the experiments later described. In the figure, the mixture **2**, consisting of the primary solid pieces **4** suspended in the liquid **3**, forms a slurry upon being agitated by the rotating propeller blade **1**. The rotating blade **1** creates shear-type strains within the slurry. Heat is supplied by an electric resistance furnace **7**; and the propeller **1** is driven by a shaft **8** powered by a variable-speed electric motor assembly **9**. In one separation process, after the agitated slurry has reached a desired temperature (and so desired fraction solid), a tap hole **5** in the crucible is opened.

Vigorous agitation is maintained especially in the vicinity of the tap hole **5** and the slurry **2** then flows out this hole. On leaving the hole, it is struck by a number of converging gas jets provided by nozzles **6** to "atomize" it as is done in standard commercial practice with fully liquid melts. In this case, the gas jets atomize the fully liquid portion of the melt to very fine particles which subsequently solidify. These fine particles, having formed from fully liquid material, possess the average composition of the liquid (usually enriched in impurity). The gas jet cannot, of course, atomize the solid particles of metal entrained in the stream. These particles, larger than those of the atomized liquid, are low in impurity. Refining is now accomplished simply by separating (as by screening) the larger size fractions from the smaller.

In the second separation process, the metal is mixed and cooled, as described above, to obtain a partially-solid slurry. A tap hole is again opened in the crucible, but this time, once the slurry is formed, the rotation speed of the blades **1** is lowered so the metal in the immediate vicinity of the tap hole **5** is not so vigorously agitated. Now, a metal stream again flows through the tap hole **5**, but this time only the liquid portion goes through the hole; the bulk of the solid phase stays behind. It is believed that the reason for this behavior is that the solid particles give the slurry a "thixotropic" nature.

A semirigid skeleton of the solid grains forms in regions of the not so vigorously agitated melt. The rigid skeleton then effectively acts as a fine filter, holding back the particles from the more vigorously agitated portion of the melt. While the second process of separation by filtration may be performed by using the solid grains as the sole filtering medium, as just indicated, a ceramic filter may additionally be used to aid in performing the filtering step of the present process. When a ceramic filter is used, the filtering action may be accelerated if pressure is also applied to the slurry.

Although two primary methods of separation have been discussed above, a third method of separation involves the use of centrifugal force to effect the separation.

Example 1: *Atomization of Liquid-Solid Lead-Tin*—With mixing initiated at a temperature above liquidus the furnace was turned off and mixing continued as the temperature dropped. Mixing equipment consisted of a variable-speed motor and three-bladed propeller. The propeller was rotated in a manner so as to push material downward toward the center bottom of the crucible and upward at the sides.

When the temperature had fallen to the point where the alloy was 40% solid, the rotational speed of the propeller was increased to 200 rpm and maintained at that value; a ¼" hole in the crucible bottom was opened to allow the slurry to emit as a stream; and atomization was begun. Nitrogen at 200 psi was applied and this caused the liquid in the stream to be blasted away from the primary solid. The liquid then solidified into fine powder. The powder was then screened. The larger particles (10 to 30 mesh) in this process should containe 2-3% lead and the smaller particles (100 mesh and up) should contain 23% lead. The first effort, however, resulted in about 12% for the larger particles and about 22-23% for the small, still a substantial improvement.

Example 2: *Draining of Liquid-Solid Lead-Tin*—The same initial procedure as in Example 1 was used, the temperature of the slurry was lowered until the alloy was 50% solid, at which point the rotation rate was slowed from about 200 rpm to about 50 rpm and the ¼" hole in the crucible bottom was opened. Substantial amounts of liquid material gravity drained from the slurry with very few solid particles being pulled down. The drained liquid contained about 22% lead and the residue in the crucible about 12% lead.

Example 3: *Draining of Liquid-Solid Aluminum-Silicon-Iron-Titanium Alloy*—The same procedure was employed for this alloy as in Example 2, except that rotation speed was not changed when crucible was tapped. The mixing speed was carefully watched and kept at 200 rpm. Results shown in the table below are for two runs at 750° and 600°C, respectively.

	Silicon	Iron	Titanium
	....	.(%)	
	. . 750°C Temperature When Tapped. . .		
Initial	29.8	6.10	1.08
Drained	22.6	4.64	0.43
	. . .600°C Temperature When Tapped. . .		
Initial	30.0	6.19	0.94
Drained	13.1	1.77	0.15

In a separation method like that employed in Example 1, both the liquid and the solid phases are forced through a hole located in the base of the crucible and the resulting mixture is atomized by a gas jet in the manner employed in usual liquid atomization processes, as mentioned above. The high flow velocity gas jet breaks up the liquid into small spherical powders thereby separating the liquid from the initally-solidified, primary solid. By varying gas pressure, i.e., velocity, and/or nozzle geometry, the size of the atomized liquid powders can be varied over a wide size range. It will be appreciated that the angular velocity of the mixer, temperature of the melt, size of the drain hole **5**, will vary with the alloy being purified in either separation process.

In the separation method employed in Examples 2 and 3, the rotational speed of the propeller-like blade **1** and/or location or direction of rotation thereof can be varied so that only the liquid phase is drained through the hole in the crucible. In this situation, the solid particles have a tendency to bond together, thereby producing a natural filter for drainage of the liquid. This type of materials separation can be applied to purification of any multicomponent alloy system because the different alloy elements normally segregate during solidification.

In these examples, apparatus like the apparatus shown schematically in Figure 8.2 was used, and this apparatus has one rotatable propeller and one drain hole in the bottom of the crucible. However, more than one propeller and more than one hole can be employed in other more sophisticated apparatus and mixers may be employed in some circumstances.

APPLICATIONS AND MISCELLANEOUS TECHNIQUES

Ductile and Stable Particle Superconductors

H. Winter and D.N. Sethna; U.S. Patent 4,050,147; September 27, 1977 have developed a method of producing a ductile and stable particle superconductor from particles of the formula NbN, $NbN_{1-x}C_x$ where x is less than or equal to 0.5, Nb_3Al, Nb_3Si, Nb_3Ge, Nb_3Sn, V_3Ga or V_3Si from a reaction in a hydrogen plasma. The particles are incorporated into a metallic matrix or coprecipitated with a metal powder.

A preferred method for the production of superconducting compounds with the required size is to introduce the halogen compounds of Nb or V into a hydrogen plasma together with N and/or a hydrocarbon like methane. The hydrogen plasma with a temperature of about 3000°C is generated by an electric arc burning between a water-cooled tungsten cathode and a similarly cooled copper anode; the copper anode has a central boring through which the hydrogen is introduced. The electrical values for the arc are for instance 200 A and 120 V. The particles resulting from the reaction fall out and are gathered on the floor of the enclosing container. This method can be employed for the mass production of A_3 compounds with βW/A15 structure. A stands for Nb or V and B for Al, Ge, Si, Ga or Sn. Some of the reaction mechanisms are presented below.

$$NbCl_5 + 0.5N_2 + 2.5H_2 \rightarrow NbN + 5HCl$$

$$2NbCl_5 + 0.5N_2 + CH_4 + 3H_2 \rightarrow 2Nb(N_{0.5}C_{0.5}) + 10HCl$$

$$3NbCl_5 + SnCl_4 + 9.5H_2 \rightarrow Nb_3Sn + 19HCl$$

$$3VCl_5 + SiCl_4 + 9.5H_2 \rightarrow V_3Si + 19HCl$$

It became possible to bring down the size of the resulting particles to 70 A by increasing the current and the voltage of the arc to 500 A and 150 V. The particles were carefully stored under argon or helium with a reduced pressure.

According to the process, the incorporation of these fine superconducting particles with a medium diameter of up to about 500 A into a ductile metallic matrix was accomplished by three different methods.

The first method is based on conventional powder-metallurgy. A carefully prepared mixture of compounds such as Nb_3Sn or V_3Si with a medium particle diameter below 500 A and a metal like copper with about the same particle size was compacted, sheathed in a container of the same metal, evacuated and closed; subsequently the compacts were extruded and further drawn into wires with a diameter of about 1 mm using Nb_3Sn particles with a medium diameter of 300 A and copper particles with about 400 A and a volume of 35% Nb_3Sn and 65% Cu, the resulting superconducting and mechanical values of the 1 mm diameter wire after an annealing treatment of 700°C for 1 hour were:

T_C	21°K
H_{C_2}	470 kG
J_C	7×10^5 A/cm^2
Tensile strength	74 kp/mm^2
Elongation	9 %

In the above table, T_C is the critical temperature, J_C is the critical current density, and H_{C_2} is the critical magnetic field.

The copper particles were produced by vacuum evaporation and stored under vacuum; in contact with air they proved to be pyrophoric. By comparison, using conventionally prepared Nb_3Sn powders made by ball milling the small est size reached was about 20,000 A or 2 μm.

The second method chosen according to the process for the incorporation of the minute superconducting particles into a metallic matrix avoided the fire hazards and the tendency of contamination of the fine metal powders with the atmosphere; furthermore only 20 volume percent of superconducting particles in the metal matrix were necessary for optimum results. About 300 g of Nb_3Sn particles with a medium diameter of 200 A were suspended in an acid $CuSO_4$ electrolyte of about 1 liter (1,000 cm^3) with the help of a wetting agent like monolaurylsulfuric acid ester; by galvanic deposition of the suspension electrolyte on a moving belt or a rotating drum, both with a stainless-steel surface, the resulting particles with a size of about 200 μm were continously scraped off and dried; they proved to be dispersion powders, that means, a fine dispersion of regularly spaced Nb_3Sn particles in a copper matrix; these powders were compacted, sheathed in a copper can, extruded and worked into semifinished material like plate, tube or wire.

After an annealing treatment at 650°C for 1.5 hours, a tube with an outer diameter of 20 mm and a wall thickness of 1 mm, suitable for conducting the cooling medium like liquid helium, showed the following superconducting and mechanical values employing a mixture of 20 volume percent of Nb_3Sn and 80 volume percent of Cu:

T_c	23°K
H_{c_2}	570 kG
J_c	9×10^5 A/cm^2
Tensile strength	78 kp/mm^2
Elongation	7 %

The resulting tubes can be cooled with liquid hydrogen, leaving sufficient superconducting properties for technical use. The third method showed the finest distribution of superconducting particles in a metal matrix and employed so called duplex powders; the superconducting particles falling out of the hydrogen plasma reaction zone were covered at the same time with a layer of the metal by introducing a halogenide of the metal, like copper, into the reaction zone.

The resulting particles bearing the required layer thickness—for the 100 A particles of Nb_3Sn or V_3Si about 40 A—were strictly kept under argon or helium to avoid contamination with the atmosphere, compacted, canned into thin copper tubes, evacuated and extruded into semifinished material like rods, tubes and wire. Using Nb_3Sn particles of about 100 A diameter and a copper layer of about 40 A, the resulting wire showed after an annealing treatment for 2 hours and 700°C the following superconducting and mechanical values:

T_c	24.7°K
H_{c2}	710 kG
J_c	9×10^5 A/cm^2
Tensile strength	82 kp/mm^2
Elongation	7 %

Using 100 A particles of V_3Si covered according to the process with a layer of about 40 A of copper, the resulting wire showed superconducting fluctuations already at 29°K; here a broad transition range started. The maximum upper critical magnetic H_{C_2} at 4.2°K nearly reached 1 Megagauss.

Dummy Bullets

N. Dautzenberg, J. Hewing, M. Michalke and H. Weber; U.S. Patent 3,951,035; April 20, 1976; assigned to Nederlandsche Wapen-en Munitiefabriek De Kruithoorn NV, Netherlands and Mannesmann AG, Germany provide a method for the construction of disintegrating bullets. According to this method an iron powder with particle dimensions between approximately 0.2-1 mm is used for the fabrication of the bullet core. The iron powder is obtained by atomizing a nonalloyed molten mass of steel having a low carbon content, by means of pressurized water. The resulting iron powder is subsequently reducibly annealed, i.e., it is softened by subjecting the powder to heat at a temperature ranging from 900°-1050°C. Thereafter, the substance is crushed and subjected to a scrubbing treatment, both of which steps are preferably accomplished in a hammer mill or swing hammer pulverizer.

The thus treated powder is subjected to a filtering process to retain those particles having a standard grain size of 0.2-1 mm, but preferably with a size of 0.4-1 mm. The powder is then used in a press die or the like to obtain bullet cores, which are then jacketed, e.g., in plastic. The latter two steps are standard procedure and are per se not part of this process.

It is advisable to use an iron powder which, after being broken up and subjected to the above scrubbing and filtering process, has a density of 3.7-3.9 g/cm^3. In order to facilitate compression of the powder into a bullet core, zinc stearate in a quantity of, for example, 0.5% may be added to the powder. The powder is then compressed to obtain density in excess of 7 g/cm^3.

Example: A molten mass of steel having a temperature of 1650°C is poured from a ladle, and the falling stream of molten steel is subjected to jets of pressurized water of 30 atm. The steel is thereby atomized resulting in an iron powder with particle dimensions below 1.5 mm.

The atomized or pulverized iron, subsequently, is dehydrated and dried and is reducibly annealed in a reducing gas. The temperature involved in softening the powder in the annealing step, in this case, is 1000°C. Following this treatment, the iron is broken up and subjected to a scrubbing process for such a duration that the iron powder, after sieving to obtain powder at particle dimensions of 0.4-1 mm, has attained a density of 3.7-3.9 g powder per cm^3.

As indicated above, the iron particles, preferably, are broken up and subjected to the beating and scrubbing treatment in a hammer mill or swing hammer pulverizer. The powder so made is then compressed under pressure of 6 to 8 metric tons per cm^2 to obtain densities in excess of 7 g/cm^3. This compression is part of the bullet core making process which proceeds otherwise along conventional lines.

Fluid Nozzle with Plenum-Forming Sections

J. H. Lafferty; U.S. Patent 3,901,492; August 26, 1975; assigned to Carpenter Technology Corp. describes a fluid nozzle structure in an apparatus for atomizing molten metal of the type in which the molten metal to be atomized is fed as a hollow stream. The fluid nozzle structure comprises plenum-forming sections threaded in spaced relation on a feed pipe coaxial with the hollow metal stream. The plenum-forming sections form a continuous radially outward presented fluid nozzle through which the atomizing fluid is directed. One of the plenum-forming sections is readily adjusted relative to the other to vary the size of the fluid nozzle.

Referring to Figure 8.3 a nozzle assembly **7** is shown for illustrative purposes in association with a tundish generally illustrated at **5**, and there is a ceramic frusto-conical distributing element **6**. The bottom end of the distributing element **6** is recessed to receive and locate the atomizing fluid nozzle assembly **7** on which the former is supported. The atomizing fluid nozzle assembly **7**

Figure 8.3: Atomizing Apparatus for Making Metal Powder

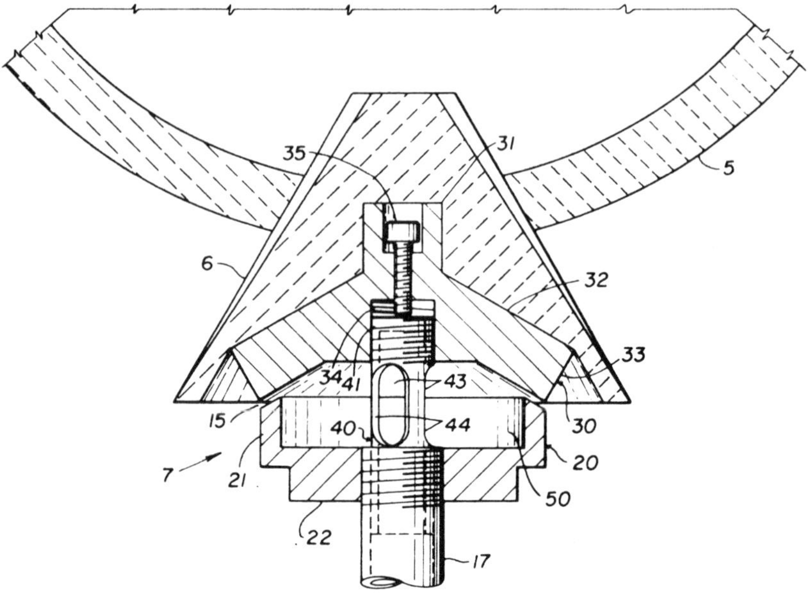

Source: U.S. Patent 3,901,492

is in turn supported by the feed pipe **17** which thereby supports the entire atomizing assembly.

The atomizing fluid nozzle assembly comprises two sections, a lower section or body **20** and an upper section or cap **30**, and a distributing support pipe **40**.

The body **20** comprises a base **22** and an upstanding cylindrical wall **21**, the former being threaded onto the gas feed pipe **17**.

The cap **30** comprises a top stud or plug **31**, a frusto-conical intermediate portion **32**, and an inclined lower portion **33**. The cap **30** has a threaded recess **34** to receive the upper threaded end **41** of distributing pipe **40** which can be an integral extension of pipe **17** or, as shown, is welded to pipe **17**. A lock screw **35** which functions as a stop is provided in the cap **30** and engages the closed end **41** of the pipe **40** and thus acts as an adjustment device for fixing the cap at a desired position on pipe **40** relative to the lower section or body **20**.

The inclined lower portion **33** is thus maintained a predetermined distance from the upwardly presented outwardly and downwardly inclined surface of wall **21** to form an annular slit or nozzle **15**. Thus, by varying the depth to which the lock screw **35** allows the pipe end **41** into the recess **34**, the distance across the slit **15** can be precisely adjusted.

Intermediate its ends, the distributing pipe **40** has three elongated holes **43** formed therein separated by three narrow support spiders **44**. Although thus shown, it will be readily apparent that a larger number of holes arranged in many possible configurations in the middle of the pipe **40** could be used. The only restrictions are that the holes be distributed about the pipe to give uniform jet velocity throughout the entire 360° extent of the slit **15**, and that enough solid pipe is left to support the cap **30** and the molten metal distributing element **6**.

The array of three elongated holes **43** is believed to be the best arrangement for maximum and uniform atomizing fluid flow with adequate support for the upper sections. The interior of distributing pipe **40** communicates with the slit **15** through its holes **43** and a plenum **50** formed between the cap **30** and the body **20**. Any undesired turbulence in the fluid flow that may be caused by the support spiders is substantially entirely dissipated in the plenum **50** and in passing through slit **15** so that the jet flow is generally laminar. Thus, a feature of this preferred embodiment resides in the arrangement of the support structure for the cap **30** which forms the upper side of the slit **15**, so that the support structure is at substantially maximum spacing from the slit **15**.

The angle of the slit or fluid nozzle **15** is not critical, but preferably it is such that it converges with the plane of the outer surface of the distributing element **6**. The distributing element **6** is supported in operative relation to the tundish **5** on the fluid nozzle assembly **7** which is in turn supported on the feed pipe **17**. The feed pipe **17**, in operation is connected to the source of atomizing fluid which can be a liquid such as water or a gas, preferably an inert gas such as argon under a pressure of at least about 300 psi to ensure desired atomization of the metal.

Powdered Metal Filter by Water Atomization

J.G. Bewley; U.S. Patent 4,166,736; September 4, 1979; assigned to Metallurgical International, Inc. has developed a process for producing a powdered metal filter composition consisting essentially of a ferrous based alloy having by weight from approximately 0.60% to approximately 4.0% silicon and from approximately 0.08% to approximately 0.18% carbon and exhibiting an apparent density of approximately 1.6 to 2.3 g/cm^3 by water atomization. The composition also exhibits compressibility and particle shape as a function of the composition chemistry and in particular the silicon, carbon and manganese content as well as the angle of intersection of the atomizing water jets which is approximately from 25° to 45°C.

The composition provides an optimization of filtration efficiency, pressure drop across the filter pack, resistance to pressure deformation of the composition particles and the useful life of both the filter pack and the composition and is produced with well-known apparatus and gives optimum yields of useful, particle-size fractions.

The preferred chemistry of the powdered metal composition is as follows:

CONSTITUENT	% BY WEIGHT
Carbon	0.14%
Manganese	0.60%
Silicon	1.0%
Nickel	12.0%
Chromium	19.0%
Molybdenum	2.2%
Iron	balance

The preferred apparent density for the preferred composition is 2.0 g/cm^3 and the preferred compressibility is 0.14″ for a particle size distribution of 50 to 100 mesh. However, with the more coarse fraction of 30 to 60 mesh, the preferred apparent density is 1.85 g/cm^3, and the preferred compressibility is 0.150″. If one were to attempt to duplicate the preferred apparent density and compressibility designated above for the 50 to 100 mesh fraction in the 30 to 60 mesh fraction, the yield would be extremely small. In this respect, it should be borne in mind that with any given composition, and under any given set of atomization operating parameters, a broad range of particle sizes is produced.

The desirable cut is then removed from the material produced and the oversized and undersized particles are recycled. With the above preferred composition, and with the use of optimum operating parameters, the yield of 50 to 100 mesh cut having the desired apparent density and compressibility will be maximized.

In the apparatus for water atomization the water pressure in the manifold should be maintained between approximately 200 to 400 lb/in^2 when generating the coarse cut of 30 to 60 mesh and between approximately 400 to 700 lb/in^2 when generating the relatively fine cut of 50 to 100 mesh for maximum yield. Preferably when producing the coarse cut a pressure of approximately 300 lb/in^2 should be employed and when producing the fine cut a pressure of approximately 600 lb/in^2 maintained.

Valve Metal Base Powder

G.S. Ronneau, P.D. Debacker and H.L. Carpentier; U.S. Patent 3,867,129; February 18, 1975; assigned to Metallurgie Hoboken-Overpelt, Belgium have developed a metal in powder form, whose particles are made up of an anodically oxidizable metal (also recognized as valve metal or metal converting alternating current to direct current, such as tantalum or niobium powder) containing molybdenum and/or vanadium and/or tungsten and/or hafnium in contents comprised between 0.06 and 2% on an additive metal to tantalum or niobium metal basis. The additive metal contents refer to each additive metal separately or to the additive metals as a whole.

The particles of another powder according to the process are made up of a nucleus or anodically oxidizable metal covered with a surface layer of the same metal, such layer containing molybdenum and/or vanadium and/or tungsten and/or hafnium in contents comprised between 0.01 and 2% on an additive metal to tantalum or niobium metal basis, these additive metal contents referring severally to each additive metal separately or to the additive metals as a whole. The thickness of the surface layer will preferentially be less than 3000 Angstroms.

Still another powder is one whose particles are made up of a nucleus of anodically oxidizable metal covered with either a continuous or a noncontinuous surface layer, or one or more areas on the surface, made up of molybdenum and/or vanadium and/or tungsten and/or hafnium. The thickness of the surface layer will preferentially be less than 3000 Angstroms.

The powders can be manufactured by a number of well known methods such as: cofusion and embrittlement, sodiothermic reduction, fused electrolysis, hydrogen reduction on a fluidized bed.

One suitable powder which can be processed according to this method is a tantalum powder obtained by electron-beam fusion and hydrogen embrittlement. Such high-purity powder is thoroughly blended with molybdenum powder in proportion of 100 to 0.2 by weight. The powder blend is compacted in an isostatic press and the resulting bars are melted in an electron beam furnace. The ingot is hydrided to a hydrogen content of about 0.4% by weight, then it is crushed and pulverized and the powder obtained is degassed and submitted to chemical assay. The desired molybdenum content is traced back in the powder with more or less 10% accuracy.

Another type of powder which can be prepared according to the process is made by charging a quantity of potassium tantalum double fluoride into a sodiothermic reduction vessel. The reduction occurs at a temperature of about 800°C in a fused salt solution containing sodium chloride: for instance 300 kg of potassium tantalum double fluoride in 10 kg sodium chloride. A stoichiometric quantity of sodium in liquid form is slowly added to the fused salt mass; a dendritic tantalum powder is rapidly formed. When the reaction signs (exothermicity) tend to disappear, indicating that the chemical reaction has been completed, a quantity of 260 g of vanadium trichloride is charged into the sodiothermic reduction vessel, then 120 g of sodium are added.

After crushing the sponge containing the tantalum, and washing the crushed sponge, a tantalum powder is obtained which contains about 600 ppm of vanadium, this latter metal being located on the surface of the powder particles. Still another type of powder which can be prepared in accordance with the process is made by an operating process similar to the one described just above. The particles of this powder show a surface layer containing both the matrix metal, which is tantalum or niobium, and the doping element.

To this end, the reaction, scheduled for reducing 300 kg of potassium tantalum double fluoride is stopped when 90% of the required quantity of sodium has been added; then a reducible compound, as in the process described above, is injected into the reaction vessel. Then the addition of sodium is resumed and the reduction together of the remaining 10% of the matrix metal and of the additive element occurs simultaneously on the surface of the powder particles obtained during the first phase of the reaction.

Powders made according to the process improve the characteristics of the dielectric (electrolytic oxide) built up on the surface of the particles of which these powders are made up. Such improved characteristics give better performance to the capacitor manufactured from these powders. Capacitors manufactured from powders made according to the process show a sizeable decrease of the capacitance variance as a function of operating temperature; in addition, such capacitors withstand reverse voltages higher than those that can be withstood by capacitors manufactured from a powder that has not been treated according to the process and the performances under life test are better.

Production of Metal Powder from Wire Stock

J.T. Blucher and D.D. Dalrymple; U.S. Patent 3,830,603; August 20, 1974; assigned to Industrial Materials Technology, Inc. produce small particles of metal, such as those used in making powder metal compacts, by feeding the end of a metal wire or rod against the edge of a rotating disc and causing a direct electrical current to flow through the wire and disc. This melts the end of the wire and also creates a magnetic field about the wire. The rotation of the disc breaks the electrical contact and forms an arc which causes additional melting of the wire.

Contact between the wire and disc takes place within a second electromagnetic field. Continuously advancing the end of the wire causes the intermittent making and breaking of electrical contact. The interaction of the two magnetic fields causes the molten particles to be removed from the area of contact. Means are provided for cooling and collecting the metal particles and for preventing the accumulation of solidifed metal particles on the electromagnet or the rotating disc. Vacuum conditions may be used, thus producing a powder of high purity.

Referring to Figure 8.4, a deeply grooved disc **10** is placed in the gap **12** of electromagnet **14** energized by coil **16**. The disc is mounted for rotation on water-cooled shaft **18** driven through pulley **19** by motor **20**. Each end of shaft **18** is provided with vacuum-tight rotary joints (not shown). Electrical current is supplied to disc **10** by means of slip ring **22** mounted on shaft **18** in contact with carbon brush **24** which in turn is connected by conductor **25** to the negative terminal of a source of direct current (not shown).

A wire reel **26** is mounted for rotation in frame **28**. Metal wire **30** having a chemical composition of that desired in the final powder product is drawn from the reel by a conventional wire feeder **32** powered by a variable speed motor **34** controlled by electronic controller **36** adjustable through knob **37**.

Figure 8.4: Apparatus for Production of Metal Powder from Wire Stock

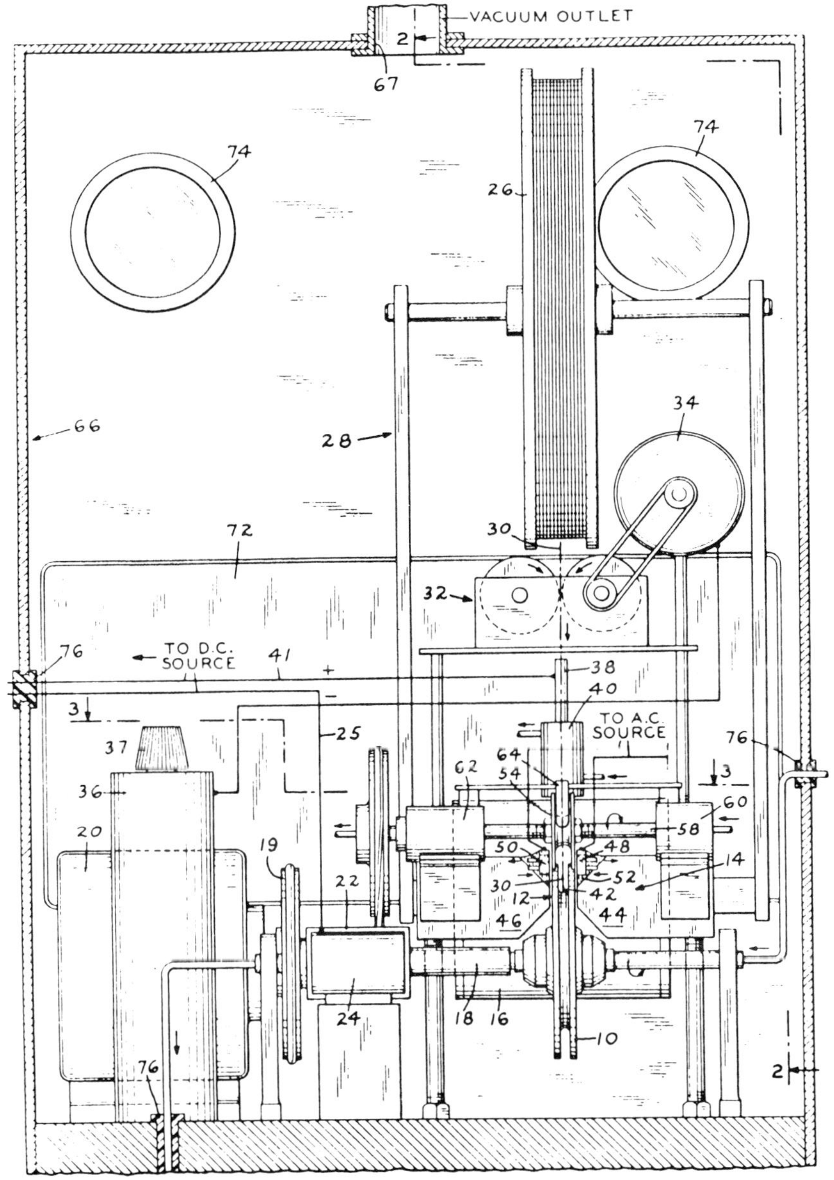

Source: U.S. Patent 3,830,603

Wire **30** is fed downwardly through guide tube **38** water cooled by housing **40**. Electrical current is supplied to wire **30** through guide tube **38** which is connected by conductor **41** to the positive terminal of the source of direct current that supplies current to disc **10**.

Thus, when the tip **42** of wire **30** makes contact with the disc **10**, a closed electrical circuit is formed from the positive terminal of the direct current source through conductor **41** to guide tube **38** to wire **30** through tip **42** to disc **10** to shaft **18** to slip ring **22** to carbon brush **24** and through conductor **25** to the negative terminal of the direct current source.

When the wire **30** makes contact with the disc **10** the current flow between the advancing tip **42** and disc **10** causes the tip **42** to melt. This melting action together with rotation of the disc **10** breaks the contact between wire **30** and disc **10** and causes the formation of an arc. The arc creates a high temperature which melts an additional portion of the continuously advancing wire.

The end of the continuously advancing wire again makes contact with the rotating disc, extinguishing the arc. The tip of the wire melts, and again, the combined melting action and the rotation of the disc **10** causes the breaking of the contact and reformation of the arc. This making and breaking of the arc with the attendant formation of molten metal particles continues at a relatively rapid rate.

The flow of electrical current through the wire **30** produces a magnetic field around the wire in accordance with well known principles of electricity and magnetism. The direction of feed of the wire is substantially perpendicular to the axis of rotation of disc **10** and in addition is substantially perpendicular to the lines of magnetic force (not shown) in the gap **12** between pole pieces **44** and **46** of electromagnet **14**.

The metal particles formed are driven out of the area of contact between the metal wire and the disc due to the interaction between the magnetic field produced by the passage of an electrical current in the wire and the second magnetic field transverse thereto produced by the electromagnet.

In the preferred embodiment disc **10** is made of graphite but it could also be made of metal of the same chemical analysis as the final powder product. The entire apparatus may be mounted within a vacuum chamber **66** connected to vacuum outlet **67**. The chamber may be a conventional steel tank with hinged and sealable doors **72** at each end. Sight viewing ports **74** may be installed in several locations along with vacuum-tight electrical and cooling water connections **76**.

In operation particles of metal are produced from metal wire **30** by continuously feeding the wire toward rotating disc **10**, the axis of rotation of which is substantially perpendicular to the direction of feed of the wire. The advancing end **42** of the wire is brought into contact with the rotating disc **10** within the field of an electromagnet **14** having lines of force with components substantially perpendicular to the direction of feed of the wire and substantially parallel to the axis of rotation of disc **10**.

A direct electrical potential is applied between the wire and the disc causing direct current to flow through the wire and the disc, completing an electrical circuit and causing a magnetic field to form around the advancing wire. This magnetic field has lines of force substantially perpendicular to the lines of force of the electromagnetic field of electromagnet **14**. The current is of a magnitude sufficient to cause the end of the wire in contact with the disc to melt. Rotation of the disc together with the melting action cause the breaking of electrical contact and the formation of an arc between the end of the wire and the disc.

Discrete particles of the metal are driven out of the area of contact between the wire and the disc by the interaction between the magnetic field produced by the passage of current in the continuously advancing wire and the magnetic field transverse thereto produced by the electromagnet.

Continuously feeding the end of the wire against the rotating disc causes the reestablishment of contact between the end of the wire and the disc. As the end of the wire reestablishes contact with the disc, additional portions of the wire are melted and the rotation of the disc once again causes formation of an arc. The making and breaking of electrical contact with the related formation of the arc are repeated at a relatively rapid rate, resulting in the production of a continuous supply of molten metal and the commensurate discharge of particles from the contact area as described above. The particles are then cooled and collected. The entire operation may be carried out in a vacuum thus avoiding contamination of the formed metal particles.

Nickel-Manganese Microballoons

G.D. Cremer; U.S. Patent 4,162,914; July 31, 1979; assigned to International Harvester Co. describes a process for making hollow metal microballoons from alloys which include a base material capable of absorbing a gas when molten and then subsequently desorbing the gas during solidification and one or more glass formers. The alloy is melted and atomized in an inert atmosphere to form a clean, dense, unoxidized powder which is screened and then subjected to a reactive plasma, melting the particles and generating a gas which is abosrbed in and inflates the particles. The gas desorbs from the microballoons as they solidify; and impervious, glassy films form on their surfaces.

The microballoons can be made from a variety of metallic alloys. The essential characteristics of the alloy are that it must include at least one glass former and a base material which is capable of absorbing hydrogen when the alloy is molten and of desorbing and expelling hydrogen as the alloy cools and solidifies. Also, the alloy is preferably one which undergoes minimal volume change as it solidifies; and the alloy must preferably have a short temperature solidification range and possess high fluidity and a high liquid film strength. It is also thought that the glass former should have a free energy of formation per gram-atomic weight of oxygen which is lower than that of hydrogen per gram-atomic weight of oxygen.

Preferred glass formers include boron and silicon. Other glass formers that may be used are phosphorus, arsenic, and germanium compounds. Mixtures of the foregoing can also be used.

Typically alloys have a glass content of 3.5 to 4.2% based on the total weight of the alloy. This range can be varied, however, depending upon the remaining constituents of the alloy. The critical factor is that the glass former or formers be present in sufficient amount to produce a thin (e.g., 100 Angstroms) but integral film or coating on the surface of the hollow metallic spheres produced in the process.

The preferred base materials are nickel-manganese alloys, especially those containing from 68-76% nickel and from 21-27% manganese. However, manganese is not believed to be an essential constituent of the base material although it is thought to desirably influence the hydrogen absorbing capability of the base material.

From a trace to 0.5% of columbium is also preferably included in the alloy with the base material and the glass former(s). This makes the alloy self-scavenging with respect to carbon and nitrogen under the process conditions.

Other elements may also be included in the feedstock alloy to enhance its capacity for forming microballoons with the wanted characteristics. Among these are iron, cobalt, chromium, aluminum, beryllium, and molybdenum.

Rare earth metals such as cerium or misch metal can also be added to the feedstock alloy to improve its self-scavenging capability, but they tend to scum up and decrease the fluidity and liquid/solid wetting characteristics of the feedstock alloy.

The preferred feedstock alloys are those having the following formula:

Element	Percent by Weight
Ni	68–76
Mn	21–27
Si	2.4–3.2
B	1.1–1.4
Cb	trace–0.5

The proportion of boron employed in the feedstock alloy is perhaps the most critical of those tabulated above as amounts of greater than the indicated maximum of 1.4% produce an unsatisfactory level of brittleness in the alloy.

Alloys with less than the specified minimum amount of boron tend to have unsatisfactorily wide solidus/liquidus spreads with correspondingly sluggish flows.

In the particular nickel-manganese based alloys covered in the tabulation, lower than specified amounts of manganese also result in undesirably wide solidus/liquidus spreads and are therefore preferably avoided. Reducing the silicon content below the specified minimum level does not appear to appreciably change the solidus/liquidus spread but nevertheless undesirably affects the flow characteristics of the alloy. Similarly, larger than indicated amounts of the columbium scavenger result in feedstock alloys with sluggish flows.

Of the alloys embraced within the tabulated formulation, those of the nominal compositions 21.0Mn-2.4Si-1.4B-trace Cb-balance Ni and 24.0Mn-2.8Si-1.4B-trace Cb-balance Ni are preferred. Both have high liquid film strength, excellent fluidity, and the ability to inflate rapidly; and neither undergoes more than a minimum volume change when it solidifies which appears to be uniquely favorable for microballoon stabilization. Also, both alloys have low microporosity and a high self-scavenging capability with respect to carbon, nitrogen, oxygen and sulfur, the chief contaminants in feedstock alloys of the character described above.

Figure 8.5 shows schematically the steps employed in making hollow, metallic microballoons in accord with the principles of the process.

Figure 8.5: Process Flow Diagram for Nickel-Manganese Microballoons

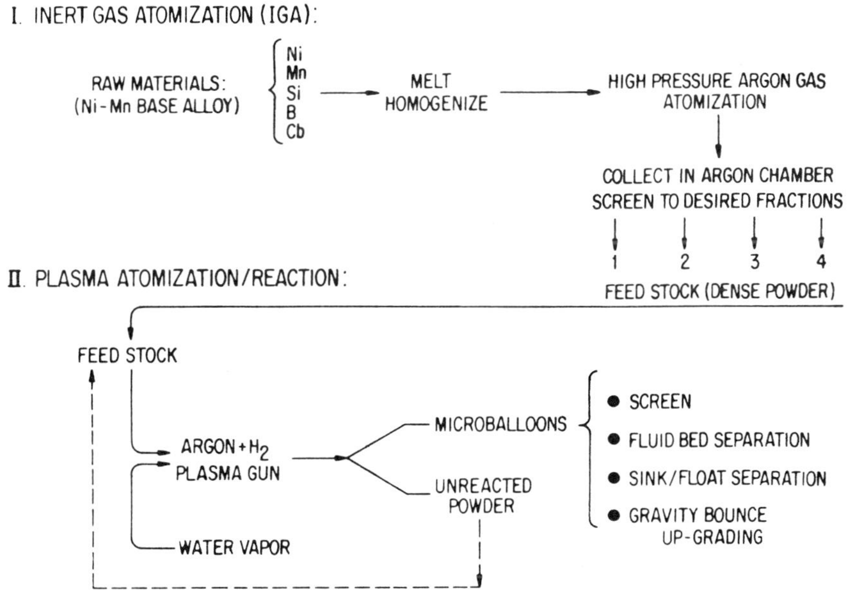

Source: U.S. Patent 4,162,914

These microballoons have uniform 1 to 3 μ thick shells with smooth, shiny external surfaces, a high degree of sphericity, and comparatively high strength for their size and shell thickness. The diameters are in the range of 50 to 1,000 μ. Such microballoons can be filled with deuterium and tritium and used as laser and electron beam targets for fusion reactions.

Trimming of Compacted Metal Powder Strip

M. Donnelly; U.S. Patent 4,172,111; October 23, 1979; assigned to Davy-Loewy Limited, England has developed a process for manufacturing a compact in the form of strip from metal powder in which material trimmed from the compact does not have to be comminuted into powder by a separate process.

Metal powder is introduced into the gap between a pair of rotating rolls to be compacted therebetween into a strip and the compact issuing from the rolls is trimmed by removing unwanted longitudinally extending edge portions thereof, the edge portions being removed by reducing them to powder of substantially the same particle size as that of the powder from which the compact is formed.

Thus, the compact is trimmed by having its low density longitudinal edges removed and at the same time the powder so formed can be compacted again to form a further strip with little or no further treatment.

The two longitudinally extending edge portions of the strip may be removed simultaneously. The edge portions are conveniently reduced to powder by rotating members which engage the edge portions to impact and break off individual particles.

The process also includes a pair of spaced apart parallel rotatable rolls, means for introducing metal powder into the gap between the rolls and means for trimming the compact by reducing unwanted longitudinally extending edge portions thereof to powder of substantially the same particle size as that of the powder from which the compact is formed.

The trimming means may comprise a pair of rotary members in the form of cutter discs which are rotatable about axes substantially normal to the plane of the compact.

After trimming, the compact is conveniently heat treated by passing it through a sintering furnace but alternatively the compact may be passed through a furnace and then between a pair of rolls which may be heated.

In order that this process may be readily understood it will now be described, by way of example only, with reference to the accompanying drawings in which Figure 8.6a is a diagrammatic view of a powder compact apparatus taken in the direction normal to the plane of the compact, Figure 8.6b is a side

Figure 8.6: Apparatus for Trimming Compacted Metal Powder Strip

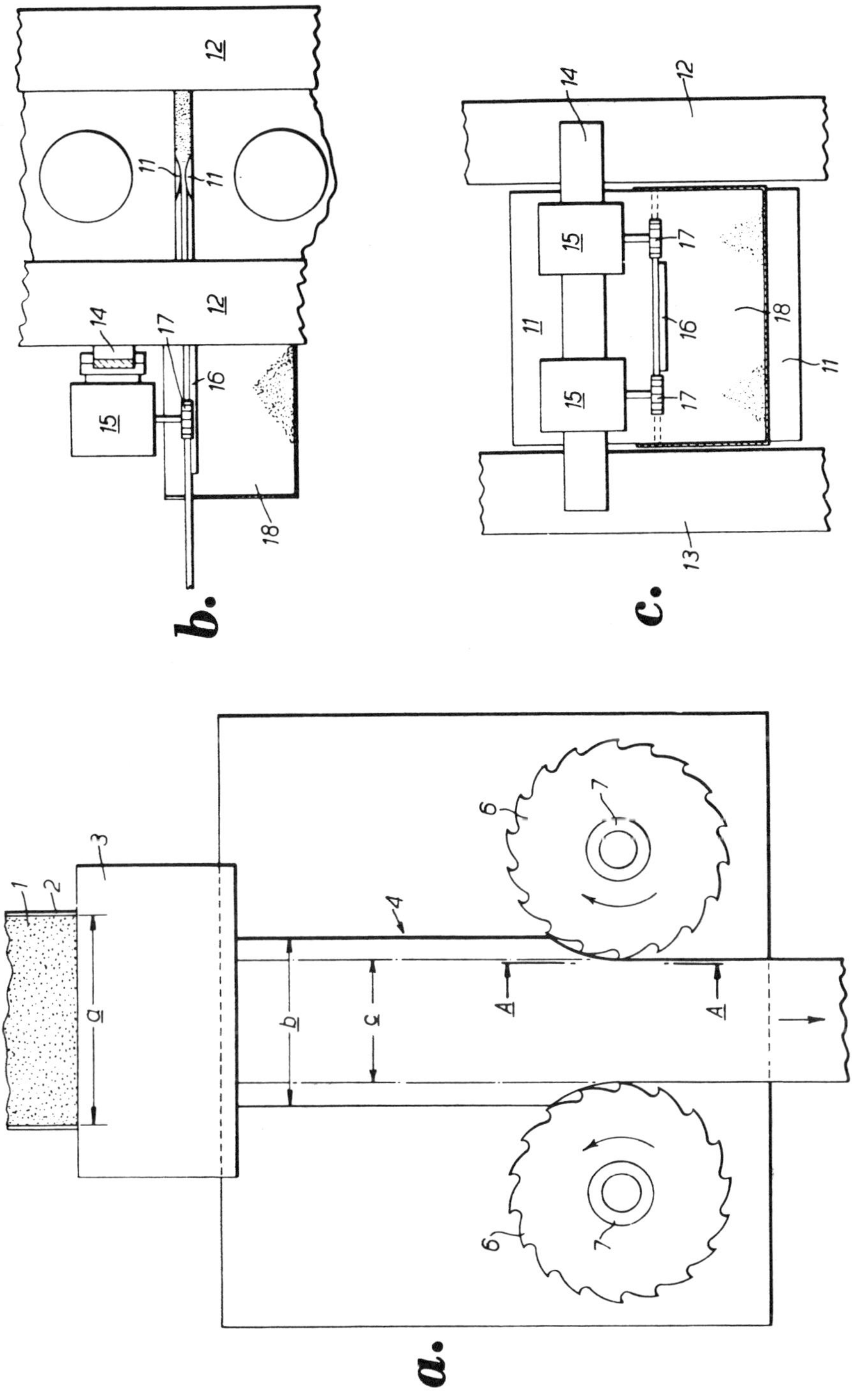

Source: U.S. Patent 4,172,111

elevation of compact forming and trimming apparatus and Figure 8.6c is an end view of the apparatus shown in Figure 8.6b.

In the manufacture of a compact in the form of a strip from metal powder, the compact being also known as a green strip, metal powder **1** is loaded into a hopper **2** from where it is fed into the gap between a pair of rotating rolls **3**. Between the rolls the powder is compacted and it issues from between the rolls in the form of a compact or strip **4**. If the width of the outlet of the hopper **2** is **a** then the width of the compact emerging from between the rolls is **b**, where **b** is slightly less than **a**. Over the central portion **c** of the compact, the density is substantially uniform but the density outside of this central portion falls off rapidly to the longitudinal edges of the strip.

For the production of high quality strip it is desirable to have a minimal density variation across the width of the strip and only the central portion of the strip of width **c** is satisfactory for further treatment. It is desirable therefor to trim the strip along both longitudinally extending edges to remove the edge portions and produce a parallel edged strip of width **c**, this strip having substantially uniform density across its width.

The edge portions are removed from the central portion of the strip by comminuting the edge portions into powder of substantially the same particle size as that of the powder from which the strip is formed. A pair of milling cutters **6** are rotatable about a pair of parallel axes **7** extending normal to the plane of the strip. The thickness of the cutters is greater than the thickness of the strip and in use the cutters are rotated at such a speed that when the teeth of the cutters contact the strip the powder particles are not cut from the strip but are individually broken off by the impact of the teeth of the milling cutter.

Referring now to Figures 8.6b and 8.6c, a powder compaction mill has a pair of rolls **11** arranged with their longitudinal axes horizontal and rotatable in a pair of housings **12** and **13**. A horizontal bracket **14** is secured to the housings **12** and **13** on the outgoing side of the mill and the brackets support a pair of electric drive motors **15**. The motors are arranged with their drive spindles extending downwardly and the motors can be moved towards and away from each other along the bracket **14** and then secured in position by means of clamps.

A compact in the form of a strip produced in the mill by compacting metal powder leaves the mill supported on a horizontal table **16**. The table is of a width less than that of the strip and the longitudinal edges of the strip project outwardly of the edges of the support table. Cutter discs **17** having teeth projecting from their peripheral edges are mounted one on each of the spindles of the drive motors **15** and the thickness of the discs is greater than the thickness of the strip. The cutters are positioned so that they engage the longitudinal edges of the strip and the cutters are spaced apart by an amount equal to the required useful thickness of the compact after the longitudinal edges have been removed.

The motors are energized to rotate the cutters so that the portion of the cutters engaging the strip is rotating in a direction opposite to the direction of movement of the strip. The number of teeth on each cutter and the speed of rotation of the cutters and the speed of movement of the strip are all chosen so that the cut per tooth is not greater than the diameter of the largest powder particles.

The particles removed from the strip fall into a bin **18** positioned below the support **16**. The powder regenerated in this way may be recycled directly back to the hopper of the compacting mill but it may be necessary to blend it with powder which has not already been compacted and it may be necessary to provide it with a degree of annealing.

The collection of the powder may be facilitated by diverting the powder trimmed from the strip below the plane of the strip and into the bin. This may be brought about by using cutters with teeth angled in a helix relative to the axis of the cutter. Alternatively, the teeth on the cutter may be straight but the axis of rotation of the cutter may be slightly inclined to the plane of the strip so that the particles trimmed from the strip are directed downwardly below the level of the strip and into the bin.

A particular example of the operation of the apparatus shown in Figures 8.6b and 8.6c is as follows. Assuming that the strip compact has been formed from powder which has been sieved through a sixty mesh screen, then the strip will contain particles of up to 0.25 mm sieve size. To ensure that particles which are not greater than this size are removed from the compact, the cut per tooth should not be greater than 0.25 mm. The drive motors **15** are thus rotated at 1,200 rpm and the strip is produced at 100 mm/sec. The strip is thus fed into the cutter at 5 mm per revolution and if there are twenty teeth on the cutter the feed will be 0.25 mm per tooth.

The diameter of the cutter should be at least twice the depth of the cut and in practice a 75 mm diameter cutter having twenty-four teeth may be used to trim 11 mm off of a strip made from sixty mesh powder at 100 mm/sec.

DENTAL AMALGAMS

Silver, Copper, Tin and Zinc

R. Kropp; U.S. Patent 4,008,073; February 15, 1977; assigned to Deutsche Gold- und Silber-Scheideanstalt vormals Roessler, Germany has produced alloy powders suitable for the production of dental amalgams containing 75-92% silver, 8-23% copper, 1-10% tin and 0-2% zinc.

Amalgams produced from these powders not only are free of γ_2 phase (a practically silver free phase of the composition $Sn_{7-8}Hg$ which in the literature is designated γ_2), but also have a very low creep behavior. The plastic deformation of a hardened amalgam under constant load in these alloys is about an entire power of 10 lower than the customary amalgams with higher tin and lower silver contents. This has the advantage that fillings of an amalgam produced with the alloy powder of the process are substantially less deformed by the constant chewing pressure in the mouth in the course of time than fillings of the previously used amalgams. Thereby there is substantially less danger of taking off the marginal filling edges and their subsequent breaking. Additionally the setting expansion of amalgams produced with alloy powders of the process lies within the pertinent specifications.

As examples of alloys of the process which after pulverization, mixing with mercury and hardening have lower values for the setting expansion and the creep compositions are listed in the table below. Unless otherwise indicated all parts and percentages are by weight.

SILVER (WEIGHT %)	COPPER (WEIGHT %)	TIN (WEIGHT %)	ZINC (WEIGHT %)
90	8	2	0
85	10	5	0
82	10	8	0
80	15	5	0
76	18	5	1

The components of these mixtures can be produced, for example, by milling of correspondingly composed bars or by atomizing of a corresponding melt. However, it is also possible to produce only one of the components by milling and the other by atomization.

Silver, Tin, Copper and Indium

H. Kaji and N. Suzuki; U.S. Patent 4,164,419; August 14, 1979; assigned to Shofu Dental Manufacturing Co., Ltd., Japan have developed a dental alloy capable of combination with mercury to form a dental amalgam.

The powdered alloys have, in weight percentages, 20% and above silver, 20 to 50% tin, 9.5 to 50% copper, and 0.5% and above indium. A preferred copper content is between 10 and 35% and especially between 11 and 20%. The most preferred copper content is 13±0.5%. A preferred silver content is at least 40% by weight and especially at least 50% by weight. A preferred tin content is 20 to 40% by weight and especially 20 to 30% by weight. Finally, the preferred indium content is 0.5 to 10% by weight and especially between 3 and 6% by weight.

The powdered alloys are triturated with mercury to form an amalgam. Preferred dental amalgams consist by weight percentages of 18 to 35% silver, 11 to 16% tin, 6 to 15% copper, 1 to 5% indium and the balance mercury at the time of trituration. The trituration is effected at a weight ratio of 1 part alloy powder to 0.82-1.0 part mercury.

In the above composition, silver content less than 20% makes amalgamation with mercury difficult, accompanied by slow hardening speed. Tin content less than 20% makes amalgamation with mercury difficult with too rapid hardening speed, while more than 50% tin results in large aging contraction and slow hardening speed. Copper content of 9.1% or less is insufficient to prevent the tin-mercury phase in the amalgam, while more than 50% copper makes amalgamation with mercury difficult with slow hardening speed due to consequent reduction in the silver and tin contents. Indium content of less than 0.5% gives insufficient protection against discoloration.

Mercury Trituration

In order to provide a metal powder mixture for use in producing dental amalgams which has improved shelf life and does not deteriorate during storage prior to amalgamation, enabling the attainment of dental restorations of continuous high strength and excellent mechanical properties, *K. Asgar and S.H. Reichman; U.S. Patent 3,980,472; September 14, 1976; assigned to Special Metals Corporation* have produced an alloy powder blend by trituration with from about 40 to about 60% mercury.

The powder particles are of an average particle size less than about 100 μ, and preferably from about 5 μ to about 40 μ in size. The alloy powder blend comprises a mechanical mixture comprised of from about 55 to about 90% of a first powder composed of a silver-base amalgamatable alloy, and from about 10 up to about 45% of a second powder composed of a substantially nonamalgamatable silver-containing alloy which includes a controlled quantity of tin.

The amalgamatable powder constituent consists essentially of about 75% silver and about 25% tin, with the silver-tin alloy being replaceable by up to about 5% copper and up to about 2% zinc. The nonamalgamatable powder consists of a silver-containing alloy containing about 1 to about 20% tin, a second alloying constituent individually selected from the group consisting of about 5% cadmium, about 5 to about 50% zinc, about 5 to about 50% aluminum, copper in an amount to provide a silver-to-copper ratio of about 2.6:1, up to 30% indium in combination with copper in an amount to provide a silver-to-copper ratio of about 2.6:1, and with the balance consisting essentially of silver.

Preferably the nonamalgamatable powder consists of a eutectic silver-copper alloy having a silver-to-copper ratio of about 2.6:1 and containing from about 2% to about 10% tin. Upon trituration with mercury in an amount from about 40 up to about 60%, and preferably from about 46 to about 50% mercury, a high strength amalgam is produced in which the nonamalgamatable particles are substantially uniformly dispersed through a substantially continuous matrix and are chemically bonded therein forming an integral high strength mass of excellent mechanical properties.

Example: Three silver-base alloy powders having an average particle size of 25 μ were prepared by gas atomization. The nominal composition of these powder samples, designated as A, B and C, are set forth in Table 1.

Table 1

Sample	Alloy Powder Composition Silver	Tin	Copper
A	70.5	27.0	2.5
B	72.0	0	28.0
C	69.0	4.4	26.6

The composition of Sample A is typical of an amalgamatable powder composition. The composition of Sample B is typical of a nonamalgamatable powder in accordance with prior art teachings. The composition of Sample C is typical of a nonamalgamatable powder containing a controlled quantity of tin in a silver-copper eutectic alloy in accordance with a preferred practice of the process.

A powder mixture was prepared employing two parts by weight of Sample A and one part by weight of Sample B, designated as Mixture D, which corresponds to a prior art powder blend as set forth in U.S. Patent 3,305,356. A second powder blend, designated as Blend E, was prepared employing two parts by weight of Sample A and one part by weight of Sample C, corresponding to the improved powder blend in accordance with the teachings of the process. While in a new or fresh condition, powder Blends D and E were triturated with one part mercury for each part alloyed powder and a resultant mixture was shaped into a test specimen and allowed to set for a period of seven days.

The resultant test specimens were subjected to compressive strength evaluations at a testing speed of 0.008 in/min. An aliquot portion of Blends D and E were allowed to age under storage for an 18-month period, whereafter they similarly were triturated with an equal quantity of mercury and formed into test specimens and were tested after setting for a seven-day period. For comparative purposes, an amalgamated test specimen was also prepared employing equal quantities of mercury and the powder of Sample A by itself. The resultant compressive strength of the several test specimens in a new as well as aged condition are set forth in Table 2.

Table 2

Sample	Seven-Day Compressive Strength of Amalgamated Powder: Condition of Powder	Compressive Strength (psi)	Standard Dev.
A	—	57,000	(2,000)
D	New	67,000	(3,000)
E	New	67,200	(1,500)
D	18 Month	59,200	(1,900)
E	18 Month	67,000	(1,500)

Data as set forth in Table 2 are based on a series of determinations and the standard deviation for the results obtained is also included in Table 2. As will be noted, the compressive strength of Blends D and E in a new condition are substantially identical. However, after aging for 18 months, Blend D corresponding to the prior art composition, decreased approximately 8,000 psi in compressive strength approaching the compressive strength obtained on Sample A. Blend E, on the other hand, after an 18-month aging period prior to amalgamation exhibited a compressive strength substantially identical to that obtained on the freshly-prepared amalgam employing the same material. The foregoing test data clearly substantiate the improved shelf life and stability of alloy powder mixtures of this process.

While the tensile strength of amalgams have been shown to have little clinical correlation, tensile strength nevertheless is a good indicator of the quality of the bond between the discrete particle phases and the continuous matrix phase. Comparative tensile strength data of amalgams prepared from Blends D and E in a new as well as an aged condition are set forth in Table 3.

Table 3

Sample	Seven-Day Tensile Strength of Amalgamated Powder Samples: Condition of Powder	Tensile Strength (psi)	Standard Dev.
D	New	7,200	(510)
E	New	8,100	(340)
D	18 Month	6,600	(600)
E	18 Month	7,900	(350)

The test data as set forth in Table 3 comprise an average of five tests and the standard deviations for the numerical values given are set forth in the last column thereof. The tensile testing was conducted at a test speed of 0.008 in/min on test specimens which had been allowed to set for seven days to attain maximum strength. The higher tensile strength of the amalgam prepared from Blend E in comparison to the amalgam prepared from Blend D in a new condition clearly evidences the improved bonding between the discrete phases and the continuous matrix of the amalgam of the process. This difference is further amplified by the results obtained on the amalgams prepared from the aged powder blends further substantiating the improved shelf life and resistance to deterioration of powder mixtures prepared according to this process.

Company Index

The company names listed below are given exactly as they appear in the patents, despite name changes, mergers and acquisitions which have, at times, resulted in the revision of a company name.

Inventor Index

U.S. Patent Number Index

Notice

Nothing contained in this Review shall be construed to constitute a permission or recommendation to practice any invention covered by any patent without a license from the patent owners. Further, neither the author nor the publisher assumes any liability with respect to the use of, or for damages resulting from the use of, any information, apparatus, method or process described in this Review.

HAZARDOUS AND TOXIC EFFECTS OF INDUSTRIAL CHEMICALS 1979

by Marshall Sittig

There is a continuing need to assess the status of potentially dangerous substances including those now available and those that may reach commercial availability in the future. This should be done with a predictive view to avoid or at least to ameliorate catastrophic episodes similar to those that have occurred with methyl mercury, polychlorinated biphenyls, vinyl chloride monomer, dioxin and a number of pesticides.

This handbook is intended to be a working guide for the industrial hygienist and other interested persons. Most of the individual chemical listings contain a number of literature references.

This book is intended to be a guide to these hazardous chemicals, to give first warning signals, and to serve as an introduction to the published literature and to government agencies offering detailed guidance.

Information (where available) under each chemical listing includes:

Description. Derivation, chemical structure and systematic chemical name.
Synonyms. Other chemical names, generic names and trademarks.
Potential Occupational Exposures.
Permissible Exposure Limits.
Routes of Entry.
Harmful Effects. Local, systemic.
Medical Surveillance.
Special Tests.
Personal Protective Methods.
Bibliography. References to articles in journals and government publications.

The book consists of about 250 individual monographs arranged alphabetically by the common name of each substance described. It is impossible to indicate here all of the chemical entries in this book. In order to give an indication of the vast scope of information dealt with here, the names of most entries in the beginning of the book are reproduced here.

Acetaldehyde
Acetates
Acetic Acid
Acetic Anhydride
Acetone (See Ketones)
Acetonitrile
Acetylaminofluorene
Acetylene
Acridine
Acrolein
Acrylamide
Acrylonitrile
Alkanes
Allyl Alcohol
Allyl Chloride
Aluminum & Compounds
Aminodiphenyl
Ammonia
Amyl Alcohol
Aniline
Antimony & Compounds
Arsenic
Arsine
Asbestos
Asphalt Fumes
Barium & Compounds
Benzene
Benzidine & Salts
Benzoyl Peroxide
Benzyl Chloride
Beryllium & Compounds
Bis(Chloromethyl) Ether
Bismuth & Compounds
Boron & Compounds
Boron Hydrides
Boron Trifluoride
Brass
Bromine
1,3-Butadiene
Butyl Alcohol
Butylamine
Cadmium & Compounds
Calcium Cyanamide
Calcium Oxide
Carbaryl
Carbon Dioxide
Carbon Disulfide
Carbon Monoxide
Carbon Tetrachloride
Carbonyls
Cement Dust
Cerium & Compounds
Chlorinated Benzenes
Chlorinated Lime
Chlorinated Naphthalenes
Chlorine
o-Chlorobenzylidene Malonitrile
Chlorodiphenyls & Derivs.
Chloroform
Chloromethyl Methyl Ether
Chloroprene
Chromium & Compounds
Coal Tar Products
Cobalt & Compounds
Copper & Compounds
Cotton Dust
Creosote
Cresol
Cyclohexanone
Diacetone Alcohol
2,4-Diaminoanisole
4,4'-Diaminodiphenylmethane
Dibromochloropropane
1,2-Dibromoethane
3,3'-Dichlorobenzidine
1,2-Dichloroethane
1,2-Dichloroethylene
Dichloroethyl Ether
Diisobutyl Ketone
N,N-Dimethylacetamide
4-Dimethylaminoazobenzene
N,N-Dimethylformamide
Dimethyl Sulfate
Dinitrobenzene
Dinitro-o-Cresol
Dinitrophenol
Dinitrotoluene
Dioxane
Diphenyl
Epichlorohydrin
Ethanolamines
Ethyl Alcohol
Ethylbenzene
plus 160 other chemicals

ISBN 0-8155-0731-3 **460 pages**

EXTRACTIVE METALLURGY 1977

Recent Advances

by Edward J. Stevenson

Chemical Technology Review No. 93

The efficient extraction and recovery of metals found in the earth's crust or on the bottom of her oceans has become increasingly important as modern industrial societies consume vast quantities of these materials.

With the increased awareness on the part of both the public and the metals-winning industry of the ecological dangers that can arise from continued customary mining procedures and the increased problems of pollution caused by obsolete refining methods, the search for clean low-energy processes has become paramount.

This book describes over 230 processes from the recent U.S. patent literature. The world-wide search for new and efficient metal recovery methods is nowhere more evident than is shown here by the scope of these processes.

Many of these processes encompass the extraction and isolation of several metals. However, for continuity and clarity, the organization of this book is focused on the major metallic component most commonly sought in a given process. Improved uses of raw materials and of human and energy resources are emphasized as is any step that might lead to a cleaner, less energy-consuming process.

A partial and condensed table of contents follows here. Chapter headings and important subtitles are given. Numbers in parentheses indicate the number of topics covered.

ISBN 0-8155-0668-6 **255 pages**

CORROSION INHIBITORS 1976
Manufacture and Technology

by Maurice William Ranney

Chemical Technology Review No. 60

Modern corrosion inhibitors with antidegradant properties not only inhibit the corrosion of metals and other materials, but also stabilize and readjust the medium or environment in which they are applied.

A customary corrosion inhibitor, for instance, will prevent acid attack in an automotive crank case by building up a protective barrier on the surface of the metal substrate by physical adsorption, chemisorption or other reaction with the substrate, but a modern antidegradant corrosion inhibitor will try to forestall the formation of acids in the crank case oil in the first place, while forming a protective coating on the metal as a second step.

Naturally, such corrosion inhibitors must be designed individually for given applications: Aluminum is attacked by alkalis, salts, and certain acids. Concrete is attacked by acids. Iron and its various alloys are very sensitive toward oxygen-containing compounds.

How to design, manufacture and apply such antidegradant corrosion inhibitors is taught in this book which describes over 260 processes relating to the protection of materials susceptible to corrosive chemical action.

A partial and very condensed table of contents follows here. Numbers in parentheses indicate the number of processes per topic or chapter. Chapter headings and some of the more important subtitles are given here.

ISBN 0-8155-0606-6 **338 pages**